US Petrochemicals

US Petrochemicals

Technologies, Markets, and Economics

ARTHUR M. BROWNSTEIN, Editor

THE PETROLEUM PUBLISHING COMPANY ■ Tulsa

The Petroleum Publishing Company
211 South Cheyenne
Tulsa, Oklahoma 74102, USA

Library of Congress Catalog Card Number: 77-184571
International Standard Book Number: 0-87814-008-5

Printed in USA

To Rosalie, Jeffrey and Laura

The Authors

ARTHUR M. BROWNSTEIN, Editor
Chem Systems Inc.
Director of Process Evaluation
and Research Planning

Contributing Authors

MYRON T. FOVEAUX
Monsanto Company

AUBREY M. KIRBY, JR.
Sun Oil Company

FRANK W. LONG
Arco Chemical Company

ELLIOT M. MAGER
Amerada Hess Corp.

NELSON E. OCKERBLOOM
Sun Oil Company

Contents

Preface

THE US chemical industry has moved rapidly from coal to a petroleum-based economy during the last two decades. In recognition of the change in origin of raw materials, the term petrochemical has come into common usage. Although a precise definition of petrochemical has yet to gain universal acceptance, it commonly refers to chemicals isolated from refinery processes and to their first generation derivatives.

The US petrochemical industry is a complex matrix in which interrelated chemicals compete for overlapping markets via increasingly competitive technologies. Woven throughout this scheme is a web of economic and political factors whose prime concern is the efficient production of gasoline and other fuels. Hence, the basic economics of the chemical industry are largely related to those of petroleum refining. Upon examination of these factors, one learns that such seemingly diverse subjects as polyester fibers and gasoline share a common base. For example, it is indeed possible to demonstrate that the removal of lead from gasoline could increase the price of polyesters.

This manuscript is based upon lecture notes from a unique course on *The US Petrochemical Industry,* which was twice presented by the authors at the University of Pennsylvania during 1971. It was offered under the auspices of the Continuing Engineering Studies program of the Department of Chemical Engineering. Its intent was to convey an understanding of the salient factors which govern the US petrochemical industry. A comprehension of these factors is essential for effective planning of research and marketing. This is the primary objective of this book.

Arthur M. Brownstein
Cherry Hill, New Jersey
December 21, 1971

US Petrochemicals

1

AUBREY M. KIRBY, JR.

The chemical industry and the oil-import program

THE chemical industry and the oil-import program may seem strange bedfellows with which to start an examination of petrochemicals. Recognition of the importance of oil to the chemical industry, however, merely reflects economic reality with respect to the chemical industry's future growth and the scientist's usual penchant to start at the beginning. The world's seemingly insatiable appetite for plastics and other chemical products can only be satisfied by utilizing as chemical feedstocks an ever-increasing proportion of petroleum products processed each year.

To place things in perspective, petroleum feedstock imports certified by the Oil Import Administration as eligible for quota allocations in 1970 amounted to the equivalent of 838,000 bbl/day. This figure represents almost 6% of an estimated total of 14,745,000 bbl/day of petroleum demand in the US during 1970.

Projecting petrochemical feedstock demands to 1980 at a 10% growth rate would suggest requirements equivalent to almost 2,200,000 bbl/day, approaching 10% of forecast US petroleum demand of 22,400,000 bbl/day. Some have even gone so far as to extend this type of projection to the year 2000, placing petrochemical feedstock requirements in the order of 20 to 25% of total US crude oil demand. By inference, it is conceivable that future restrictions on the domestic use of oil for energy might be considered to reserve such resources for chemical usage.

But all this is misleading in some respects since it is all conjecture. Currently, only a fraction of the 838,000 bbl/day certified petrochemical inputs is actually petroleum derived. The bulk of domestic petrochemical feedstocks come from natural gas.

Prior to World War II, the infant chemical industry was nursed along with a feedstock diet derived from coal. After World War II, a more substantial feedstock diet was introduced in the form of refinery dry gas and olefins resulting from the growth of catalytic cracking installations for gasoline manufacture in refinery operations.

This junior food was in turn replaced starting in the early 1950's by a low priced LP-gas by-product (ethane and propane) from a rapidly expanding natural-gas industry based on the Gulf Coast. And thus matters stood when the oil import program was instigated in the late 1950's. The chemical industry paid little heed at the time, since its feedstocks were in ample supply, at low cost, and independent of crude-oil supply and demand.

THE MANDATORY OIL-IMPORT PROGRAM

The beginning of the mandatory oil-import program can be traced as far back as July, 1954. At that time, growing concern over increasing imports for foreign crude oil and the implications of such imports for domestic oil exploration prompted the President to establish an Advisory Committee on Energy Supplies and Policy Resources. The committee was charged to make a study of energy supplies and resources in the United States, "with the aim of strengthening the national defense, providing orderly growth, and assuring supplies for our expanding national economy and for any future emergency."

The committee concluded that foreign oil imports should be restricted in the interest of national security and, in February, 1955, recommended voluntary restraint by importers to hold imports at the same proportion of future domestic crude oil production as existed in 1954. The director of the Office of Defense Mobilization was directed to monitor the situation.

Foreign oil imports continued to increase, however, leading to the appointment of another committee in April, 1957—the Special Committee to Investigate Crude Oil Imports. This committee also concluded that crude-oil imports should be limited in the interest of national security. In July, 1957, it recommended a plan for voluntary limitation of imports into the area east of the Rocky Mountains, setting the level at about 12% of crude oil production in that area. The President approved the recommendations of the committee and selected the Department of the Interior to administer the Voluntary Oil Import Program.

By the end of 1958, it had become apparent that the Voluntary Oil Import Program was also failing to restrict imports, despite continuous monitoring, admonitions to violators, and attempts to modify the program. A new series of investigations was started, which culminated in a recommendation in March, 1959, for mandatory controls by the Special Committee to investigate Crude Oil Imports.

On March 10, 1959, the President issued Proclamation 3279 which established the Mandatory Oil Import Program. The original proclamation has been amended many times since by subsequent proclamations, but is still the basis for the Oil Import Regulations currently administered by the Department of the Interior.

The Mandatory Oil Import Program divides the country into five districts for administrative purposes as shown in Fig. 1.1. The licensed level of imports, except for residual fuel oil, in Districts I–IV (east of the Rockies) is fixed at 12.2% of the estimated crude oil and natural-gas-liquid production in these districts, less estimated overland imports (Proclamation 3509, November 30, 1962).

Other deductions are also made from this total to provide amounts for historic importers of record, finished products, reserves for the Oil Imports Appeals Board, and special allocations. Unfinished oils are limited to 15% of this total; finished products are fixed at the level of such imports into these districts in 1957.

Finished products are defined as petroleum products, such as gasoline, jet fuel, naphtha, lubricating oil, asphalt, etc., which are to be used without further processing except blending by mechanical means. Unfinished oils are defined as any petroleum product which is to be further processed other than by mechanical means.

District V (west of the Rockies) is recognized separately as being a crude-deficient area. Import quotas for District V are set at a level equal to the difference between estimated demand and production,

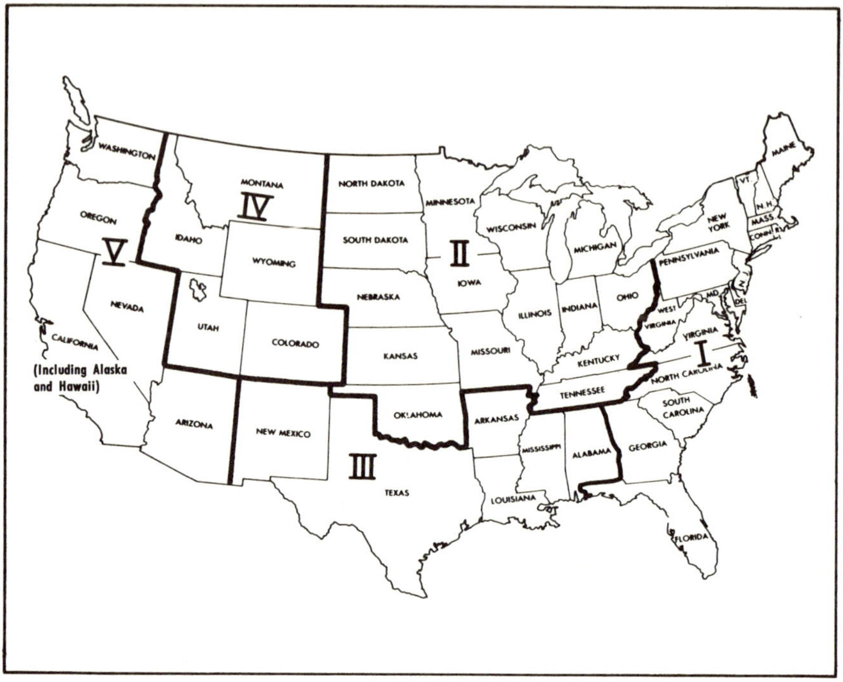

FIG. 1.1 Petroleum administration for defense (PAD) districts

less overland imports. Unfinished oils are limited to 25 percent of total imports; finished products are also limited to the historic import level in 1957.

The initial distributions of authorized crude oil imports into Districts I–IV for the years 1967 and 1970 are shown in Table 1.1, both for comparison and to illustrate the typical division among various categories. Residual fuel oil is not included in the figures presented since importation of this product into District I is not controlled.

Import allocations are granted for a specific period (ordinarily annually as of January 1) both to those companies having refinery capacity and "inputs," and petrochemical plants with qualified "inputs." Refinery allocations are based on a graduated scale related to refinery capacity so as to benefit the small refiner. Petrochemical allocations are based on a fixed percentage of qualified "inputs" (11.2% for District I–IV in 1970) regardless of plant size. Table 1.2 shows the crude oil input scale for refinery allocations in 1970.

TABLE 1.1

Distribution of crude oil imports for Districts I–IV

(Bbl/day)

	1967	*1970*
Bureau of Mines		
Estimated production	8,724,000	9,879,000
Total quotas – 12.2%	1,064,328	1,205,238
Less: *Overland exempt*		
Mexico	30,000	30,500
Canada – crude	225,000	332,000
– products	–	30,000
Light oil shipments (Puerto Rico)		
Phillips	3,750	24,800
Commonwealth		20,000
Total deductions	258,750	437,300
Maximum imports for allocation	805,578	767,938
Allocations		
Finished products		
Hess (Virgin Islands)	–	15,000
Oil-import appeals board	5,000	13,279
1957 historical quotas	71,636	48,357
Total finished products	76,636	76,636
Crude and finished oils		
Oil-import appeals board	5,275	2,737
Court orders set aside	–	4,146
Petrochemical	40,084	90,100 (est.)
Refiners – historic	228,096	89,163
– input	455,485	472,326
Sec. 25 Starters	–	26,586
(refinery & petrochemical)		
Not allocated	2	4,244
Total crude and unfinished	728,942	691,302
Grand total	805,578	767,938

TABLE 1.2

Refinery input scale used for quota allocation – 1970

Average bbl/day input	*Percentage allocation*
0 – 10,000	19.5
10,000 – 30,000	11.0
30,000 – 100,000	7.0
100,000 plus	3.0

THE CHEMICAL INDUSTRY'S FEEDSTOCK SITUATION

The chemical industry came of age during the decade of the 1960's. The history of ethylene consumption shown in Fig. 1.2 illustrates the rapid growth culminating in the huge petrochemical business we have today. This rapid growth of over 10% per year presented enough problems that it is not surprising that feedstock supply and adequacy received little attention in the early 1960's. Towards the

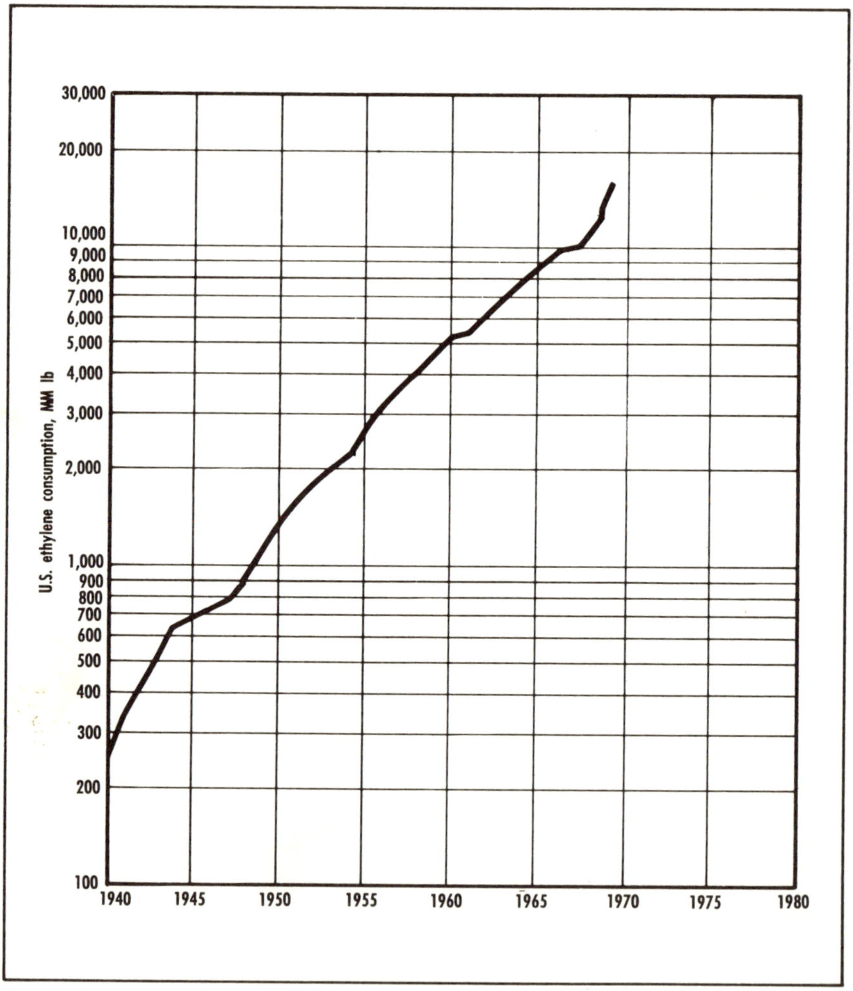

FIG. 1.2 Ethylene-consumption history

middle of the decade, however, the discrepancy had become obvious between demand and supply with petrochemical growth at twice the rate of the source of its feedstock supply, natural gas. Fig. 1.3 shows the growth of ethylene, representing the petrochemical industry, compared with LP-gas feedstock to illustrate the forecast gap in feedstock availability.

Richard C. Perry of Union Carbide was one of the first to comment publically on this in a paper, "International Competition in Petrochemicals," delivered at an AICHE meeting in Houston, February, 1967:

> "Based on ethylene projections, our figures indicate that due to limited LP-gas availability after 1975, most new US capacity will have to be based on a more fundamental hydrocarbon in the form of crude oil-derived liquids, if US producers are to continue to participate significantly in the growing world petrochemical business."

Stated thus, devoid of economic considerations, the solution to the problem seems simple. All we have to do is to start using light oil distillates as feedstock to manufacture ethylene and other olefins. But the real world isn't that simple, and economics have to be considered.

The only economical alternative use for LP-gas is fuel. This alternate fuel value would place a price of 0.42¢/lb (1.31¢/gal) on ethane and 0.41¢/lb (1.75¢/gal) on propane, in comparison with methane valued at 20¢/Mcf. The natural-gas industry has found it advantageous, therefore, to sell these products for chemical usage at approximately 0.7¢/lb (ethane at about 2¢/gal; propane at about 3¢/gal).

Light liquid distillates, however, have an alternate use and value as motor gasoline. A typical barrel of crude oil usually contains only about 20% light distillates in a boiling range suitable for use as gasoline. All of this must be processed at some expense, and, in order to satisfy our automobile oriented society, another 25% of the higher-boiling crude-oil distillates must be converted into the gasoline boiling range by catalytic cracking and other refinery processing. With crude oil valued at about 7.5¢/gal ($3.15/bbl), the final gasoline product, representing about 45% of the barrel, must be valued at close to 11 to 12¢/gal. Light-distillate fractions suitable either for manufacturing petrochemicals or processing into gasoline carry an intermediate value of about 8½¢/gal, or 1.4¢/lb.

Moreover, only 1.25 lb of ethane are required to make 1 lb of ethylene, whereas a minimum of 2.85 lbs of light distillate is required/lb of ethylene, with concurrent production of a number of byproducts. With

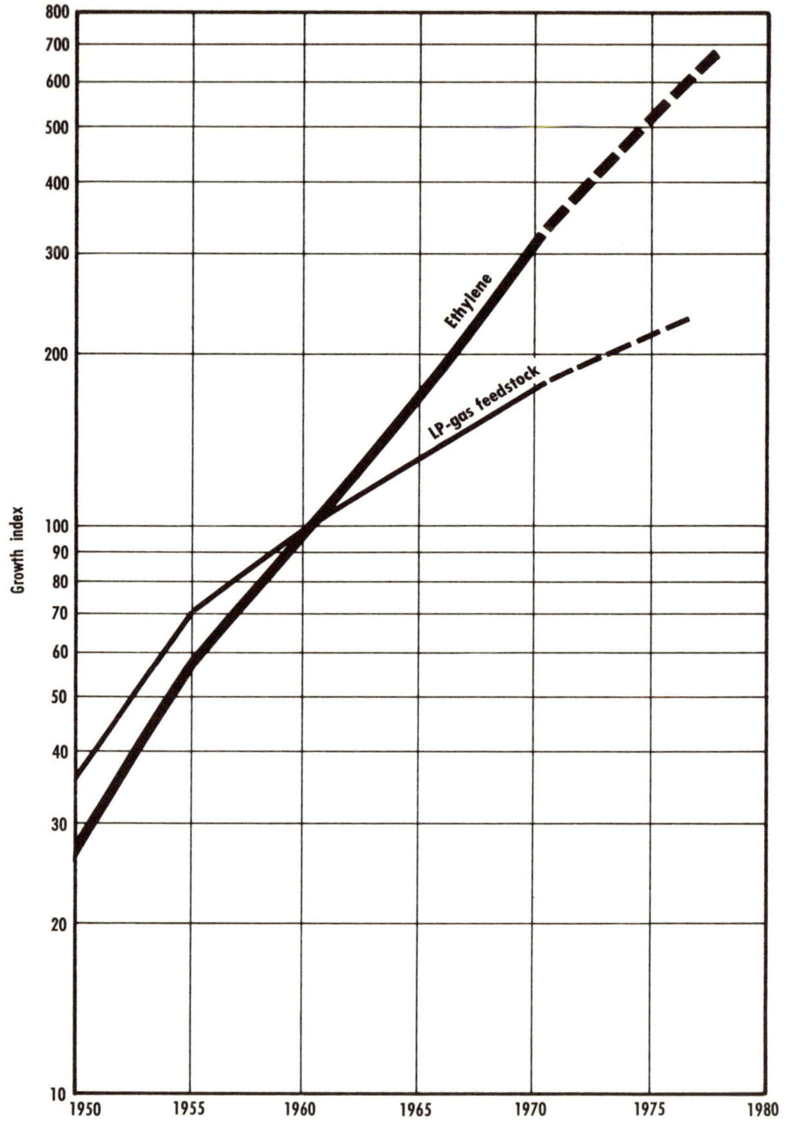

FIG. 1.3 Historic and projected growth rates for LP-gas feedstock and ethylene

such an increase in feedstock requirements, these light-distillate co-products must also be sold at no less than feedstock value to limit an ethylene feedstock cost to no more than double, or 0.7¢/lb increase. Such a cost increase would be about 25% of the total 3¢/lb ethylene price prevailing for Gulf Coast markets during the past several years.

The situation in Europe is different. With a fuel-oriented economy, only about 25% of the crude-oil barrel is converted into light distillates. Even this amount is more than necessary to supply the lower automobile population of smaller cars, so that light distillate, or naphtha, has been in excess supply for a number of years. Recent forecasts have suggested no change, or if any, a probable increase in such naphtha surplus.

Naphtha values in Europe with a lower foreign crude price have averaged 5 to 5½¢/gal, or about 0.83 to 0.92¢/lb. In the absence of a natural-gas industry to provide by product LP-gas, such as existed in the US naphtha has served as the replacement for coal-derived feedstocks for petrochemicals at a slightly higher cost to European chemical manufacturers over their US counterparts.

The obvious solution which suggested itself for economic expansion of US petrochemicals was to obtain access to foreign light distillates. This solution was not overlooked. Several approaches were tried.

FOREIGN TRADE ZONES

Foreign Trade Zones were established initially by Act of Congress (Public Law 397, 71st Congress), June, 1934, "to expedite and encourage foreign commerce." The Zones are special areas, usually near ports of entry where foreign and domestic merchandise may be brought without payment of custom duties and without other restrictions applying to the movement of foreign materials to US shores. As amended in 1950 (Public Law 566, 81st Congress), the Act permits manufacturing and exhibiting, as well as the sorting, drying, cleaning, grading, and other activities authorized in the original legislation.

Foreign Trade Zones offer, among other advantages, the opportunity for duty-free and quota-free manufacturing for export, utilizing foreign components. Any products or merchandise intended for US markets become subject to applicable duties only upon entry into US territory from such zones.

All the information necessary to establishment and operation of a foreign trade zone is included in a pamphlet, "Laws, Regulations, and Other Information Relating to Foreign Trade Zones in the United

States," available from the Superintendent of Documents, Washington, D.C. The Act established a board for authorization and administration of foreign trade zones, consisting of the Secretary of Commerce, as chairman and executive officer, and the Secretaries of the Treasury and the Army.

Any public or private corporation may apply for a zone, but private corporations require authorization before approval by an act of the state legislature concerned. The Collector of Customs in whose district a zone may be located is charged with direct supervision as resident representative of the Board.

Foreign trade zones are currently in operation in such places as New York, New Orleans, Toledo, San Francisco, and Mayaguez, Puerto Rico. Union Carbide Corp. operates a special subzone in Penuelas, Puerto Rico, for petrochemicals manufacture using foreign naphtha.

In late 1964, Union Carbide applied for a similar subzone in Taft, La., to process a proposed 10,000 to 15,000 bbl/day of Caribbean naphtha. Dow Chemical Co. also announced its intentions to apply for a foreign trade zone for its petrochemical operations at Bay City, Michigan. The oil industry, of course, opposed both proposals on the grounds that widespread utilization of such zones would either displace or drastically reduce prices for traditional, domestic petrochemical feedstocks.

All of the arguments with respect to petrochemical manufacture in foreign trade zones became academic, however, in late 1965. Presidential Proclamation 3693 was issued December 10, 1965, to modify the Mandatory Oil Import Program established by Proclamation 3279, March 10, 1959. The movement of quota materials such as crude, unfinished, and finished oils into foreign trade zones was prohibited except under a prior quota granted by the Oil Imports Administration.

Although Dow Chemical Co. did subsequently receive both authorization for a foreign trade zone and an oil-import allocation for Bay City, Michigan, use of foreign trade zones as a general method to obtain widespread access to foreign feedstock for petrochemicals was effectively eliminated. All was not lost, however, since Proclamation 3693 also provided authorization for inclusion of petrochemical allocations in the oil-import program.

PETROCHEMICAL IMPORT QUOTAS

The approach which has proved most successful has been direct appeal to the Oil Import Administration and the Department of the

Interior for a share of the import quota. The opening salvo in this campaign was fired on August 28, 1965. Arthur D. Little, Inc. presented to the Departments of Commerce and the Interior a study sponsored by Monsanto and Dow Chemical Co. entitled, "Effect of Naphtha Imports on the Competitive Position of the US Petrochemical Industry."

This study presented arguments mainly as to the need for access to low-cost foreign naphtha for maintenance of a competitive petrochemical situation in foreign markets. About a year later, in August, 1966, Arthur D. Little presented a follow-up study, "European Naphtha Demand," sponsored this time by Dow Chemical Co. and Union Carbide Corp. This second study provided arguments demonstrating a plentiful supply of foreign naphtha at low prices.

Meanwhile, the groundwork had already been laid for chemical entree to the import program by Presidential Proclamation 3693, December 10, 1965. This proclamation primarily intended to permit utilization of import allocations to provide for economic development of Puerto Rico with Phillips Petroleum Co. receiving the first allocation. As previously mentioned, however, it simultaneously restricted the movement of oil into foreign trade zones, but permitted the inclusion of petrochemical plants in the import program.

In early September, 1966, the Oil Import Administration announced petrochemical allocations of about 32,000 bbl/day, retroactive to cover the full calendar year. In December 1966, the petrochemical quotas were increased to about 44,000 bbl/day for allocation year 1967. The continuing success enjoyed by the chemical industry subsequent to 1967 is documented in Table 1.3, showing the growth of petrochemical quotas both in volume and as a percentage of petrochemical feedstock input.

Not all of the increase in certified inputs has resulted from chemical industry growth alone, nor have the same companies always shared in the quota allocations, since changes have occurred in the regulations from time to time. For example, the chemical industry complained that oil refineries manufacturing olefins enjoyed a "double dip" by receiving both a refinery quota and a petrochemical allocation for the same olefins.

This double dip was eliminated during allocation year 1968 by classifying ethylene, propylene, and butylene as unfinished oils ineligible for petrochemical quotas except as certified inputs for derivatives manufacture. Thus was created the "nothing plant"—a merchant olefins facility which qualifies neither for a refinery input quota nor a petrochemical allocation.

TABLE 1.3

History of certified plant inputs and petrochemical quota allocations

	Certified plant inputs (Bbl/day)			*Crude and unfinished oil allocation (Bbl/day)*				
Year	*Districts I–IV*	*District V*	*Total*	*Districts I–IV*	*% of inputs*	*District V*	*% of inputs*	*Total*
1966	447,421	25,340	500,761	29,975	6.7	1,990	7.85	31,965
1967	530,057	27,131	557,188	40,215	7.6	4,295	15.7	44,510
1968 (1st half)	622,605	29,888	652,493	51,984	8.35	3,169	10.6	55,153
1968 (2nd half)	639,155	24,196	663,351	72,224	11.3	2,988	12.35	75,212
1969	693,636	25,178	718,814	77,690	11.2	2,998	11.9	80,688
1970	809,994	27,495	837,489	90,721	11.2	3,271	11.9	93,992

In the meantime, the petroleum industry was not silent in its opposition to the A. D. Little studies, especially after the Department of Commerce initiated a survey to evaluate the impact of increased petrochemical quotas on balance of trade. The Department issued specific instructions to assume naphtha availability at 5½¢/gal.

Illustrative of the earlier petroleum-industry responses was a report to the Oil Import Administration in July 1967, by J. H. Lichtblau, Director of the Petroleum Industry Research Foundation, Inc. This report, "Current and Proposed Import Allocations to the Petrochemical Industry," countered the A. D. Little studies with an estimate of 7.0¢/gal as the minimum delivered price of foreign naphtha to US markets.

July, 1967, also marked the beginning of the battle in earnest with the presentation of the original Chemco Plan, "Petrochemical Industry Proposal to the Department of the Interior," by eight chemical concerns. DuPont, as a member of the original group, has since withdrawn to provide the makeup of the current Chemco Group as shown in Table 1.4. The Chemco Plan in essence advocated a phasing out of existing allocations to allow free import of foreign feedstock for petrochemical manufacture by 1972.

TABLE 1.4
The Chemco Group

Celanese Corp.
The Dow Chemical Co.
Eastman Kodak Co.
Monsanto Co.
National Distillers and Chemical Corp.
Olin Corp.
Publicker Industries, Inc.
Union Carbide Corp.

Reaction was strong and immediate in the form of a study, "Petrochemical Feedstocks and the US Balance of Payments," by the Stanford Research Institute. Nine petroleum companies, as listed in Table 1.5, sponsored this study which was first presented to government representatives on August 23, 1967. The study concluded that adequate domestic feedstocks would be available to 1975 at slowly increasing prices, and that petrochemical exports would continue growing, regardless of the level of foreign feedstock imports.

TABLE 1.5
Sponsors of the Stanford Research Institute study "Petrochemical Feedstocks and the US Balance of Payments"

Ashland Oil & Refining Co.
Atlantic-Richfield Co.
Cities Service Co.
Kerr-McGee Corp.
Marathon Oil Co.
Mobil Oil Corp.
Standard Oil Co. (Indiana)
Standard Oil Co. (Ohio)
Union Oil Co. of California

The fall of 1967 also revealed for the first time a split in the chemical industry. In November a group of smaller chemical companies, listed in Table 1.6, released a statement to the Departments of Commerce and the Interior in opposition to the Chemco Plan. These companies represented non-integrated petrochemical manufacturers who did not use petroleum feedstocks, but utilized, the olefins products, and who were in favor of continued and growing import allocations. Thus was laid the groundwork for a separate organization from the Chemco Group, which culminated in the current Petrochem Group as listed in Table 1.7.

TABLE 1.6
Chemical dissenters to the 1967 Chemco Plan

Cabot Corp.
Copolymer Rubber & Chemical Corp.
El Paso Products Co.
Foster Grant Co., Inc.
J. M. Huber Corp.
Marbon Chemical Division of Borg-Warner Corp.
Rexall Chemical Company (Dart Industries, Inc.)
Sid Richardson Carbon Co.

The end result of all this activity during 1967 came in the form of a joint announcement in December by the Secretaries of the Departments of the Interior and Commerce, Stewart L. Udall and Alexander B. Trowbridge, of their intentions to revise petrochemical import allocations to implement an export-for-import program. Shortly thereafter, Secretary of Commerce Trowbridge suffered a heart attack and nothing further transpired during 1968, beyond continued publications of rebuttals and counter-rebuttals by the various groups.

TABLE 1.7
The Petrochem Group

Cabot Corp.
Chemplex Co.
Copolymer Rubber & Chemical Corp.
Dart Industries, Inc.
Eastman Kodak Co.
E. I. duPont de Nemours & Co.
El Paso Products Co.
Ethyl Corp.
Firestone Tire & Rubber Co.
Foster Grant Co., Inc.
Goodyear Tire & Rubber Co.
B. F. Goodrich Chemical Co.
Hercules Inc.
J. M. Huber Corp.
Koppers Co., Inc.
Marbon Chemical Division of Borg-Warner Corp.
National Distillers and Chemical Corp.
Northern Petrochemical Co.
Sid Richardson Carbon Co.
Uniroyal, Inc.

After the national elections in the fall of 1968 and the change in administration, activity by the petrochemical industry to promote independent action on petrochemical allocations was forestalled by President Nixon's appointment of a Cabinet Task Force on Oil Import Control to review the entire oil-import situation. This committee, formed in March, 1969, under the chairmanship of Secretary of Labor, George C. Shultz, requested submissions from all interested parties in reply to a lengthy questionnaire covering all aspects of petroleum imports.

Two replies were submitted from the chemical industry, one from the Chemco Group and another from the AllChem Group which included sixteen companies (the forerunners of the current Petrochem Group) in addition to the Chemco Group. The AllChem Group recommended continuation of petrochemical quotas, with increased allocations supplemented by an import-for-export program. The Chemco Group endorsed this recommendation as an interim measure, but also recommended a Stage D program culminating in quota-free access for petrochemical feedstocks.

The Cabinet Task Force issued its report in February, 1970, with a majority report recommending a major change in the oil-import program by substituting tariffs for direct-quota controls. A minority report differed from this majority view. With respect to petrochemicals, the majority report recognized "a case for granting a controlled product exception for petrochemical feedstocks imported or exchanged on a like-for-like basis for actual use in a petrochemical plant, and that the present crude allocations to petrochemical plants should be phased out." The minority report recommended two options: Option A provided allocations of 20% of the volume of feedstocks converted into petrochemicals for domestic sale, and 100% of the volume converted for exports. Option B provided pound-for-pound imports of feedstocks converted into petrochemicals in operations covered by the option.

In late February President Nixon appointed an Oil Policy Committee under the direction of Gen. George A. Lincoln to formulate policy for immediate oil problems, including petrochemical imports. Several months later, the recommendation of the Task Force regarding tariffs was clearly rejected.

In late May, the Chemco and Petrochem Groups jointly submitted a memorandum to the Oil Policy Committee entitled "Proposed Petrochemical Regulations," which broadly followed the lead suggested by the minority Task Force report. The joint "Proposed Petrochemical Regulations" provided for a petrochemical facility to choose between two options:

Option A: A 20% allocation based on feedstock inputs with 100% import-for-export privileges. The 20% allocation should be reviewed by the Oil Policy Committee in 1973 or, in the absence of review, should escalate to 27% for allocation year 1974.

Option B: Free access on a feedstock-input weight-equivalent basis for all feedstocks converted into petrochemicals or noncontrolled products (including plant fuel) in a "petrochemical access" facility. In order to accommodate plants which cannot use foreign feedstocks directly, eligible feedstocks may be certified to upstream suppliers.

Both options require changes in the current definition of petrochemical plants and petrochemicals.

Current definitions

Sec. 22 (n) "petrochemical plant" means a facility to which petrochemical plant inputs are charged, and in which more than 50% (by weight) of such inputs are converted by chemical reaction into petrochemicals, or in which over 75% (by weight) of recovered product output consist of petrochemicals."

Sec. 22 (p) "petrochemicals" means organic compounds or chemical elements other than unfinished oils or finished products, produced from petrochemical plant inputs by chemical reaction."

The proposed changes would also eliminate the 50% conversion of feedstock input requirement and substitute other alternatives for the 75% petrochemical output requirement:

> Option A – petrochemical plants: At least 40% by weight of product outputs should be petrochemicals or unfinished oils intended for petrochemicals, and no more than 25% by weight of total output should be finished products (other than fuel used in the same plant or residual fuel) or unfinished oils not intended for petrochemicals.
>
> Option B – petrochemical access facility: No less than 30% by weight of inputs should be converted into petrochemicals or unfinished oils intended for petrochemicals, provided that less than 30% by weight of inputs be finished products (other than fuel used in that facility or residual fuel) or unfinished oils not intended for petrochemicals.

Changes in other definitions would accomodate the new petrochemical-plant definitions; would provide a redefinition for petrochemical access facilities of benzene, toluene, ethylbenzene, and xylene as unfinished oils regardless of grade; and would define "a like for like exchange" recommended under the proposal.

On Aug. 5, 1970, the Office of Emergency Preparedness under Gen. Lincoln invited interested parties to answer eight general questions regarding petrochemical feedstocks, "in order to build a good fact foundation for policy evaluation of the program." Dead line for replies was indicated to be 30 days after publication of the questions in the "Federal Register," or Sept. 4, 1970. It is not known if any significant new information was revealed in any of the replies to these questions.

PETROCHEMICAL QUOTAS PRO AND CON

The chemical industry's claim to foreign feedstocks is based upon one major premise – the need to maintain economic competitiveness with foreign manufacturers who have free access to foreign oil at low world prices. Otherwise, it is claimed, the domestic industry will no longer be able to maintain an export trade. Lower-price foreign chemical products will take over third-country markets and eventually an ever increasing share of the domestic market, to the detriment of our balance of trade.

The only solution suggested if foreign feedstocks aren't made available is a threatened relocation of petrochemicals plants overseas. The end result would be a loss of domestic investment, jobs, tax revenues, and a weakening of national security at least as bad as a shortage of domestic oil production.

Major assumptions are implicit in this basic premise of necessity to match foreign competitors with its implications:

1. Low-cost foreign crude oil is the only, or at least the most significant, difference in foreign manufacturing economics.

2. Low-cost crude oil is synonymous with low-cost petrochemical feedstocks and low-cost petrochemicals.

3. Low-cost feedstock is the major factor in petrochemical plant-site selection.

4. Low-cost feedstock is the critical factor controlling chemical imports and exports.

These assumptions, perhaps, should be kept in mind for confirmation in our review of the various arguments surrounding this situation.

Oil-industry spokesmen may well agree and sympathize with the chemical industry's desire to maintain a competitive advantage over foreign manufacturers. At the same time, however, they have never failed to point out that the chemical industry, while claiming the need for foreign *imports,* have actually always asked for *import tickets*. Such tickets, they maintain, would in effect provide a subsidy valued at about $1.25 per barrel when traded for the same domestic feedstocks already used in petrochemical manufacture.

Obviously, like the three blind men with the proverbial elephant, there is a lack of communication somewhere which may never be resolved. Let us attempt, nevertheless, to shed some light on the differing viewpoints by examining some of the main arguments on both sides.

Economics

Ethylene, which will be used as an example for an economic review, is made by the decomposition, or cracking, of higher hydrocarbons at high temperature. The primary feedstock in the US is ethane and propane (LP-gas.) The primary feed in Europe is light naphtha or gas oil distillates, not crude oil.

The yield of ethylene varies, not only with the feedstock used, but also with the severity (time and temperature of reaction) of the cracking operation. Typical feedstock requirements and co-product yields from cracking operations are shown in Fig. 1.4 reproduced from an article, "Ethylene," by D. P. Burke and R. Miller published in *Chemical Week,* October 23, 1965. The diversity and amount of co-products should be noted. These co-products play an important role in manufacturing economics.

Ethylene co-products may carry a value ranging from a minimum

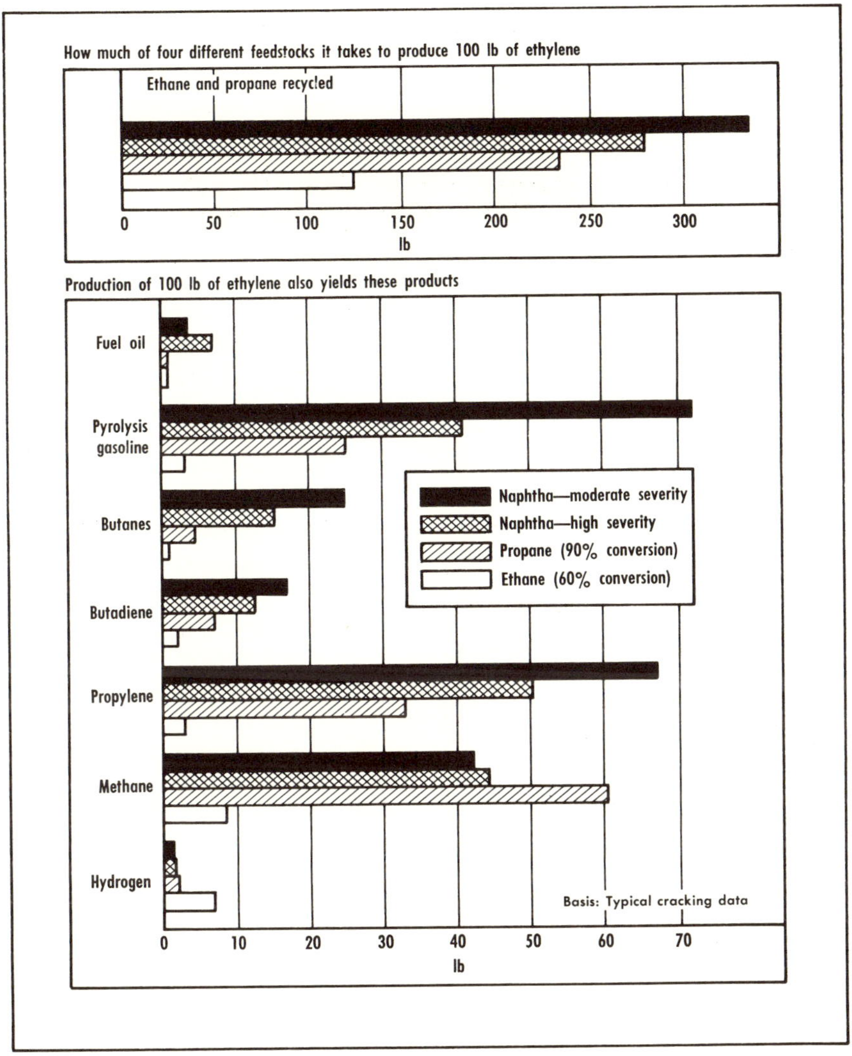

FIG. 1.4 Typical feedstock requirements and co-product yields for ethylene production

of feed value to a maximum of gasoline or chemical value, depending on the co-product, plant location, and market situation. Fig. 1.5, also from Burke and Miller's cited article, shows their calculation of the impact of US co-product values. Also shown is their estimate of the effect of co-product credits on ethylene manufacture for a European

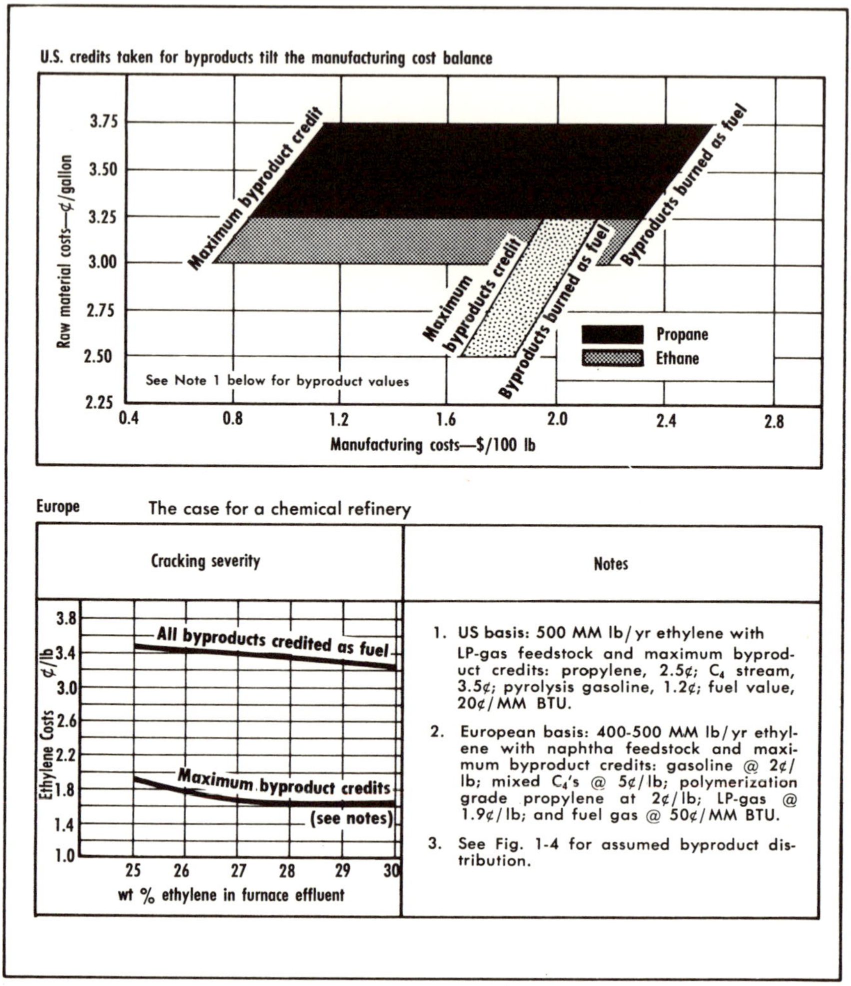

FIG. 1.5 Effect of co-product values on ethylene-production cost

situation using co-product values as a parameter of ethylene yield (cracking severity).

The 1.5 to 1.6¢/lb differential in ethylene manufacturing cost shown in Fig. 1.5 has received support from another paper by J. G. Freilling, C. C. King and J. Newman, "Ethylene Raw Materials and Production Economics," presented at a joint meeting of the American Chemical Society and the Chemical Institute of Canada on May 28, 1970. Fig. 1.6 shows their evaluation of the effect of cracking severity

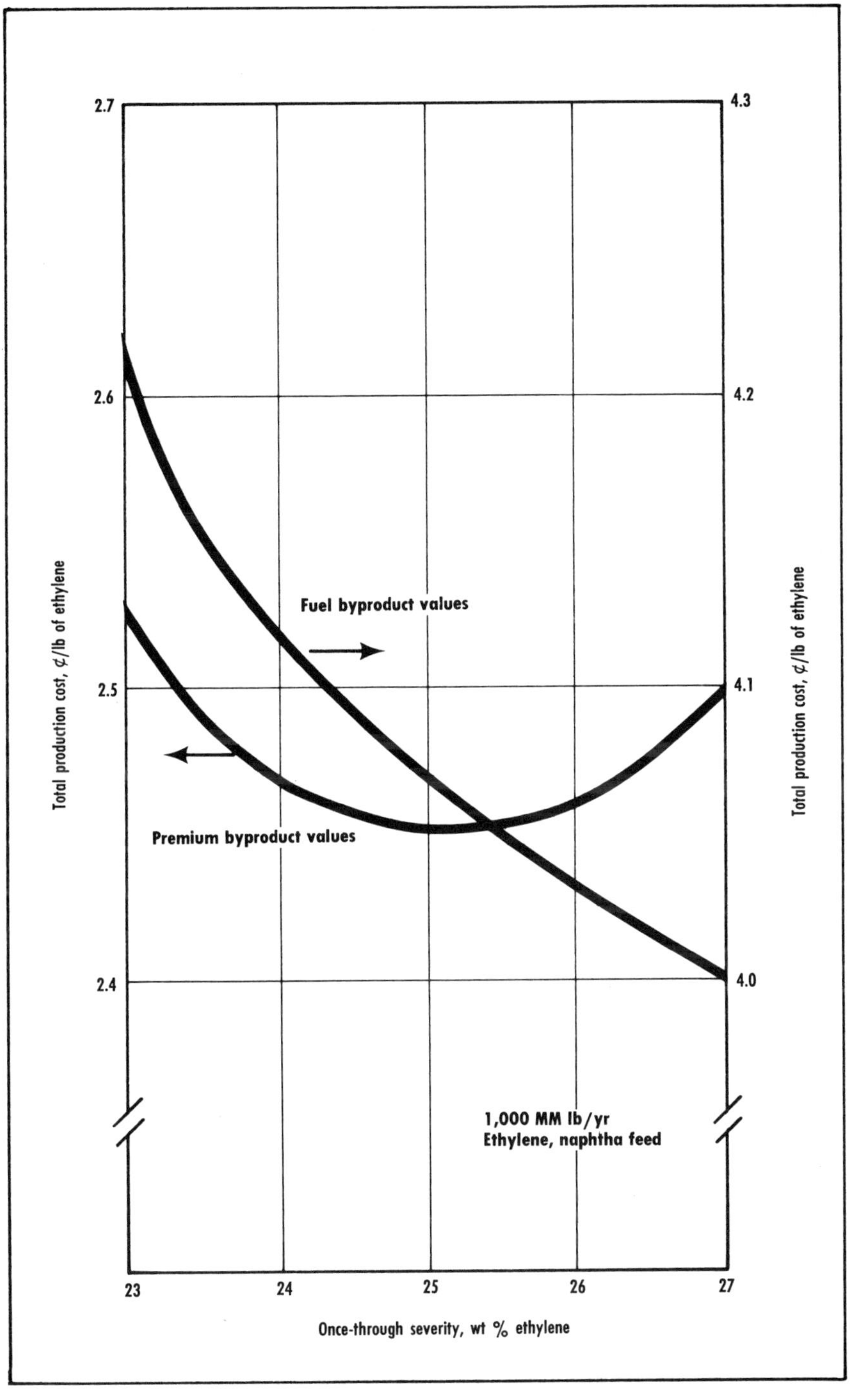

FIG. 1.6 Impact of co-product credits on ethylene manufacturing costs for Europe

on European ethylene manufacturing cost, with co-product valuation as a parameter. Their spread ranges from 1.5 to 1.7¢/lb.

Going beyond this Freilling et. al., also presented Figs. 1.7 and 1.8 showing the effects of feedstock price with premium co-product credits on European and US ethylene manufacturing costs respectively.

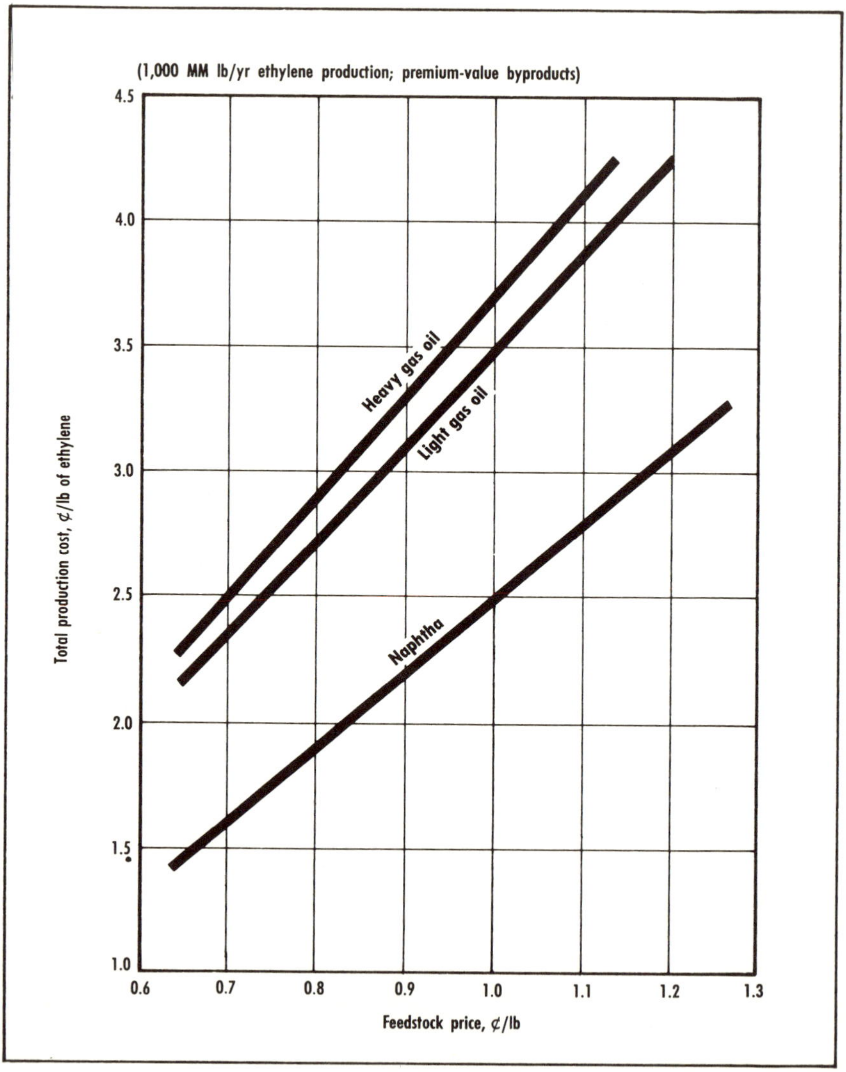

FIG. 1.7 Effect of feedstock price on European ethylene production costs

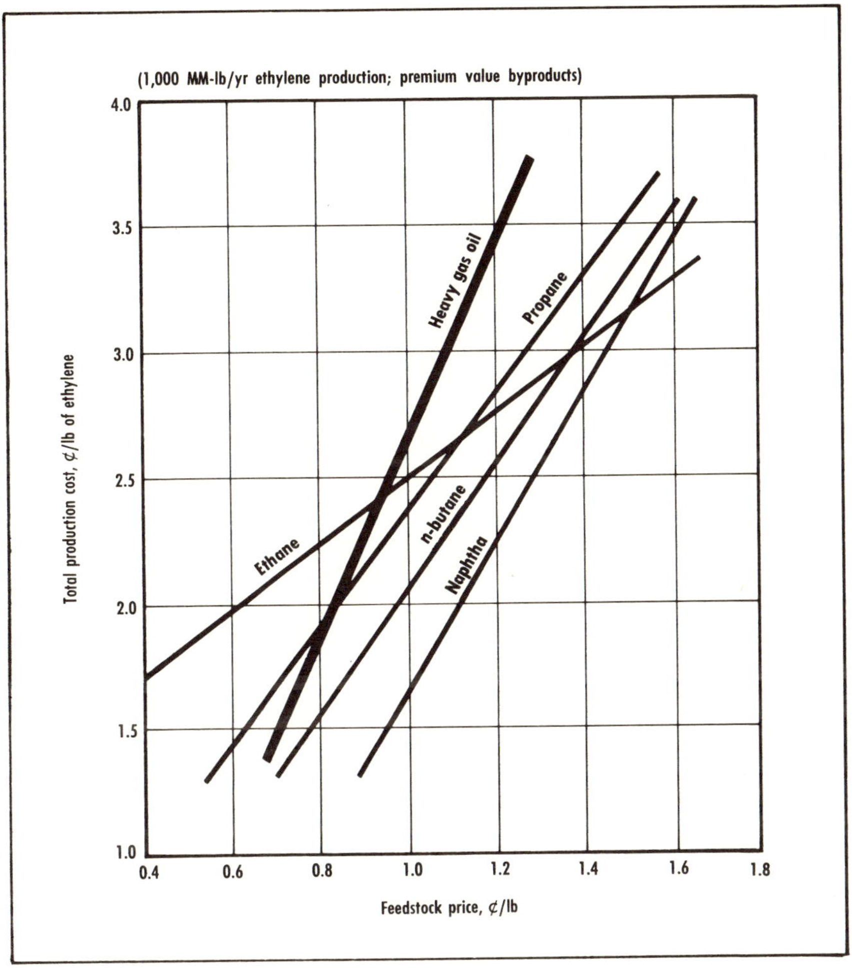

FIG. 1.8 Effect of feedstock price on US ethylene-production costs

Table 1.8 and 1.9 present details of the assumed feed and co-product values.

Fig. 1.7 and 1.8 explain why naphtha is the preferred feedstock in Europe, with butane and propane preferred in the US at current prices. If we now combine the naphtha curves from Figs. 1.7 and 1.8, as shown in Fig. 1.9, the data suggest US naphtha values of about 0.3¢/lb, or 1.8¢/gal, higher to be directly equivalent with foreign naphtha for ethylene manufacture.

It must be concluded that the economic facts of foreign vs. do-

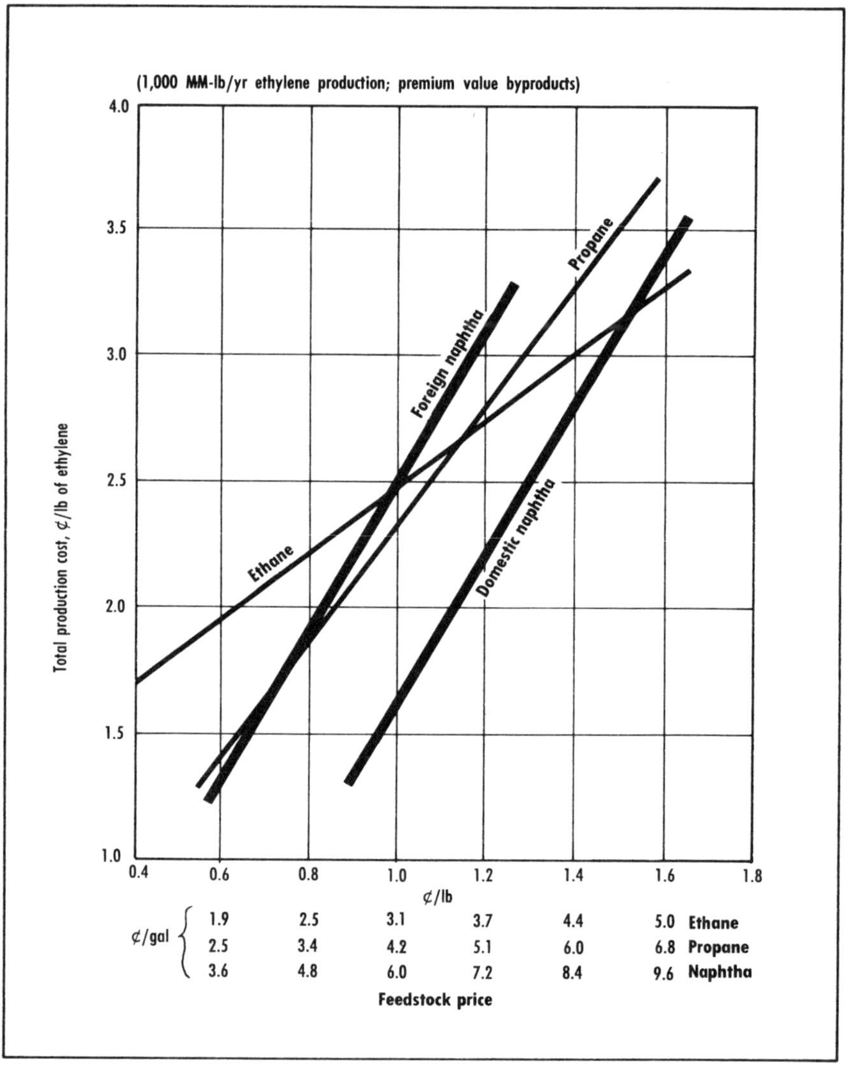

FIG. 1.9 Comparison of effect of feedstock price on US–European ethylene-production costs

mestic ethylene manufacture are complicated by the differences in the kind, volumes, and values of co-products, as well as the feedstock. Considerable latitude exists for estimators to exercise personal judgment.

TABLE 1.8

Assumed feed and byproduct values for Figures 1.7 and 1.8

Feedstock	*Ethane*	*Propane*	*n-Butane*	*Full-range naphtha*	*Heavy gas oil*	*Full-range naphtha*	*Light gas oil*	*Heavy gas oil*
Location	USA					Europe		
Unit cost, ¢/lb	1.0	1.0	1.0	1.6	1.1	1.0	1.1	0.75

By-products	*USA*		*Europe*	
Unit values, ¢/lb	*Fuel value*	*Premium or chemical value*	*Fuel value*	*Premium or chemical value*
Off gas	0.48–0.74	0.73–2.85	1.05–1.06	1.40–1.45
Propylene (chemical grade)	0.41	2.75	0.89	2.0
C_4 fraction				
Butadiene	0.40	5.0	0.86	4.0
Butylenes/butanes	0.41	1.30–2.20	0.87	1.0
C_5/400°F. + gas oil	0.38	1.30–2.10	0.81	1.05–1.30
400°F. + fuel oil	0.35	0.50	0.59	0.59

TABLE 1.9
Further details on product unit values (for Figs. 1.7 and 1.8)

	USA	*Europe*
1. *Off gas*		
A. Taken at fuel value, ¢/MMBTU(LHV)	21	45
B. Taken at premium value		
Contained H_2, ¢/MMBTU(LHV)	30	40
Methane & other, ¢/MMBTU(LHV)	21	45
2. *Butylenes/butanes*		
A. Taken at fuel value, ¢/lb	0.41	0.87
B. Taken at premium value		
Butylenes, ¢/lb	2.3	1.0
Butanes, ¢/lb	1.0	0.87
3. *Butadiene*	– unextracted state –	
4. *C_5/400°F. naphtha*		
A. Taken at fuel value, ¢/lb	0.38	0.81
B. Taken at premium value		
Untreated and before separation, based following values as finished products:		
Benzene, ¢/gal	25	85% of
Toluene, ¢/gal	17	US values
C_8 Aromatics, ¢/gal	19	
Non aromatics	–valued as naphtha–	

The chemical industry view of petrochemical feedstock economics is highlighted by these excerpts from testimony on "Oil Imports and the Petrochemical Industry" before the Antitrust and Monopoly Subcommittee of the Senate Judiciary Committee, July 22, 1969. The statement was made by Kenneth H. Hannan, Vice Chairman of the Board of Union Carbide, on behalf of the Chemco Group.

There are two major types of raw materials – feedstocks – for petrochemical production.

One is natural-gas liquids, which are by-products – originally wasted – of natural gas production.

By 1975, virtually every industry expert expects that the price of natural-gas liquids will increase substantially, making them non-competitive with foreign feedstocks for the production of most petrochemicals.

The alternative for natural gas liquids is the second major type of petrochemical feedstock – heavy liquids distilled directly from crude oil (or crude oil itself). The price of heavy liquids is determined by the price of crude oil, and the domestic price of these materials today is at least 60% higher than the foreign price because of the Oil Import Program.

Thus, unless the Oil Import Program is changed, all petrochemical raw materials in the United States will be uneconomic when compared to lower cost foreign crude oil and its derivatives.

The effect of the Oil Import Program on feedstock prices has another competitive aspect. Because it requires the industry to rely on domestic raw materials, of which only natural-gas liquids are competitively priced today, it tends to restrict most of the industry's production capability to those chemicals that can be produced from natural-gas liquids. Crude-oil-based feedstocks, however, result in the manufacture of a greater variety of petrochemicals than do natural-gas liquids. Many of these chemicals are key, high-volume intermediates used in the production of synthetic fibers and rubbers and other important products.

The economics of ethylene manufacture was covered in the Appendices to the Joint Submission by the Chemco Group to the Cabinet Task Force on Oil Import Controls," June 16, 1969, starting on page 2 of Appendix B to Question 28:

"Economics of Ethylene Manufacture"

The Stanford Research Institute report, cited earlier, contains detailed information on the manufacture of ethylene at a rate of 1 billion lb/yr by thermal cracking of various feedstocks. Detailed data covering the use of ethane and gas oil as feedstocks are reported here.

When cracking ethane, 82% of the feedstock is converted to ethylene. In contrast, when cracking gas oil, only 11% of the feedstock goes to ethylene. Thus the accounting methodology also becomes important to the understanding of the feedstock contribution to total costs. For the heavier feedstocks such as gas oil, a co-product accounting system, where feedstock costs are prorated among all products, is usually used. (SRI Report, Volume I, Page 97, notes: "At the moment no adequate cost accounting methods have been developed for chemical refineries of the kind represented by naphtha and gas oil cracking facilities." The SRI Report also warns one on the "fallaciousness of a costing methodology based on a subtraction of co-product credits from total manufacturing costs.) The treatment used here has been to consider residual gas value as a by-product credit reducing gross raw material costs. Then all other materials are treated as co-products of ethylene manufacture.

Details of this evaluation are given in Table I on costs of ethylene manufacture. It demonstrates that feedstock costs comprise 53% of total costs in cracking ethane and 85% of total costs of petrochemicals when cracking gas oil. Table II shows the material balances for these two cracking operations.

Tables I and II referred to above are reproduced here as Tables 1.10 and 1.11. It should be noted that no credit has been taken for pyrolysis gasoline or other energy products; all products have been assumed to be petrochemicals except fuel gas.

The oil industry has tended to ignore this aspect of the controversy. At most, a few industry spokesmen have attempted to point

TABLE 1.10
Costs of ethylene manufacture

(Data: SRI Report, Volume II, page 77)

	Ethane		Gas oil	
Feedstock, ¢/gal	3		8	
	$ million/yr	%	*$ million/yr*	%
Feedstock	12.000		91.900	
Less residual gas	1.140		4.400	
Net feedstock	10.860	53	87.500	85
Direct costs				
Utilities	4.500		8.000	
Catalyst & chemicals	0.631		1.000	
Labor & supervision	0.572		0.700	
Maintenance	0.770		1.000	
	6.473	32	10.700	10
Indirect costs				
Depreciation	2.370		3.000	
General overhead	0.474		1.500	
Taxes & insurance	0.237		0.300	
	3.081	15	4.800	5
Total costs	20.414	100	103.000	100
Total MM lb products (Table 1.11)	1075		7750	
Unit product cost, ¢/lb	1.9[1]		1.33[1]	

[1] These data do not include sales, research, administration or other corporate expenses, nor income taxes or profit on the investment.

TABLE 1.11
Cracking products

	Ethane[1]		Gas oil[2]	
	MM lb/yr	%	*MM lb/yr*	%
Ethylene	1,000	93	1,000	13
Propylene & higher	75	7		
Propylene	–		1,000	13
Butenes & butadiene	–		750	9
Pentanes & higher	–		5,000	65
	1,075	100	7,750	100

[1] As residue gas was 150 MM lb, about 82% of the feedstock was converted to ethylene.

[2] Residue gas 1,000 MM lb; 11% feedstock to ethylene.

out the relationship shown in Fig. 1.10. These flow charts suggest that the higher cost of domestic naphtha for chemicals might well be offset by the higher value of the larger quantity of energy products. This would tend to equalize the domestic and foreign manufacturing cost of the by-product chemicals.

The oil industry has tended rather to emphasize the future availability of plentiful supplies of domestic feedstocks, leading to the next argument.

Domestic feedstock availability and price

The Stanford Research Institute study, "Petrochemical Feedstock Imports and their Effects on the US Balance of Payments," stated as the first of three major conclusions:

> 1. Adequate supplies of domestic petrochemical feedstocks will be available in the United States to 1975 at slowly increasing – but internationally competitive – prices.

The chemical industry denies this emphatically, continuing to insist that the difference in domestic and foreign liquid feedstock costs is the difference in the price of crude oil, or approximately $1.25/bbl (3.0¢/gal). Thus, US feedstock costs for petrochemicals is at least 60% higher (5.0¢/gal foreign naphtha, plus 3¢/gal) than foreign costs. On this basis, crude-oil import allocations are essential to equalize the economic competitiveness of domestic and foreign manufacture.

The real issue then may boil down to the actual future availability and price of domestic naphtha. But since the publication of the SRI study, another factor has been introduced to complicate this picture even more. Acceleration of the ecological goal to remove lead from gasoline might have greater and further reaching impact in many areas than anyone had ever anticipated.

Bonner & Moore Associates, Inc., among others, have pointed out in various studies that the current gaoline pool contains 10 to 15% highly paraffinic, low-octane components. These include straight run gasoline, natural gasoline, and paraffinic raffinate resulting from the manufacture of aromatics. Such materials require so much lead to raise their octane level to satisfy engine requirements as to force their removal from even a low-lead gasoline pool of the future.

Some forecasts suggest that as much as 500,000 bbl/day might be rejected from the gasoline pool over the next few years in providing the no-lead gasoline goal currently mentioned for 1975. Since these reject streams are ideal feedstocks for petrochemical manufacture,

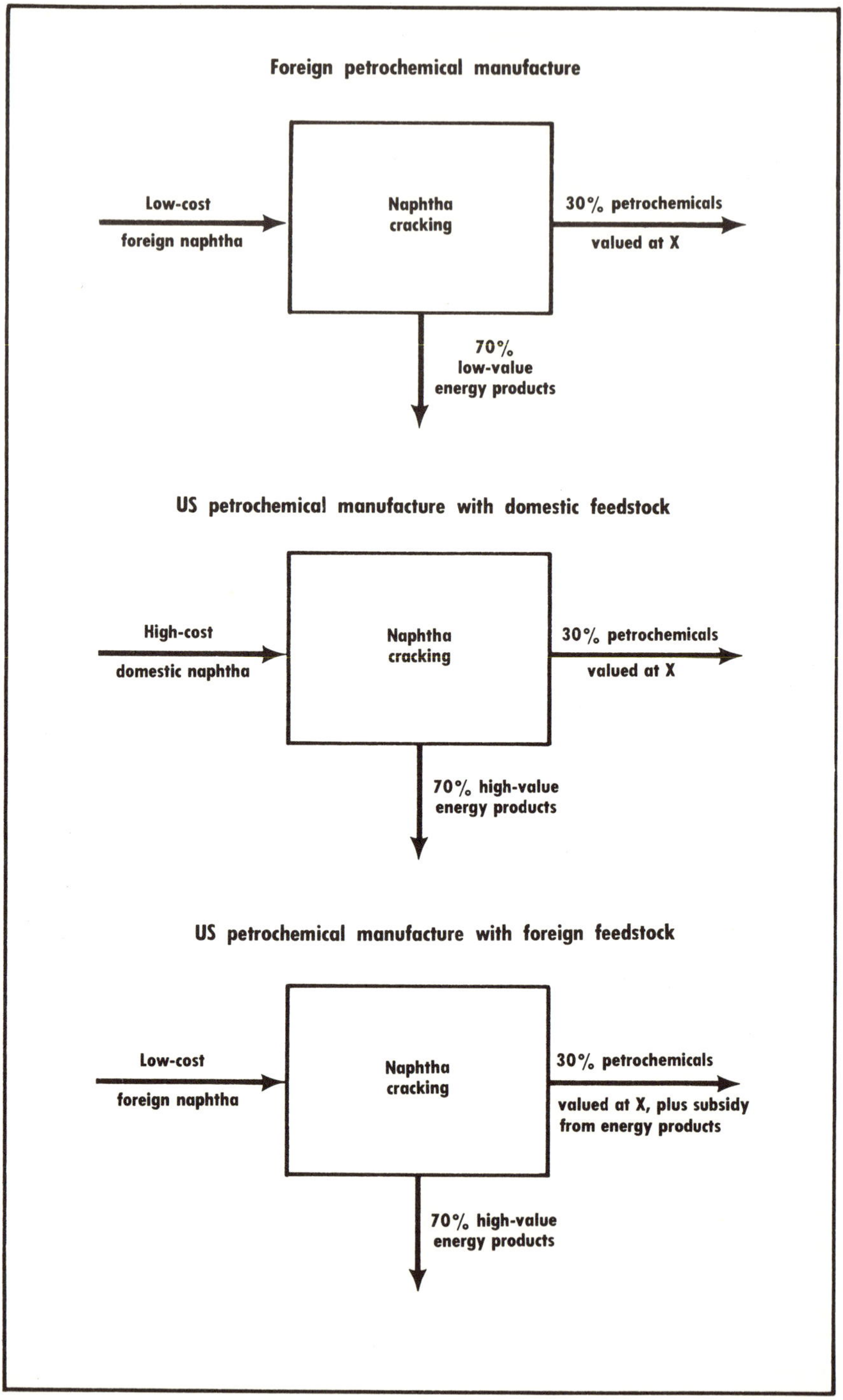

FIG. 1.10 A possible view of the relationship of foreign and domestic petrochemical manufacture

such a development would not only more than satisfy projected chemical demands through the late 1970's, but also might result in sufficient surplus to push domestic naphtha prices down to the level of the landed price of quota-free foreign naphtha.

The chemical industry, on the other hand, has suggested that such a high volume cannot possibly be spared from the gasoline pool, but must be used to manufacture high-octane aromatics and olefin alkylates for return to the pool. In fact, it is stated, the demand for aromatics and olefins for such octane improvement will be so great as to force price increases in these petrochemical feedstocks even further, degrading their economic competitiveness with foreign chemical manufacturers.

Insufficient time has elapsed to allow accumulation of all the data and the necessary studies to permit definitive conclusions in this area at this time. The results of such studies could provide the key to petrochemical aspirations and should be watched closely to perceive the earliest indications of future petrochemical quota developments.

Balance of trade

The chemical industry in this area maintains that it is the unintended victim of an oil-import program which resulted from concern over adequate domestic supplies of oil for energy purposes. As a result of the oil-import program, the oil industry operates in a protected environment. Imports of foreign petroleum products, as well as raw materials, are controlled, even though domestic petroleum products cannot compete in world markets because of higher prices. In contrast, the chemical industry must compete in world markets. Petrochemical imports are not restricted, although imports of petrochemical feedstocks are controlled.

The chemical industry submissions to the Cabinet Task Force on Oil Import Control included another study by Arthur D. Little, Inc., "Trade Trends in Petrochemicals," June, 1969. This study details an imposing array of statistics to document the leveling off of the US balance of trade in petrochemical imports since 1964. As shown in Fig. 1.11, abstracted from this study, US petrochemical imports have risen much faster than exports in recent years. Fig. 1.12 shows the declining US share of the total world export market from 25 to 20% during this same period. The study concluded that unless US industry can remain competitive with foreign plants, the US trade balance will continue to decline. The chemical industry, of course, blames feedstock cost.

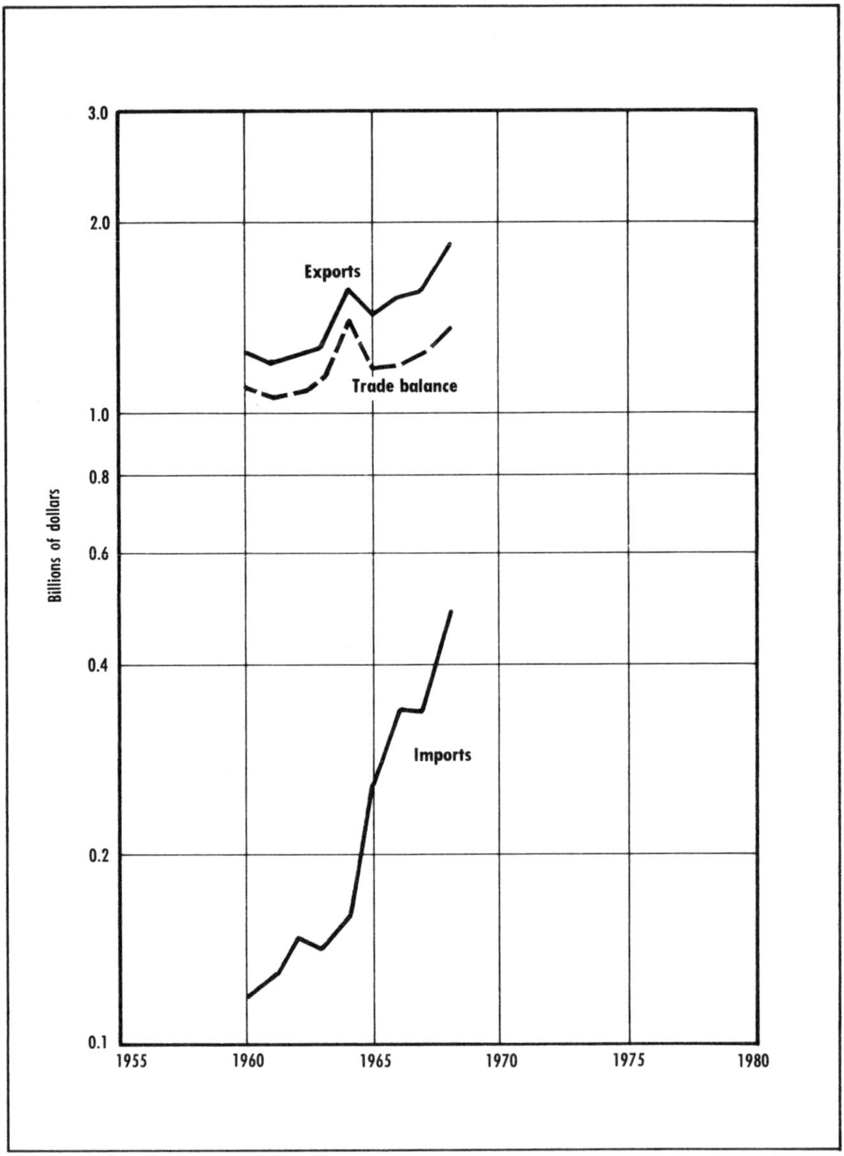

FIG. 1.11 US balance of trade in petrochemicals

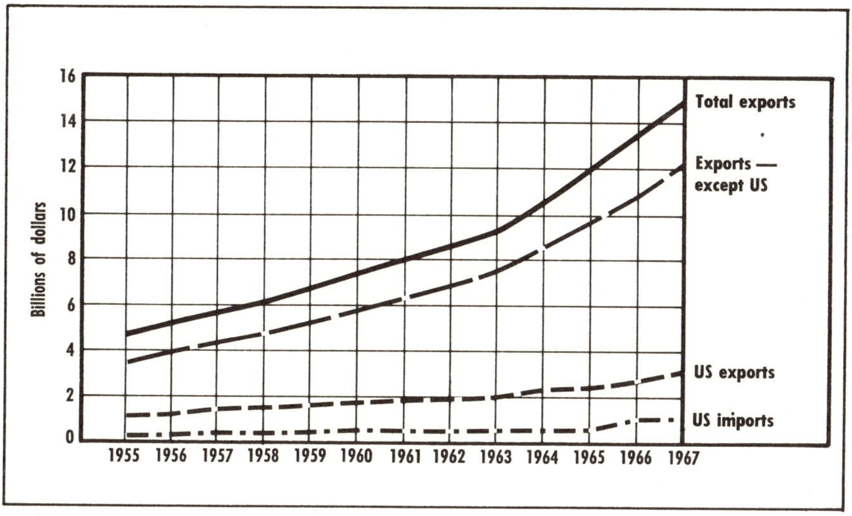

FIG. 1.12 The declining share of world chemical exports represented by US exports

One oil industry response to this type of argument was presented by Sun Oil Co. in its "Rebuttal Statement to the Cabinet Task Force on Oil Import Control," August 15, 1969:

Benzene is an excellent example to use for detailed examination of the alleged problem of low-cost foreign feedstock. Benzene is the most important member of a family of petrochemicals called aromatics.

Petroleum-based aromatics, including benzene, can be made only from naphtha. Considerable quantities of aromatics, including benzene, are made in the United States using domestic naphtha valued at 8–9¢/gal. The price differential between domestic and foreign naphtha has ranged between 3–5¢/gal during most of the past decade of oil import control. In addition, benzene is one of the few petrochemicals which can be imported duty-free into the US for chemical manufacture.

So, if the argument is valid that the use of low-cost feedstock results in low-cost petrochemicals, then the dire pattern projected for all petrochemicals in the future should already have shown up in benzene. Foreign benzene prices should have rapidly declined due to lower naphtha prices abroad, and any US exports of benzene or benzene derivatives should have rapidly disappeared in the early 1960's. Subsequently, during the mid to late 1960's, the US should have been flooded with foreign benzene or benzene derivatives to the detriment of domestic benzene manufacturers. Tables I and II show otherwise.

Table I presents a history of European benzene prices, as reported in "European Chemical News," in comparison with list prices for the US from "Oil, Paint and Drug Reporter," as well as the unit value of benzene sales from the

TABLE 1.12

Benzene price history USA versus Western Europe

	US Tariff Commission unit values		*USA OPD price*	*European Chemical News price ¢/lb*				
Year	*¢/gal*	*¢/lb*	*¢/lb*	*UK*	*France*	*Germany*	*Italy*	*Belgium*
1963								
J			3.42	5.03[1]	4.07	2.83	3.99[2]	3.01
A	23	3.14	3.42	3.40[1]	4.0	2.8	4.0[2]	2.8
J			3.42	3.4[1]	4.0	3.4	3.6[2]	2.8
O			3.42	3.4[1]	4.0	4.5	3.6[2]	2.8
1964								
J			3.42	4.1[3]	4.0	4.5	3.6	3.3
A	23	3.14	3.42	4.1[3]	4.0	4.5	3.6	3.3
J			3.42	4.1[3]	4.0	4.5	3.6	3.3
O			3.42	3.3[3]	4.0	4.5	3.6	3.3
1965								
J			3.42	3.3[3]	4.0	4.5	3.3	3.3
A	24	3.28	3.42	3.3[4]	4.0	3.7	3.3	3.3
J			3.54	3.7[4]	4.0	4.3	3.4	3.3
O			3.67	3.7[4]	4.0	3.6	3.6	3.6
1966								
J			3.67	3.7[4]	4.0	4.0	3.8	3.6
A	24	3.28	3.67	3.7[4]	4.0	3.8	4.0	3.6
J			3.67	4.2[4]	4.0	4.0	4.2	3.6
O			3.67	4.2[5]	4.0	3.8	4.2	3.6
1967								
J			3.67	4.2[5]	4.0	4.0	4.2	3.6
A	24	3.28	3.67	4.2[5]	4.0	4.1	4.2	3.6
J			3.66	4.2[5]	4.0	4.2	4.2	3.6
O			3.66	4.2[5]	4.0	4.1	4.2	3.6
1968								
J			3.66	3.63[6]	4.0	4.0	3.8	3.6
A	–	–	3.66	3.63[6]	4.0	3.7	3.6	3.6
J			3.66	3.86[7]	4.0	3.7	3.6	3.4
O			3.66	3.86[7]	4.0	3.6	3.2	3.4
1969								
J			3.11	3.86[8]	4.0	3.6	3.2	3.3
A	–	–	3.11	3.86[8]	4.0	3.5	3.2	3.3
J			3.11	3.86[9]	4.0	3.5	3.4[10]	3.3
O			–	–	–	–	–	–

OPD–Oil, Paint & Drug Reporter.

[1] UK price excludes tax and includes rebate.

[2] Italian price does not include 3.3% sales tax.

[3] UK price excludes hydrocarbon duty of 2s 9d/gal.

US Tariff Commission. European benzene prices have historically been the same as or slightly higher than US prices.

Table II shows the historical import-export records for benzene and several important benzene derivatives, based mainly on US Department of Commerce figures. From a net import balance in 1959, exports of these benzene chemicals have increased steadily to equal, in benzene content, almost one-third of the total domestic benzene production in 1968. This table includes only some of the first derivative benzene compounds but excludes many important benzene-based end products such as nylon, styrene-based rubbers, and plastics. Domestic benzene production during this period increased by nearly 300% from 2.48 billion lb/yr to 7.11 lb/yr."

Tables I and II have been reproduced here as Tables 1.12 and 1.13.

The implication, of course, is that some other cause than feedstock costs must be responsible for the increasing imports of petrochemicals and the leveling off in balance of trade.

The previously cited SRI study also included these major conclusions:

2. The net petrochemical export position of the United States will increase —irrespective of the level of feedstock imports—from a current level of about $1.3 billion to a high of $2.3 billion by 1975.

3. Changes in petrochemical feedstock import quotas from 1967–75 will not measurably affect the present positive trade balance for the US petrochemical industry.

These comments on feedstock prices were included in the summary of the report.

Many factors come into play in the export of commodities from one nation or group of nations to another. Feedstock prices—one of the many factors—do not now, nor will they in the future, have a significant effect on petrochemical exports, since (1) average feedstock costs are and will be in the same range in the various developed countries of the free world; and (2) feedstock costs are

---

[4] UK price excludes hydrocarbon duty of 3s 3d/gal.

[5] UK price excludes hydrocarbon duty of 3s 7d/gal.

[6] UK price excludes hydrocarbon duty of 3s 7d/gal, plus 2d surcharge.

[7] UK price excludes the 3s 11d/gal hydrocarbon duty.

[8] UK price excludes the 4s 3½d/gal hydrocarbon duty not levied on chemical feedstocks.

[9] UK price excludes the 4s 6d/gal hydrocarbon duty not levied on chemical feedstocks.

[10] Includes local taxes.

Basis: January, April, July, October issues of European Chemical News.

actually a small part of the total production and entry cost of an imported petrochemical. The most important factors affecting the exportability of petrochemicals are changes in supply and demand, tariff and non-tariff barriers, trading blocs, costs of production and transportation, and investment in overseas production facilities. A comparison of typical feedstock costs with other exportability factors for a number of US exportable ethylene-derived petrochemicals is shown in the tabulation on the following page. This tabulation indicates that differences in raw materials' prices of even 10–25% produce effects much smaller than other factors, such as tariff and transportation.

TABLE 1.13

US foreign trade of benzene and benzene derivatives

(MM lb)

	Benzene		*Cyclohexane*		*Cumene*	
	imports	*exports*	*imports*	*exports*	*imports*	*exports*
1959	403	51	Nil	NA	NA	NA
1960	278	176	Nil	46(43)[1,2]	NA	NA
1961	139	344	Nil	98(91)[1,2]	NA	NA
1962	154	300	Nil	130(121)[1,2]	NA	NA
1963	95	476	Nil	390(363)[1,2]	NA	NA
1964	117	637	Nil	520(484)[1,2]	NA	NA
1965	183	329	Nil	720(670)[1]	6(4)[1]	NA
1966	168	710	Nil	738(686)[1]	29(20)[1]	NA
1967	168	732	Nil	843(784)[1]	22(15)[1]	95(66)[1,3]
1968	241	608	Nil	822(764)[1]	6(4)[1]	95(66)[1,3]

	Styrene		*Ethylbenzene*		*Phenol*	
	imports	*exports*	*imports*	*exports*	*imports*	*exports*
1959	NA	80(70)[1]	NA	NA	NA	39(39)[1]
1960	Nil	158(137)[1]	NA	NA	Nil	48(48)[1]
1961	Nil	138(120)[1]	NA	NA	Nil	41(41)[1]
1962	Nil	123(107)[1]	NA	NA	.2(.2)[1]	12(12)[1]
1963	Nil	120(104)[1]	Nil	287(221)[1,4]	Nil	35(35)[1]
1964	Nil	342(298)[1]	NA	NA	Nil	40(40)[1]
1965	Nil	367(319)[1]	Nil	307(236)[1,4]	.03(.03)[1]	12(12)[1]
1966	9(8)[1]	360(313)[1]	1(Nil)[1]	250(192)[1,4]	9(9)[1]	8(8)[1]
1967	17(15)[1]	436(379)[1]	Nil	200(154)[1,4]	4(4)[1]	13(13)[1]
1968	11(10)[1]	539(469)[1]	NA	NA	NA	28(28)[1]

	Detergent alkylate		DDT		Total benzene equivalent		Benzene equivalent exports as % domestic benzene production	
	imports	*exports*	*imports*	*exports*	*imports*	*exports*	*actual*	*net*
1959	Nil	66(26)[1]	Nil	62(45)[1]	403	231	9.3%	–
1960	Nil	81(32)[1]	Nil	81(59)[1]	278	495	14.8%	6.5%
1961	Nil	94(38)[1]	Nil	78(57)[1]	139	691	17.3%	13.8%
1962	Nil	103(41)[1]	Nil	102(74)[1]	154	655	16.4%	12.5%
1963	Nil	82(33)[1]	Nil	88(64)[1]	95	1296	27.4%	25.4%
1964	Nil	82(33)[1]	Nil	85(62)[1]	117	1554	29.1%	26.9%
1965	Nil	137(55)[1]	Nil	90(66)[1]	187	1687	27.9%	24.8%
1966	Nil	164(66)[1]	Nil	91(66)[1]	205	2041	29.2%	26.3%
1967	Nil	175(70)[1]	Nil	82(60)[1]	202	2258	32.0%	29.2%
1968	Nil	288(115)[1]	Nil	109(80)[1]	255	2130	30.0%	26.4%

[1] Key to dual numbers: Figure in parenthesis is the actual quantity of benzene required to manufacture the derivative weight shown.
[2] Estimate from Stanford Research Institute.
[3] Estimate from Sun Oil Company.
[4] Estimate from the PACE Company.
[5] NA = Not Available.
Source: US Department of Commerce except as noted above.

The tabulation referred to is shown in Table 1.14.

TABLE 1.14

	Unit value (¢/lb)	*Feedstock requirement (¢/lb)*	*EEC* tariff (¢/lb)*	*Transportation† (¢/lb)*
Polyethylene	17–18	1.0	7.6‡	1.5
Ethyl alcohol	6	0.64	7.2	0.6
Ethylene glycol	9	0.7	1.8	0.6
Polyvinyl chloride	17	0.47	4.0	1.5

*European Economic Community.
†New York to Amsterdam.
‡4.0¢/lb if of non-US origin.

The chemical industry response to this was another Arthur D. Little, Inc., study, "Oil Import Quotas and the US Balance of Payments in Petrochemicals, 1965–1980," June, 1969. This was included as part of the Chemco submission to the Cabinet Task Force on Oil Import Control. This study concluded that free access to foreign petrochemical feedstocks would provide as much as a $1 billion benefit by 1980 in the US petrochemical trade balance, as documented in Table 1.15. Further details of this projected improvement were provided by the study as shown in Table 1.16.

The basis for this improvement was again reduced feedstock costs with no debit being considered, so far as could be ascertained, for foreign feedstock costs against the balance of payments. Such a debit could amount to as much as $500 million, or half the projected improvement for the 600,000-bbl/day increase due to decontrolled imports projected for 1980 by the Chemco submission.

TABLE 1.15
US balance of payments in petrochemicals
(Millions of dollars)

1975	*Most likely range*
With access	2,300–2,700
Without access	1,900–2,300
Difference due to access	400
1980	
With access	3,100–3,700
Without access	2,100–2,700
Difference due to access	1,000

Source: Arthur D. Little, Inc., estimates.

The chemical industry also commented on the SRI Tabulation of Exportability Factors of Some Major Ethylene Derivatives, Table 1.14, in response to questions from Senator Hruska to Kenneth H. Hannan at his previously mentioned appearance before the Antitrust and Monopoly Subcommittee. In a letter dated September 11, 1969, to Senator Philip A. Hart, Chairman of the Subcommittee, Hannan attached the original and a corrected version of the tabulations as shown in Table 1.17. The 31.5% feedstock cost claimed by the chemical industry is based on a 3¢/gal differential in naphtha price. The intent, of course, is to show that higher feedstock cost is significant and

TABLE 1.16
Projected US petrochemical trade flow to 1980

	1975 Most likely range		1980 Most likely range	
	Current quota	With access	Current quota	With access
Exports of				
Organic chemicals	1470–1550	1590–1670	2230–2340	2550–2650
Plastics	840–870	890–920	1390–1450	1540–1610
Fibers	180–200	200–220	250–280	270–300
Elastomers	230–250	250–270	290–320	330–360
All other	160–180	190–210	180–200	230–250
Total	2880–3050	3120–3290	4340–4590	4920–5170
Imports of				
Organic chemicals	610–650	520–560	1100–1200	800–890
Plastics	210–230	200–220	470–510	420–460
Fibers	240–260	240–260	600–630	580–610
Elastomers	70–80	60–70	140–160	120–140
All other	150–170	140–160	290–330	240–280
Total	1280–1390	1160–1270	2600–2830	2160–2380

Source: Arthur D. Little, Inc., estimates.

of the same order of magnitude as the other barriers to export trade highlighted by the SRI study. Reference must be made to Chapter 6 of the SRI study for detailed discussion of the many other factors affecting exportability.

THE CURRENT SITUATION AND OUTLOOK FOR THE FUTURE

Our review of the petrochemical industry and the oil import program has finally reached an impasse. Regardless of one's individual bias for or against increased petrochemical quotas or even free access, the next move is up to the Oil Policy Committee.

Only a nominal increase in petrochemical quota to a level of about 98,000 bbl/day was allowed by the Oil Import Administration for 1971, and it appeared that the chemical industry had lost the battle when the Oil Policy Committee announced in March, 1971, that similar nominal increases would continue through 1973 before any final decision would be made.

Subsequent activity behind the scenes in Washington must have been fast and furious. This decision was reversed in mid-August, 1971, by an announcement that an import for export policy would be inaugurated for 1972 to provide an increase in petrochemical quota

TABLE 1.17

Original and revised tables for SRI exportability factors

Product	Unit value ¢/lb	EEC[1] tariff ¢/lb	Transportation[2] ¢/lb	Total ¢/lb	Feedstock requirement ¢/lb	20% of feedstock requirement ¢/lb
Polyethylene	17–18	7.6[3]	1.5	26.1–27.1	1.0	0.2
Ethyl alcohol	6	7.2	0.6	13.8	0.64	0.13
Ethylene glycol	9	1.8	0.6	11.4	0.7	0.14
Polyvinyl chloride	17	4.0	1.5	22.5	0.47	0.09

[1] European Economic Community.

[2] New York to Amsterdam.

[3] 4.0 ¢/lb if of non-US origin.

Cf. Stanford Research Institute, op cit, Vol 1, p 5.

Product	Unit[1] value ¢/lb	EEC tariff ¢/lb	Transportation[2] ¢/lb	Total ¢/lb	Feedstock requirement ¢/lb	31.5%[3] feedstock requirement ¢/lb
Polyethylene	12.0	5.4	1.4	18.8	2.23	0.70
Ethyl alcohol	6.0	7.2	0.4	13.6	1.22	0.38
Ethylene glycol	7.0	1.2	0.4	8.6	1.51	0.48
Polyvinyl chloride	10.5	2.2	1.4	14.1	1.10	0.35

[1] Current domestic sales price (SRI figures reflected 1965 prices).

[2] New York to Amsterdam or Antwerp.

[3] Differential feedstock cost of domestic naphtha at 9.5 ¢/gal vs. internationally priced naphtha at 6.5 ¢/gal.

variously estimated at 35,000 to 50,000 bbl/day. In addition, it was proposed that free access to liquid feedstocks, in proportion to the amount converted into petrochemicals, would become available to all liquid-feed petrochemical plants upon completion of the first liquids plant whose construction was started after August 12, 1971. However, it seems extremely hazardous to attempt any guess as to what may happen in 1972 and the next few years.

Longer range, strange as it may seem, most forecasters seem more willing to take a stand. And the forecasts are unanimously in one direction—up. US domestic-oil production in the lower 48 states is projected to peak at somewhere near 12 MM bbl/day by the year 1975. Estimates of the total imports required to make up the deficit between demand and supply may be placed at about 6.3 MM bbl/day in 1975, and 11.3 MM bbl/day in 1980.

Contingent upon how much of these imports may be Canadian overland, waterborne imports could range anywhere from 5 MM to 10 MM bbl/day, as compared to the current volume of about 3 MM bbl/day (including residual fuel). There now seems to be little doubt but that chemicals will share in this growth to the benefit and well-being of all.

BIBLIOGRAPHY

Anderson, E. V., "Petrochemical Feedstock," *Chemical and Engineering News,* June 15, 1970, pp. 18–27.

Burke, D. P. and Miller, R., "Ethylene, CW Report Part 1," *Chemical Week,* October 23, 1966, pp. 64–74.

"Effect of Naphtha Imports on the Competitive Position of the US Petrochemical Industry," Arthur D. Little, Inc., August, 1965.

"European Naphtha Demand," Arthur D. Little, Inc., August, 1966.

Freilling, J. G., King, C. C., and Newman, J., "Ethylene Raw Materials and Production Economics," Joint Meeting of the American Chemical Society and the Chemical Institute of Canada, Toronto, Ontario, May 28, 1970.

Hannan, K. H., "Oil Imports and the Petrochemical Industry," Testimony before the Antitrust and Monopoly Subcommittee of the Senate Judiciary Committee, July 22, 1969.

"Joint Submission by the Chemco Group to the Cabinet Task Force on Oil Import Control," June 16, 1969.

"Laws, Regulations and Other Information Relating to Foreign-Trade Zones in the United States," Rev., June, 1966, US Government Printing Office.

Lichtblau, J. H., "Current and Proposed Import Allocations to the Petrochemical Industry," Memorandum to the US Oil Imports Administration, July, 1966.

Malone, J. C., Perry, R. C., and Shoenberger, R. S., "International Competition in Petrochemicals," American Institute of Chemical Engineers, 61st Annual Meeting, Houston, Texas, February 21, 1967.

"The Mandatory Oil Import Control Program and the Petrochemical Industry," Submitted to the Cabinet Task Force on Oil Import Control by the Allchem Group, July, 1969.

Oil Imports Digest, Vol. No. 1, National Petroleum Refiners Association.

"Oil Import Quotas and the US Balance of Payments in Petrochemicals," Arthur D. Little, Inc., Rev. June, 1969.

"Petrochemical Feedstock Imports and Their Effects on the US Balance of Payments," Stanford Research Institute, September, 1967.
"Petrochemical Industry Proposal to the Department of the Interior," The Chemco Group, July 14, 1967.
"Proposed Chemical Regulations," Memorandum to the Oil Policy Committee by the Chemco and PetroChem Groups, May 22, 1970.
"Rebuttal Statement to the Cabinet Task Force on Oil Import Control," Sun Oil Company, August 15, 1969.
Snedeker, T. C., "Mechanics of the Oil Import Program," National Association of Purchasing Agents Meeting, Philadelphia, Pennsylvania, February, 1967.
Spencer, F. J., "Foreign-Trade Zones – Loopholes on Gateways?," *Hydrocarbon Processing,* Vol. 46, No. 1, January, 1967, pp. 161–166.
"The Oil Import Question," The Cabinet Task Force on Oil Import Control, February, 1970.
"Trade Trend in Petrochemicals," Arthur D. Little, Inc., June, 1969.

2

MYRON T. FOVEAUX

US chemical industry in world trade

THE life style of the US petrochemical industry is changing rapidly, just as it is for other major industries. It is critically important that industry executives develop new perspectives that exceed those now thought of as sophisticated. The industry must look far beyond having the best technology, the optimum location, the biggest plant, and the best plant startup system. It must increasingly focus on actions by governments, both US and foreign, that will often outweigh other factors.

These actions cover a broad range of governmental policy areas:

- The health of the US economy. The President's economic program is only the latest and strongest of economy-changing programs of recent years.
- The US oil-quota system as it affects the comparative cost of heavy feedstocks in the US with those abroad.
- New government programs which mandate higher production costs, such as labor laws, pollution clean up, product safety and plant safety and health rules.
- Tariff and non-tariff barriers are changed significantly affecting both imports and exports.

- Incentives or lack of incentives to investment.
- Effects of mandatory removal of lead from gasoline on feedstock supply and price.
- Stronger bargaining position on oil pricing by foreign governments.

All of these influences on the petrochemical industry are current and active over the past several years. Some are still hanging in the balance. It has become important for petrochemical companies to be completely conversant with customs laws, administration of the OIA petrochemical feedstock allocations, both primary and secondary effects of the Kennedy Round tariff cutting, and the Presidentially imposed import surcharge. It follows that it is important to maintain well-informed and effective Washington representation to assure a flow of accurate information and equitable treatment.

DEFINING PETROCHEMICALS

At the outset, the problem of defining the industry must be considered. The category of chemicals called petrochemicals is understood differently in different places. Added to this is the problem of loose usage. One should always question the meaning of the word as it is used in reporting data or facts about the industry.

In the government policy area, petrochemicals include a broad range of products including, surprisingly, man-made fibers. This is more an accident of history than a definition which is best for all purposes. However, it does emphasize for petrochemical people that products such as man-made fibers are major end products for what the industry considers to be petrochemicals. This should be turned around to say that the health of the petrochemical industry depends on the health of the major product sectors which it supplies. The depressed state of the man-made fibers industry from massive import competition and over-capacity has had a significant, adverse effect on the petrochemical industry.

The industry itself has a range of definition of petrochemicals. Sometimes the first-line raw materials and monomers are called petrochemicals. On other occasions, it is broadened to include polymers and plastics. Unfortunately, the data collected on the "petrochemical" industry are not on a base common to other data.

For those intending to pursue petrochemicals into the US Tariff Schedules, the term includes both straight-chain and aromatic chemicals. The aromatics are starting materials for products classified as benzenoids which, if imported, are classified under the American Selling Price section of the US Tariff Schedules.

References will be made to both the chemical industry and the petrochemical industry, and the terms will not be used interchangeably.

A PROFILE OF THE INDUSTRY

A profile of the chemical industry which highlights trends and comparisons is worth studying. Table 2.1 includes data for the chemical and allied products industry. Although the major sectors of the industry show different trends, the state of the petrochemical industry is indisputably related to the parent industry.

TABLE 2.1
Chemical and allied products industry
1965–1970

	1965	*1966*	*1967*	*1968*	*1969*	*1970*	*Average annual growth*
Shipments ($ billions)	37.5	40.8	42.3	46.4	48.7	49.6	5.8%
Net income ($ billions)	3.2	3.5	3.26	3.53	3.59	3.43	1.4%
Capital							
Expenditures ($ billions)							
(Domestic)	2.73	3.26	3.06	2.83	3.10	3.44	4.8%
(Foreign)	0.86	1.04	1.21	1.21	1.11	1.35	9.4%
Ratio: domestic/foreign	3.17	3.13	2.53	2.34	2.79	2.55	
Research & development							
($ billions)	1.20	1.27	1.36	1.44	1.54	1.62	6.2%

Source: Department of Commerce, Bureau of the Census, FRB, FTC, SEC, National Science Foundation.

Some of the compounded annual growth rates reveal interesting trends. Shipments of the last five years are growing at four times the pace of net income. Net income is still at the level of five years ago, and, if expressed in constant value dollars, would be significantly less. The current level of net income is insufficient for an adequate growth rate based on retained earnings and depreciation. The result has been an unusually high level of chemical-industry borrowing which, at current high interest rates, is becoming an excessive burden against earnings.

The growth rate of foreign plant and equipment expenditure by US firms (9.4%) is almost double that of domestic investment. Research and development growth has been moderately good in dollar expenditure, but has actually declined when adjusted for the decreased value of the dollar.

United States chemical exports, including a large export volume by the petrochemical industry, have been a mainstay in a deteriorating United States balance-of-trade and balance-of-payments positions of the United States. Table 2.2 shows that even with an annual growth rate of imports (13.3%) much higher than the export growth rate (9.8%), the trade balance grew impressively to $2.4 billion in 1970.

TABLE 2.2
Chemical and allied products trade balance
(Millions of dollars)

	Exports	*Imports*	*Balance*
1965	2,402	778	1,625
1966	2,672	943	1,729
1967	2,803	958	1,845
1968	3,289	1,122	2,167
1969	3,383	1,232	2,151
1970	3,833	1,454	2,379
Growth rate	9.8%	13.3%	7.9%

This chemical trade surplus amounted to all but $320 million of the entire US merchandise trade balance. However, the disproportionately high import growth rate will soon operate to cause a reduction in the chemical trade balance. As Table 2.3 shows, the US share of world chemical exports has declined in five years by 2.8%.

The comparative growth rates of the US and foreign chemical industries should be of interest to industry and government alike. The US petrochemical industry seems likely to grow at an 8 to 9% annual rate, but the Japanese and European industries will grow much faster. The US share of petrochemical investment is now about 40%. However, only 17% of the free world petrochemical investment in the next 3½ years will be made in the US.

Although the United States has enjoyed a chemical trade surplus with all other countries and trade blocs, some important trends are underway. The comfortable trade surplus with Japan probably turned to a deficit in 1971. Other trade balances may shift in the next few

TABLE 2.3
*US share world chemical exports**

	Share
1965	24.7%
1966	24.6%
1967	23.8%
1968	24.1%
1969	21.7%
1970	21.9%

*US Department of Commerce.

years. US exports of resins have averaged 8 to 10% of total sales but are predicted to drop to perhaps 5% in 1971. International petrochemical companies will adjust operations to adapt to changing situations but must watch for US government measures designed to emphasize exports and minimize imports and, perhaps, to minimize foreign plant investment.

DIFFERENCES IN GOVERNMENTAL PHILOSOPHIES IN ECONOMIC POLICY

The petrochemical industry has been increasingly affected by differences between US and foreign governmental control and influences. Major foreign countries have implemented taxation, incentive, and trade policy devices that emphasize strength in both domestic and export markets. The United States has done essentially nothing until the promulgation of the President's economic program on August 15, 1971. It remains to be seen whether the US can achieve a better balance with its commercial competitors.

The world's major goods-producing countries have recovered from World War II, accumulated capital, and developed capabilities never faced before by US industries in both domestic and foreign markets. Unfortunately, the Kennedy Round tariff cuts worsened the situation.

Foreign governments have long practiced measures which aid exports and inhibit imports. Cash grants of 40% of new capital investment have been available. Depreciation of an entire facility in one year can be obtained. The indirect tax system used in the Common Market burdens the domestic consumer at a 10% to 18% level, but the foreign purchaser of the same item is not charged this tax because the exporter is rebated or forgiven the tax. In some countries, special inducements to new plants include forgiveness of taxes for a specific period of time and worker training costs are sometimes completely paid by the government.

The United States has done none of these, but is apparently moving to new and important policies in the President's economic program. The temporary 10% surcharge offsets much of the Kennedy Round tariff cutting, but its effect along with that of currency revaluations will not be apparent for some time.

While the United States has been pursuing and leading a free trade policy on a multilateral basis, a polarization of major nations is taking place that is discriminatory to the US. The Common Market has preferences or arrangements with 27 other countries. Additionally, preferences have been extended by the Common Market to less developed countries which may be a disadvantage to the United States. The US had, in 1970, proposed that the Common Market and the US jointly offer preferential customs treatment to the LDC's, but the Common Market proceeded unilaterally. For a number of reasons the United States has not moved to implement its own plan.

The EFTA, including seven European countries, is still a tariff-preference group but not as important to United States interests as the Common Market. It will become much less a factor now that the U.K. has joined the Common Market.

Canada and Japan also find themselves isolated, but Japan has practiced a mix of intense domestic protectionism and aggressive world selling that has resulted in the most astonishing industrial growth rates in the world. The petrochemical industry has been particularly fast growing, as noted earlier. The administrative guideline system in Japan is becoming an admired model for other countries. It is quite possible that the United States will ultimately come to a similar, lower level of national planning in the petrochemical industry at least partly as a result of Japan's success.

The most striking difference in governmental philosophies lies in the procedures by which policies, including trade policy, are determined and implemented. This has resulted in very important differences in policy and effects on trade.

The United States governmental system includes a complete forum of hearings, sometimes before more than one body on the same issue, where adversaries confront each other. Foreign interests are represented strongly and effectively in hearings before such groups as the US Tariff Commission, the Committees and Subcommittees of the Senate and House, the Trade Information Committee, and others. Thus, widely different viewpoints are expressed and thoroughly studied before decisions are made.

In doing this, the US is unique because no other country in the world does it. In the United States, industry is frequently considered to be the adversary of what is considered to be in the public interest. In contrast, United States firms find no forums abroad comparable to those of the United States. There is no opportunity to give testimony to legislatures in Europe, for example, nor to send representatives to its legislators. Also, there is no effective way for US industry to present its views about policy making to the Common Market.

Added to the vastly different opportunities to promote viewpoints is the difference in appeals to administrative decisions available in the courts. The United States Customs Court carries a heavy work load of cases arising from protests of importers against the US Customs Bureau treatment of imported goods. US exporters do not find the same opportunity in other countries.

There is a gradual awakening in the US, forced by its extremely serious international economic problems, that there must be a more "commercial" and aggressive attitude in trade matters. The President's economic program's eight recommendations include three which directly impinge on other countries. The effects will be developing for some time, including the legislative activity needed for parts of the program. It does seem clear that the US will be dealing more with its self-interest in mind, and this could affect petrochemical investment in many different ways.

KENNEDY ROUND

In 1962, the US Congress authorized the most sweeping tariff reduction negotiations in history in the Trade Expansion Act of 1962 (TEA). The Executive carried out the negotiations under the direction of a Special Representative for Trade Negotiations who reported to the President. This post was held during most of the Kennedy Round by the distinguished former Secretary of State, Christian A. Herter. After his untimely death and for the last months of the Kennedy Round, the Honorable William M. Roth served as the STR.

Members of the chemical industry were not in complete agreement on the tariff cutting proposed in the TEA. Some companies, sensitized by serious competition for certain products, were strong in predicting a lack of reciprocity in agreements made in other countries and an overall adverse effect. Other companies were neutral and unconcerned.

There followed four years of unbelievably complicated, "stop-and-go" bargaining. The writer was at the Geneva, Switzerland, scene of bargaining for the last six weeks of the negotiations for consultation with the US negotiators. After many crises, a final agreement was signed by all parties on June 30, 1967. The sweeping concessions exceeded all expectations. They involved over 60,000 different products and affected $40 billion in international trade.

Most industrial goods received a 50% tariff cut. The only exceptions were labor-intensive products such as footwear, watch movements, transistor radios, textiles, and apparel. Essentially nothing was done on non-tariff barriers with the exception that a new anti-dumping code was agreed to. It has since been implemented to a considerable degree by the countries involved, including the US.

The United States made its tariff cuts in five equal steps over a four-year span completed on January 1, 1972. With two of these cuts implemented by 1969, the distribution of petrochemical tariffs (man-made fibers excluded) showed that 40% of the tariffs were lower by 15% or more. The average duty was 17.2%, which was influenced both by $42 million of duty-free imports and by $19 million imports having 26% or greater tariffs (Table 2.4).

TABLE 2.4
*Distribution by tariff level of dutiable petrochemical products imported into the United States in 1969**

Tariff level			*Average duty for group*
Range	*% of total*		*%*
0–5%	4%		2.1
6–10	32%		6.8
11–15%	4%		14.4
		40%	
16–20%	21%		18.7
21–25%	23%		22.3
26% +	16%		34.0
		60%	
Total		100%	17.2

* Based on Department of Commerce figures for the amount of duty paid on 87% of dutiable petrochemical imports (excluding synthetic fibers).

Although there is a lag in the effect of reduced tariffs on import levels, US tariff reductions have become important to a number of chemical-industry customers in both beneficial and adverse ways. Chemical industry sales to the automobile industry and parts suppliers are affected by the increasing share of foreign automobiles (estimated at 16 to 18% in 1971) on US roads. Heavy volumes of man-made fiber textiles obviously affect sales of domestically produced petrochemicals. On the other hand, customers also are afforded cheaper raw materials.

The most contentious part of the Kennedy Round settlement was the chemical agreement. It was in two parts. The first was the general tariff-cutting agreement authorized by the Congress in which the US cut chemical tariffs by 50%, while Common Market countries and the UK cut tariffs by 20%. The second part of the chemical settlement was a supplementary agreement subject to the approval of the US Congress. It provided for additional 30% Common Market and UK chemical tariff cuts (to the 50% level) if the American selling price (ASP) system of valuation was eliminated.

The chemical industry strongly opposed Congressional approval of the supplementary agreement. It contended that although the 50% tariff cut would have a greater adverse effect on the industry, the loss of ASP was significantly large and unfair when added to the 50% tariff cut. The industry also criticized the original inequity of the 50% US cut for a 20% cut by the seven European countries.

At this time, ASP elimination has not been approved by the full Congress. This is due to a complex foreign trade policy situation for the US and the continuing influence of Senate Resolution 100 which warned that the US negotiators were not to bargain on ASP.

The effect of the Kennedy Round chemical tariff cuts are close to what the chemical industry predicted. The combination of the worsening of competitive factors, including comparative wage rates and productivity, have been reflected in both imports and exports. The chemical trade surplus discussed earlier has declined and is predicted to worsen, even though producers are making every effort to increase very low plant utilization through more exports.

An example of the US market effect of a typical duty reduction can be instructive. A typical tariff rate on benzenoid chemicals (comprising about 10% of total US chemical sales) imported in 1967 was 2.8¢/lb plus 18% of the US price (ASP) of the product. If there was a competitive product made and sold in the US and its selling price was 25¢, the duty would be 7.3¢/lb on the import. This duty, along with shipping and other delivery costs and profit, is added to the producer's netback on the product for the delivered price in the US.

The netback price, due to these added costs, is necessarily less than the netback to the domestic producer of the same product. For many products, the pre-Kennedy Round level of duty did not permit a satisfactory netback price to exporters. However, with the tariff on this product reduced one-half, the netback to exporters is increased by the same amount (assuming a constant US price). This increase has been significant enough to attract new imports, with benzenoid imports increasing at an average annual rate of 17.8% for 1968–70.

Product-by-product studies made by members of the industry shows that there will be a fast escalation of imports. However, the most damaging aspect of imports to US producers is not the volume of the market lost to importers. It is, rather, the damage done because of the lower price at which imports are sold and at which the domestic producer is thus forced to sell his share of the market.

The balancing side of the equation involving imports is that of exports. For a number of reasons, the chemical industry has predicted, for petrochemicals as well as other chemicals, that the growth rate of US exports would not match that of increased imports. For one reason, the 20% European tariff cuts have not significantly improved the competitive position of US chemicals. Products having export success due to uniqueness are unaffected by tariff cuts, of course. Another block to US chemical exports are the Common Market border equalization taxes, commonly called border taxes, Table 2.5. They have more than replaced the reduction of tariffs from the Kennedy Round.

TABLE 2.5
*Common Market border taxes**

Country	*Tax*
Belgium	18%
France	23
Germany	11
Italy	approx. 10%
Luxembourg	10
Netherlands	14

*Paid in addition to tariffs and on a value including duties paid.

The effect of the Kennedy Round on petrochemicals is less, perhaps, than for other sectors of the industry but is predicted to be adverse. Down-stream products, including man-made fibers, are experiencing heavy import competition. Over 10% of the US consumption of man-made fibers is now imported. Other customers have similarly been affected and the result is a lack of growth comparable to that abroad. This is one of the factors in the down turn of the growth rate of prominent petrochemicals.

US REMEDIES FOR IMPORT DAMAGE

The problem of excessive imports and adverse impact on a domestic industry is not new but is certainly at the highest level in history. The United States has developed a number of methods and procedures which have been used with varying degrees of success to mitigate the impact of imports. An understanding of these will become increasingly important to the petrochemical industry, partly because they will be important to customers.

Quotas

The United States can impose quotas under authority of the Trade Expansion Act of 1962 or under the Tariff Act of 1930. Quotas may be imposed where a foreign country imposes discriminatory restriction on US exports or engages in unfair methods of competition in the importation of articles into the US, where imports seriously injure or threaten to seriously injure US industries, or where the national security is threatened. Quotas can be of the type restricting quantities of imports at fixed levels, or assessing low tariffs to the specified level of imports and a higher tariff on all imports exceeding that amount. Quotas may be set unilaterally or they may be voluntary limitations worked out between governments.

The US imposes quantitative limitations on 67 categories, the most important of which are petroleum and products (the only products dealt with under the National Security Section 232 of the Trade Expansion Act of 1962), meat, sugar, dairy products, wheat, peanuts, steel (voluntary), cotton, and cotton-textile products. The Common Market and Japan have roughly equivalent quota barriers.

A number of industries have been actively seeking quotas on imports during the last few years. The shoe, man-made-fiber textile, steel, and other industries have failed thus far to achieve mandatory limitations on imports. They ask for quotas because a return to previous tariff levels, as a remedy, would not be sufficiently inhibiting to imports. Quotas are sought typically as a "last resort" effort because it means revealing details on the seriousness of the problem.

Without judgment here as to the propriety of the quota requests by US industries, it's clear that major customers of the petrochemical industry are experiencing problems which cause secondary but adverse sales effect on the industry.

Anti-dumping

Dumping is a world wide commercial practice occurring when producers export products at a lower price than that for which the product is sold in the domestic market. However, most governments do not take action against dumped products unless they cause a disruption ("serious injury" in the US) in a domestic industry. Between 1964 and 1970 the Treasury Department referred 60 cases to the Tariff Commission after determining that sales of products were made to the US at prices below those in the country of origin. The Tariff Commission found injury in eleven of these cases. The remedy is an anti-dumping charge assessed against the dumped product which is equal to the difference between the dump price and the home market price.

Dumping of petrochemicals in world markets, including the US, is common. However, most of the aggrieved companies cannot prove the very high degree of injury to the entire industry which is necessary for a favorable finding by the Tariff Commission. Hence, there have been relatively few charges of dumping against foreign chemicals in the US.

There are shifting trends in the Tariff Commission on finding injury in dumping of foreign goods in the United States. During 1970, the six-man Tariff Commission was finding affirmatively more frequently in dumping cases. However, in 1971, the pattern reversed to fewer affirmatives.

Countervailing duties

This is a remedy available under Section 303 of the Tariff Act of 1930 which can be used to offset subsidies on goods imported into the US. Complaints of suspected subsidies can be submitted to the Bureau of Customs, requesting an investigation. If the finding is positive, a duty equal to the subsidy is applied to the normal customs duty.

From 1934 through 1969, approximately 200 cases were processed, resulting in 34 countervailing duty orders. Seventeen are currently in effect, affecting such things as steel transmission towers from Italy and butter from Australia.

There has been little interest in countervailing duties in the chemical field. It could become important in the future if national planning in the major countries emphasize and subsidize exports even more than now. The difficulty in determining which of the many forms of export incentives constitute subsidy will not be easy to solve. The rebate, or forgiveness, of indirect tax on exports from the major European countries is a case in point. It is possible that the US will ultimately make greater use of this remedy.

National security

If imported articles are determined to be in sufficient quantity to impair national security, the Director of the Office of Emergency Preparedness can advise the President to increase tariffs or impose quotas. The authority is Section 232 of the Trade Expansion Act of 1962.

A number of industries have sought to invoke the provisions of this law, but only petroleum and petroleum products have been so treated. Quotas on importation of these products were imposed in 1959, and these were administered under the Oil Import Administration. The only regular allocation of crude and unfinished oils made outside the refining industry is given to the petrochemical companies as a uniform percentage of the qualified import feedstock.

The petrochemical industry is obviously affected by this limitation of imports. In appealing for access to, or increased quotas of, the cheaper feedstocks available to its competition abroad, a large sector of the US petrochemical industry believes that national security will not be impaired.

Escape clause

Under Sections 302, 351, and 352 of the Trade Expansion Act of 1962, tariffs can be raised, quotas can be imposed, international agreements can be negotiated, and trade-adjustment assistance can be provided to individual firms and groups of workers. Such action can be initiated by the filing of a petition with the Tariff Commission. This agency makes investigations to determine if the imports seriously injure, or threaten to seriously injure, a US industry. The Tariff Commission has a number of escape-clause applications under study. The chemical industry has rarely sought to invoke it. Of all the remedies available to those affected by imports, this provision, and that of anti-dumping, are the most likely to be used.

Unfair competition

Under Section 337 of the Tariff Act of 1930, there are provisions for relief from unfair methods of competition in the importation of items into the US. Appeals for relief to the Tariff Commission result in findings submitted to the President for final decision which could result in exclusion of the items concerned from importation. In over 20 years, there have been only two positive findings by the Commission, and the President has taken affirmative action once.

There is a changing mood, however, which may lead to a more frequent use of this device. The long-established practice of foreign cartels selling in the United States market, including petrochemicals, is coming under increasing criticism, and this may result in action.

NON-TARIFF BARRIERS

Non-tariff barriers (NTB's) can be defined as, "all hindrances to trade other than tariffs." They are frequently measures which impede imports in an unintended way. Very often incentives to domestic production serve as NTB's. The more important of the more than twenty categories of NTB's are:

Special charges applied like tariffs
Quantitative restrictions
Export subsidies
Governmental procurement
State trading enterprises
Special valuation procedures
Licensing arrangements
Industrial health and safety standards

The General Agreement of International Trade (GATT), the international organization providing the forum and machinery for resolving international trade problems, has compiled a large compendium of the world's NTB's. Most of these have been long established, but new ones are always in the development stage. They are almost without exception measures taken to solve national economic-social problems which have arisen and which existing tariff levels do not solve. For this reason, they will be more difficult to deal with and to eliminate in international negotiations of the future. Much post-Kennedy time and effort has been spent by the major countries to prepare for such négotiations under OECD auspices, although there is no present, US authorization in effect.

The chemical industry has been most concerned about NTB's since the early 1960's. Two groups of knowledgeable marketing representatives of US chemical companies presented a report on a large number of NTB's to the Department of Commerce in 1963. The Kennedy Round (1962–67) did not, however, result in NTB negotiations.

Probably the two most important non-tariff barriers to US chemical exports are the indirect tax systems in Europe (which incorporate border tax adjustments for imports) and the Japanese administrative guidance system. The 20% chemical tariff cut by the Common Market countries has been more than offset, the US industry believes, by the subsequent increases in border taxes in the Common Market countries.

Belgium, taking the most extreme case, has raised its border taxes from 4% to 18%. In the early 1960's, chemicals entered Belgium at a zero tariff and a 4% border tax. It was essentially a free trade country. Today, tariff rates average about 12% (the Common Market external tariff) and a border tax of 18% is applied to the duty-paid value of the chemical. Although internal taxation on the same chemical has also gone up, the US chemical industry believes that it has been disadvantaged in markets where border taxes have been significantly increased.

The Japanese administrative-guidance system represents a highly effective coordination between governmental trade policy and Japanese commercial interests. This amounts to probably the most advanced state of national economic planning in the world. Essential to such a system are the strong limitations put by Japan on imports and capital investment from abroad.

The US also has many non-tariff barriers. Quotas have already been discussed. The next and most contentious of the NTB's is the American Selling Price system of valuation. It is considered so because for benzenoid chemicals (amounting to 10% of US chemical sales) which are imported and competitive to a US-made product, duties are assessed on the US selling price of the competitive product and not on the usual export value.

Non-tariff barriers will be the focus of future international negotiations, and they will be more difficult to negotiate than tariffs. The prediction is that NTB's will not be decreased significantly, and may increase. There is not likely to be the free inter-country movement of petrochemicals that management sometimes pictures as the way of the future.

POLLUTION CONTROL COSTS

The most important new development that will significantly increase costs for US petrochemical producers is the drive by federal and state governments to protect the environment. New legislation has been passed by the Congress calling for very stringent measures to prevent pollution of air and water. 1971 saw the creation of the Environmental Protection Agency (EPA) and the first administration of the new laws.

The cost of anti-pollution efforts includes the installation of nonproductive equipment and expensive research and development. Companies have put teams of competent personnel on pollution control, thus making them unavailable for other pressing needs.

Chemical-industry firms are affected not only by anti-pollution efforts in its own plants but also by the effect on its customers. A McGraw-Hill survey shows that the total industrial investment for 1971 pollution-control costs was over $12.5 billion. The steel industry, for one example, made 12% of its total investment in anti-pollution equipment.

The chemical industry spent about $260 million for pollution control in 1971. Investment for this purpose is predicted to grow to 7.7% of total investment in 1972 and perhaps as much as 10 to 12% thereafter. In Dow's Midland, Michigan, plant, $8.5 million, or 23% of the total investment there, was for pollution control in 1971. There are over 4,500 vents from that installation!

To some degree, the industry will benefit from the sale of anti-pollution chemicals and of pollution-control systems and equipment. However, the net effect of the anti-pollution efforts is likely to be a new, serious cost burden.

The anti-pollution measures required in plants of the future will cause some serious thinking. Often the technology required to meet federal standards is not available. There is uncertainty that anti-pollution systems which are approved will be considered satisfactory within a few years. In many situations, the standards call for almost zero emissions from plants to air and water, including no change in temperature. It remains to be seen whether chemical plants can be built to meet these standards.

At the present time, the manner of administration of the anti-pollution laws is not affected by the degree of adverse economic impact. A number of companies have shut down plants because the required anti-pollution investment cannot be justified. There is concern about the philosophy expressed in President Nixon's economic report of February, 1971. In it, there is a section on safeguarding the environment.

"New rules," it says, "for use of the environment are bound to effect competitive relationships in and among industry, localities, and nations. As industries are forced to bear the costs of using the environment, those who have high costs will lose part of their market to those with lower costs of using the environment."

It continues in a later paragraph, "Our high level of material wealth has caused us to place a higher value on clean air and water than they are assigned in countries which have lower incomes or where clean air and water may still be abundant." It goes on to say that the resulting reallocation of production among nations should benefit all nations.

William D. Ruckelshaus, Administrator of the Environmental Protection Agency, is quoted in Fortune Magazine:

> In the Clean Air Act I'm mandated by Congress to set a standard that will protect public health. It does not say protect public health if we can afford it or if the costs are not exorbitant–it says protect public health. So to that extent, there is built into the laws the factor of ignoring the cost.

The likely result of the US anti-pollution activities will be an accelerated growth of petrochemical investment abroad. Although there is a growing concern in Japan and Europe about pollution, it is expected that there will be considerable lag in required anti-pollution measures as compared to those in the United States. It is important also that the US recovery of pollution control costs must be achieved, if at all, through the pricing system. In other countries, recovery of these new costs will probably be achieved by other means, including government support through taxation and other methods which spread the cost throughout the economy.

Cost factors mandated by the government, other than pollution control, will be those resulting from the newly passed Occupational Safety and Health bill and the cost of complying with the forthcoming Toxic Substances Act. If it is possible to recover these significant new costs in prices, the petrochemical industry will grow and thrive. If not, there will be serious difficulty in justifying capital flow into US investment.

MULTINATIONAL COMPANIES AND LABOR UNIONS

The US petrochemical industry is made up largely of multinational companies (MNC's), including oil companies and the integrated chemical companies. Foreign investment by these firms, as noted earlier, grew at an annual rate of 9.4% during the last five years. However, the share of foreign plant and equipment investment made by the chemical industry has actually fallen from 23 to 20% (1966–70) of that for all US manufacturing firms.

There are strong indications in Washington that multinational companies will experience discouragements in new governmental policy. The desire of the petrochemical multinational firm to freely shift products between its plants in different countries will probably not come to pass. Members of Congress and the administrative agencies are more frequently asking how construction of a foreign plant helps the US.

US labor unions are increasingly resentful of the internationalization of US companies and are moving strongly to block further expansion abroad. The AFL–CIO has played a leading part in attempting to restrict such expansion. The reasoning is that jobs created abroad are jobs eliminated at home. They see no benefit to American labor from any favorable balance of payment effect or beneficial effects accruing to individual companies.

The AFL–CIO presents its views on MNC's at every opportunity to Members of Congress and others. The pressure will grow. The form of restrictions which may come would involve such things as a change in present rules of taxation of foreign earnings or tighter restrictions on foreign investment.

Activity against multinational corporations is growing also at the international union and local union level. The International Chemical Workers Union passed a resolution in 1970 calling for chemical companies to place a union member on the Board of Directors to prevent foreign investment and the presumed transference of US jobs to plants abroad. Other international unions in the industry are considering similar efforts. There may be labor union efforts to include restrictions on imports in management-union contracts as they come up for renewal with individual companies.

The labor-union effort against multinational companies was evident in 1970 when they were responsible for the elimination from a Senate bill of a provision called the Domestic International Sales Corporation (DISC). This bill never came to a vote in the Senate. The DISC provision permitted a deferral of taxation of income from exports until the profits were returned to the parent company. The unions reason that this was a windfall device which would allow companies to invest more money abroad without increasing exports.

However, in December, 1971, the Administration-introduced Revenue Act of 1971 became law, and it included a DISC provision which defers taxes on 50% of export earnings. The unions were credited with forcing a cutback from the 100% deferral in the original bill.

Multinational companies should be aware not only of the possible adverse action by the US government but also be prepared for possible adverse action by foreign governments.

SUMMARY

The US government will play an increasingly important part in the degree of success of the petrochemical industry. It will do this through a broad range of policy-making, including import limitations on feedstocks, anti-pollution requirements, moves to bolster or slow the economy, and taxation incentives or disincentives. US chemical tariffs have been reduced 50% in five years, and the US will absorb a greater proportion of chemical imports. The adverse effect on pricing will be the most important effect.

The petrochemical industry must keep extremely well informed on developments in the federal government. The industry must coordinate action in Washington to achieve equity with all the other demands on the government. This calls for having the best possible representation in Washington.

3

FRANK W. LONG

Technology and markets of petrochemicals derived from synthesis gas

THE importance of synthesis gas to the petrochemical industry is much greater than one might at first realize. This can be seen by looking more closely into the broad areas of markets and technology that are influenced by the three major chemicals made from synthesis gas —ammonia, methanol, and oxo chemicals.

The goal of this chapter is to provide an overall picture of synthesis gas and these three derivatives.

We are in the middle of an "energy crisis" in the United States, with the supply of natural gas unable to keep pace with increasing demands. Accordingly, the price of natural gas has advanced sharply and is expected to continue to do so. The total picture is very complex, involving social, political, technical, and marketing factors. Therefore, no attempt will be made to predict here the exact changes to come in synthesis gas.

Whereas in the past, natural gas has been the nearly unanimous choice as feedstock for synthesis gas plants, other sources of hydrocarbons must now be considered and will undoubtedly be used in place of natural gas. This will directly affect the choice of location for new ammonia and methanol plants. The increase in price of natural gas and the use of more expensive alternate feedstocks will mean higher prices for chemical derivatives which in turn can alter the markets for these materials.

SYNTHESIS GAS

Background

The term, "synthesis gas," as used in the petrochemical industry refers to a mixture of carbon monoxide gas (CO) and hydrogen gas (H_2) in any proportion. It is further defined by a ratio, such as 2 to 1, or 1 to 1, to denote the ratio of volumes (or mols) of H_2 to CO. The 1-to-1 ratio is usually called water gas because this is the approximate ratio formed in the typical water gas generator by blowing coke with steam:

$$C + H_2O \rightarrow H_2 + CO \quad (1 \text{ to } 1)$$

Synthesis gas may also be produced by the high temperature reaction of steam on hydrocarbon feeds such as methane:

$$CH_4 + H_2O \rightarrow 3H_2 + CO \quad (3 \text{ to } 1)$$

Or by partial oxidation of a hydrocarbon such as methane:

$$CH_4 + \tfrac{1}{2}\, O_2 \rightarrow 2H_2 + CO \quad (2 \text{ to } 1)$$

Most of the synthesis gas for petrochemical production in the United States is derived from steam reforming of natural gas. While steam reforming is preferred for natural gas and naphthas, partial oxidation is preferred for the heavier feedstocks such as crude oil or residual oils.

Raw materials for synthesis gas

Natural gas is by far the most important feedstock in the United States for manufacture of synthesis gas for conversion to petrochemicals.

From a purely technical consideration, light hydrocarbons, naphthas, off-gases, and other hydrocarbons could be utilized, but the low price of natural gas makes it the most attractive in the United States. In Europe, however, steam reforming of naphtha has been the preferred method.

Many factors must be considered when choosing the hydrocarbon source. The amount of sulfur in the feedstock is important, since it must be removed to prevent poisoning of the catalyst. The ratio of carbon to hydrogen is important. A 1-to-2 ratio is optimum for methanol production, but since methane has a 1-to-4 ratio, additional carbon usually in the form of CO_2 must be added. Hence, availability of CO_2 is a consideration. On the other hand, ammonia production requires only H_2 and is a net producer of CO_2. Often methanol or urea plants are located next to ammonia plants for this reason.

The off-gas from a basic oxygen furnace contains mostly CO, with the rest CO_2 and N_2. This may be reacted with steam after iron dust is removed to form a synthesis gas suitable for methanol. The off-gas is low-cost, but the additional cost to purify and convert the synthesis gas is usually more expensive than a normal steam reforming plant. Location of markets for the end product may also be a factor, such as ammonia for fertilizer applications.

Natural gas

Natural gas is a mixture of the low-molecular-weight paraffin hydrocarbons methane, ethane, propane, and butane, but also may contain nitrogen, carbon dioxide, hydrogen sulfide, and sometimes a small amount of helium. The composition of natural gas will vary with each

TABLE 3.1
Natural gas—production
(Trillions of cu ft)

	Gross production			*Marketed Production**	*Value at well-head-¢/Mcf*
	Gas wells	*Oil wells*	*Total*		
1960	10.85	4.23	15.09	12.77	14.0
1961	11.20	4.26	15.46	13.25	15.1
1962	11.70	4.34	16.04	13.88	15.5
1963	12.61	4.37	16.97	14.75	15.8
1964	13.04	4.40	17.44	15.46	15.4
1965	13.52	4.44	17.96	16.04	15.6
1966	13.89	5.14	19.03	17.21	15.7
1967	15.35	4.91	20.25	18.17	16.0
1968	16.54	4.78	21.32	19.32	16.4
1969	17.49	5.19	22.68	20.70	16.7
1970	18.59	5.19	23.79	21.92	17.1

*Marketed production is gross production adjusted for gas used for repressuring underground reservoirs and that vented or flared.

Source: Bureau of Mines, US Department of the Interior.

location, but methane is normally the predominant gas. Processed natural gas from which the higher-molecular-weight natural-gas liquids have been removed is usually from 90 to 99% methane.

The heating value of natural gas is normally over 1,000 BTU/cu ft. About 80% of the natural gas is obtained from gas wells, while 20% is produced from oil wells concurrently with crude oil, from which it is separated. The data on gross production is adjusted for the gas used for repressuring underground reservoirs and that vented or flared. Because of the importance of natural gas for synthesis gas production, pertinent statistics are included in Tables 3.1 through 3.5.

TABLE 3.2
Natural gas—delivered to consumers

	1970	
Type	*Cu ft (Trillions)*	*Value (¢/Mcf)*
Residential	4.84	109
Commercial	2.06	82
Industrial	7.89	37
Electric utility	3.89	29
Others	.34	48
Total	19.02	59

Source: Bureau of Mines, US Department of the Interior.

TABLE 3.3
Natural gas—supply distribution and reserves
(Trillions of cu ft)

	1960	*1968*	*1969*	*1970*
Marketed production	12.77	19.32	20.70	21.92
Withdrawn from storage	.71	1.33	1.38	1.46
Imports	.16	.65	.73	.82
Exports	.01	.09	.05	.07
Stored	.84	1.42	1.50	1.86
Losses	.27	.32	.33	.23
Consumption	12.51	19.46	20.92	22.05
Proved reserves, end of year	262.3	287.3	275.1	290.7
Reserves added in year	14.2	13.8	8.4	37.2*

*Includes 26.0 new reserves in Alaska.
Source: Bureau of Mines, US Department of the Interior.

TABLE 3.4
Natural gas—consumption
(Trillions of cu ft)

	1960	*1970*
Residential	3.10	4.84
Commercial	1.02	2.06
Electric utility plants	1.72	3.89
Field uses	1.78	2.30
Petroleum refineries	.78	1.03
Pipeline fuel	.35	.72
Portland cement plants	.17	.20
Carbon black	.20	.08
Other (includes chemicals)	3.39	6.93
Total	12.51	22.05

Source: Bureau of Mines, US Department of the Interior.

Methods to manufacture synthesis gas

The two methods commonly used to manufacture synthesis gas for petrochemical operations are steam reforming and partial oxidation of hydrocarbon feedstocks.

In the United States, steam reforming of natural gas is preferred while in Western Europe steam reforming of naphtha is most important. However, partial oxidation is very versatile and can be used with any fluid hydrocarbon including high sulfur residual oils. Therefore, partial oxidation in the United States is becoming more important as the price of natural gas increases.

Steam reforming

In steam reforming, natural gas is taken from battery limits and passed through an activated carbon bed for sulfur removal. This gas is then mixed with steam, preheated, and fed to reforming tubes that contain a nickel based catalyst at 1,400 to 1,600°F. and 400 to 500 psi operating pressure. The resulting mixture contains mainly H_2 and CO in a 3-to-1 ratio.

When the synthesis gas is for ammonia production, the CO is converted to CO_2, which must then be removed with alkaline materials such as ethanolamine or K_2CO_3. When the synthesis gas is for methanol production, a 2-1 ratio is required, so that CO_2 must be added either before or after the reforming step to obtain the optimum ratio.

TABLE 3.5

Marketed production of natural gas by area

	1968			*1969*			*1970*		
	Quantity Bcf	*% of total*	*Average wellhead value ¢/Mcf*	*Quantity Bcf*	*% of total*	*Average wellhead value ¢/Mcf*	*Quantity Bcf*	*% of total*	*Average wellhead value ¢/Mcf*
Texas	7,495	(38.8%)	13.5	7,853	(37.9%)	13.7	8,358	(38.1%)	14.4
Louisiana	6,416	(33.2%)	18.9	7,228	(34.9%)	19.2	7,788	(35.5%)	19.3
Oklahoma	1,391	(7.2%)	14.2	1,524	(7.4%)	15.3	1,595	(7.3%)	15.6
New Mexico	1,164	(6.0%)	13.4	1,138	(5.5%)	13.7	1,139	(5.2%)	14.3
Kansas	836	(4.3%)	13.8	883	(4.3%)	13.9	900	(4.1%)	14.0
California	715	(3.7%)	30.9	678	(3.3%)	30.6	649	(3.0%)	32.1
Other states	1,305	(6.8%)	–	1,394	(6.8%)	–	1,491	(6.8%)	–
Total	19,322	(100%)	16.4	20,698	(100%)	16.7	21,920	(100%)	17.1

Source: Bureau of Mines, US Department of the Interior.

Partial oxidation

This process is essentially a thermal cracking of a hydrocarbon under partial oxidation conditions. Any liquid or gaseous hydrocarbon may be fed with oxygen to a combustion chamber operated at about 2,500°F. and pressures of 500 to 850 psi. The reaction product is first treated to remove H_2S and then processed in the same manner as with steam reforming.

With natural-gas feed, the synthesis-gas ratio is approximately 2-to-1, while heavy fuel oil will give a 1-to-1 ratio. Thus, one has the capability to produce either a 2-to-1 ratio for methanol production or 1-to-1 for oxo chemicals with the same equipment.

Natural gas requirements for synthesis-gas-derived chemicals

The relative importance of synthesis gas used for petrochemical production can be seen in Table 3.6, assuming that all production of synthesis gas is based on natural gas. The percentage of natural gas used for petrochemicals is very small compared to that used for fuel and energy. So, petrochemical usage can be expected to influence the overall picture of natural gas in only a small way. This can clearly be seen from the data in Table 3.6. Petrochemical producers will have to compete with the consumers of natural gas for fuel uses and pay the prices established for these end uses.

TABLE 3.6
Potential consumption natural gas—1970

	1970 production	*Natural-gas requirement*
Ammonia	13 MM short tons	390 Bcf
Methanol	750 MM gal	95 Bcf
Oxo synthesis	1,400 MM lb	5 Bcf
Total US consumption		22,050 Bcf

AMMONIA

Introduction

Ammonia is one of the world's most important commercially produced chemicals. It is used as the basic raw material for the many different forms of nitrogen-containing chemical compositions. These products and end uses include fertilizers, refrigeration, nitric acid, water treatment, synthetic plastics and fibers, animal feeds, explosives, rocket fuels, ore refining, and many others.

The importance of ammonia to the United States economy is shown by the facts that in 1967, ammonia's production value of over $750 million was greater than that for any other chemical, and total production of 12 million short tons was surpassed only by sulfuric acid.

Ammonia is a colorless gas with a characteristic pungent odor. It is very soluble in water, alcohol, and ether. Solutions of ammonia gas in water are commonly referred to as ammonia. The liquid form of ammonia more closely resembles water in its solvent properties than any other liquid. Ammonia properties are:

Chemical formula	NH_3
Molecular weight	17.0
Boiling point	−33.5°C.
Freezing point	−77.7°C.
Vapor pressure, liquid	8½ atm @ 20°C.
Critical temperature	133.0°C.
Critical pressure	1,657 psi
Nitrogen content	82.2%

The first synthesis of ammonia directly from its elements appears to be by Döbereiner in 1823. By the end of the 1800's it became clear that the world's supply of fixed nitrogen would soon be depleted, so considerable research and technical effort was put into efforts to develop a commercial synthetic ammonia process. World War I especially created problems for warring nations cut off from nitrate supplies in Chili.

Consequently, the first commercial synthetic ammonia plant using a process developed by Haber was operated by BASF in Germany in 1913, when 750 tons was produced. The principles of this process are used today in all synthetic ammonia plants (German Patent 235,421 was issued to Haber in 1908).

Manufacturing processes

Manufacture of all synthetic ammonia is carried out by passing the correct combination of nitrogen and hydrogen over an iron oxide catalyst promoted with aluminum or potassium oxide, at near 700°F. and about 250 atmospheres pressure, according to this equation:

$$N_2 + 3H_2 \rightarrow 2NH_3$$

Nitrogen is usually obtained by feeding air, but the sources of hydrogen are many and will vary considerably for each plant. In the

United States, most of the hydrogen feed comes from steam reforming of natural gas. The overall ammonia plant usually consists of four main parts—steam reforming, carbon monoxide shifting, carbon dioxide removal, and compression and conversion to ammonia. Fig. 3.1 is a typical flowsheet of an ammonia plant. The hydrogen required may come from a number of sources.

Preparation of hydrogen

Steam reforming of natural gas calls for natural gas and steam to be passed over a reforming catalyst to form a mixture of H_2 and CO. This mixture is passed through a second reformer utilizing a nickel catalyst. Air is introduced to give a 3-to-1, H to N mol ratio, and any residual methane is burned off.

This gas is then passed through two shift converters to convert the CO to CO_2 with steam. The first converter uses a chromium-promoted iron catalyst at 700°F., while the second uses a copper-zinc catalyst at 375°F.

$$CO + H_2O \rightarrow CO_2 + H_2$$

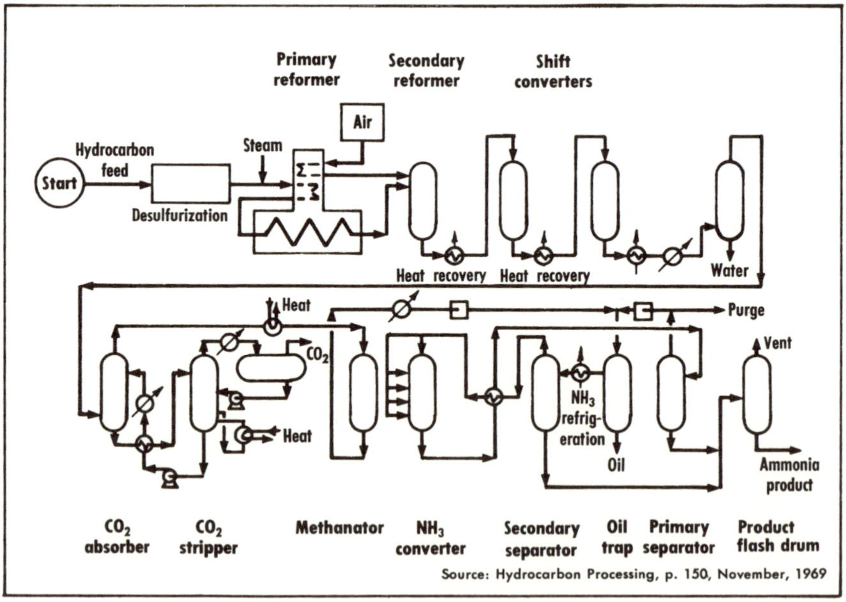

FIG. 3.1 Typical ammonia process

The synthetic gas is then treated with a solution of alkaline materials such as monoethanolamine, K_2CO_3, or others to remove the CO_2. A final step, methanation, removes any residual CO or CO_2 by reaction with hydrogen over a nickel catalyst at 600°F. to yield methane and water. A steam reforming flowsheet is also shown in Fig. 3.1.

Partial oxidation of hydrocarbons offers flexibility in the choice of feedstock as well as ability to change feeds. It is normally used on heavier hydrocarbons including crude and residual oils, coke-oven gas, and even solids, such as coal or lignite. Several companies have developed this process, which is continuous and non-catalytic under highly controlled operating conditions.

Quite often an on-site air separation plant is used to supply the oxygen as well as the nitrogen. Fig. 3.2 shows the flowsheet for Shell's process. An excellent article on this appears in *The Oil and Gas Journal*, p. 85, September 6, 1971.

Electrolysis is used to produce hydrogen for several plants, either from the electrolysis of water or from electrolytic cells for the manufacture of caustic soda and chlorine from salt brine. The hydrogen produced is then mixed with nitrogen in a 3-to-1 ratio.

Cryogenic separation is used in a few plants to recover hydrogen from coke-oven gases or hydrogen rich refinery gases.

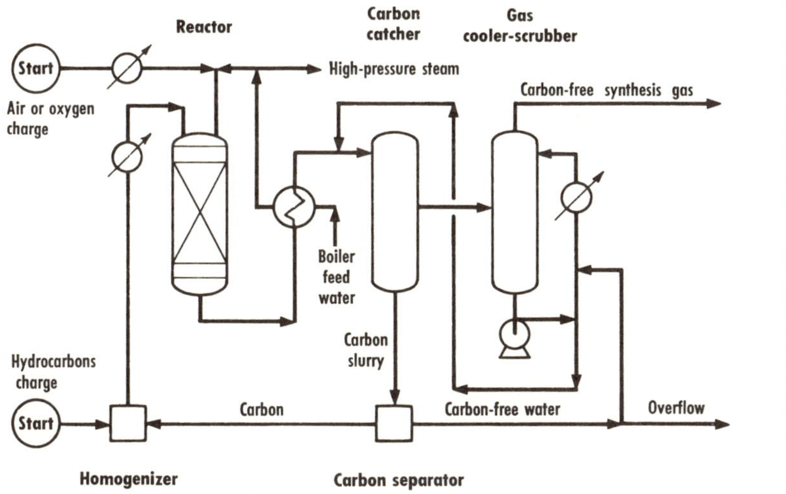

Source: Hydrocarbon Processing, p. 237, November, 1969

FIG. 3.2 Shell synthesis-gas process

Water gas, the product of the reaction of water on coke and coal or other solid fuels has been used to produce a 1-to-1 synthesis gas mixture for ammonia manufacture. However, today little, if any ammonia is made using hydrogen from this source.

Purification of hydrogen in synthesis gas

Shift conversion of CO to CO_2 is used to remove CO and recover more hydrogen for ammonia synthesis. Too, CO poisons the ammonia catalyst, and any source of hydrogen containing carbon monoxide must be treated. These objectives are accomplished in a two-step "shift conversion."

The first step converts most of the CO with steam at 650°–750°F. to CO_2 and H_2 over a chromium-promoted iron catalyst. The second step, at lower temperatures of 400°–475°F., further reduces CO to under 0.5% over a copper-zinc-chromium oxide catalyst. The hydrogen thus produced is also used to adjust the N_2 to H_2 ratio (see Fig. 3.1).

Removal of carbon dioxide produced by the shift conversion is the next step. Effluent gas is treated with solvents or aqueous solutions of potassium carbonate (with and without activators) or ethanolamine. Water scrubbing under pressure is no longer used (see Fig. 3.1).

Removal of residual CO_2 and CO, still present in trace amounts is accomplished by "methanation." This involves their reaction with H_2 at 600°F. over a nickel catalyst (see Fig. 3.1).

Ammonia synthesis

The final step in the syntheses of ammonia is to compress the prepared gas containing H_2 and N_2 in a 3-to-1 mol ratio and charge to the converter (see Fig. 3.1).

Recent developments and trends

The most important development in recent years has been the dramatic switch to very large single-train ammonia plants based on new technology. Capital and operating costs have been reduced by going to 1,000 to 1,500-tons/day plants. This has resulted in a concentration of ammonia plants in low-cost natural-gas-producing areas on the Gulf Coast, and in the need for extensive downstream storage and distribution systems.

The statistics in Table 3.11 show the building of large new plants has resulted in a large overcapacity. Accordingly, profits have been much lower or non-existent in the last couple of years. The trend to large plants is expected to continue, but no new plants have recently been announced in the US.

As production begins to catch up to capacity some hard decisions will be necessary on plant locations. It is quite possible that the shortage and rising prices of natural gas in the US will force new capacity to foreign areas rich in natural gas with the result of increased imports of ammonia. At the present time, there is no import duty on agricultural-grade ammonia.

Economics of ammonia

A close look at the costs associated with the delivering of a ton of ammonia to the customer reveals several interesting facts. One rule of thumb to use is that one-third of the selling price goes for manufacturing, one-third for transportation and one-third for distribution. A recent study showing changes in these costs from 1968 to 1971 is shown in Table 3.7.

TABLE 3.7
Makeup of ammonia cost

	Cost, $/ton	
	*1971**	*1968†*
Manufacturing	30.49	29.70
Transportation (pipeline 1,000 miles)	7.30	25.15 (*total*)
Terminal storage	5.50	
Transportation (truck 100 miles)	5.80	
Retail	30.00	49.85
Total cost to consumer	79.09	104.70

* Based on a 1,000-ton/day plant on Gulf Coast, natural-gas cost 25¢/MM BTU, consumption in farm belt.

† Based on a 600-ton/day plant in New Orleans area, natural-gas cost 20¢/MM BTU, consumption in Midwest.

Source: The Oil and Gas Journal, p. 117, August 23, 1971.

In the manufacturing area, the greatest improvement has come about in going to 1,000-ton/day plants utilizing gas-fired turbine-driven centrifugal compressors instead of electric reciprocal compressors. Some estimates indicate savings as much as $8/ton in direct operating costs between an older 200-ton/day plant and a new 1,000-

ton/day unit. Improvements have been made in transportation, including pipelines, rail, and barge carriers. Large refrigerated storage terminals are less expensive than pressure storage units. Retailing costs have been reduced by aggressive marketing and more efficient distribution systems.

The important fact to recognize is that the manufacturing process is not the only area in which cost improvements can be made.

Manufacturing costs for a 1,000-ton/day plant are given in Table 3.8. The significance of natural gas prices is very apparent, being 40% of the operating cost at 25¢/MM BTU. It is estimated the total manufacturing cost would be decreased approximately $1.30/ton in going to a 1,500-ton/day plant.

TABLE 3.8

Manufacturing cost for 1,000-ton/day plant

Capital cost		
Plant cost	$18.0 million	
Catalyst & chemicals	$ 0.6 million	
Total	$18.6 million	
Operating cost		
	Unit rate	*Cost, $/ton*
Natural gas, 25¢/MM BTU	840.9 MM BTU/hr	5.05
Utilities		
Natural gas	487.4 MM BTU/hr	2.93
Power, 7 mils/kw-hr	545 kw	0.09
Total makeup water, 10¢/1,000 gal	2,075 gpm	0.30
Export steam	In balance	
Catalyst & chemicals		0.80
Labor, 4 men/shift, $5/hr		0.48
Maintenance, 2½% of investment		1.83
Indirect charges, 16½% of investment*		8.61
Total operating cost		20.09
Pretax profit, 20% return on investment		10.40
Total manufacturing cost		30.49

*Indirect costs include 10% depreciation, 5% capital charges, and 1.5% for taxes and insurance.

Source: The Oil and Gas Journal, p. 117, August 23, 1971.

Another study summarized in Table 3.9 gives a comparison of the relative costs of ammonia production when different sources of hydrogen are used. This shows why steam reforming of natural gas is preferred in the US.

TABLE 3.9
Comparison of processes for hydrogen production

*Process**	*Raw material*	*Total energy required*	*Investment cost*	*Production cost*
Steam reforming	Natural gas	100	100	100
Steam reforming	Naphtha	122	132	120
Partial oxidation	Fuel oil	113	191	170
Water gas	Coal, coke, lignite	242	600	460
Gasification	Coal, coke, lignite	192	456	170
Steam-iron	Coke	186	432	170
Electrolysis	Water	95	107	394†

*All processes are expressed as a percentage of the steam reforming of natural gas process.

†Credit taken for by-product oxygen.

Source: Bhattacharyga, K. K. et al, Chemical Age India 17, 3 (March 1966), 169–176.

Producers and capacities

In 1960, there were no ammonia plants with capacities over 600 ton/day (TPD). With the new trend toward the very large 1,000-TPD or larger plants, by June 1969, 35 of the 93 ammonia plants reported to be operating were of the new larger capacity type. Consequently, the ammonia industry has undergone many changes in the last several years and more can be expected in the future. As the larger plants are brought on stream, many smaller ones will be shut down.

A breakdown of capacity in the six largest producing states shows how the industry is concentrating in the natural gas producing states of Louisiana and Texas (see Table 3.10).

Capacity, production, and sales of synthetic ammonia are shown in Table 3.11. Production and sales are published by the Bureau of the Census, while capacity figures were developed by Stanford Research Institute. The capacity figure for 1970 has not been adjusted for some plants that may have already been shut down. So it is a maximum capacity figure.

Consumption

The major end use for ammonia is domestic fertilizer applications which accounted for nearly three-fourths of the total consump-

TABLE 3.10
Synthetic ammonia capacity
(Thousands of short tons)

	1960	*1968*	*1970*
United States total	5,239	14,705	18,430
Louisiana	575 (11%)	3,010 (20%)	3,960 (21%)
Texas	569 (11%)	1,912 (13%)	3,314 (18%)
Mississippi	253 (5%)	1,106 (8%)	1,025 (6%)
California	567 (11%)	970 (7%)	1,180 (6%)
Arkansas	224 (4%)	830 (6%)	830 (4%)
Iowa	—	786 (5%)	1,126 (6%)

Source: Chemical Economics Handbook, Stanford Research Institute, Menlo Park, Calif.

TABLE 3.11
Synthetic ammonia
(Thousands of short tons)

	Capacity	*Production*	*Sales*
1960	5,239	4,818	2,000
1961	5,316	5,207	2,173
1962	6,068	5,810	2,418
1963	6,903	6,693	3,040
1964	8,021	7,634	3,387
1965	8,623	8,869	4,091
1966	10,597	10,605	5,343
1967	11,944	12,194	5,885
1968	14,705	12,120	6,405
1969	17,540	12,714	6,400 est.
1970	18,430	13,000 est.	6,500 est.

Source: Chemical Economics Handbook, Stanford Research Institute, Menlo Park, Calif.

tion of ammonia in 1967. Table 3.12 gives a breakdown by major industry groups. Even though none of the other groups come close to matching the large volume of ammonia used in fertilizers, the many other derivatives go into many different end uses and are very important to the petrochemical industry.

Many of the recent new developments in process technology involve ammonia derivatives. For example, the production of acrylonitrile has switched from using acetylene and HCN to propylene and ammonia. Much of this technology is covered in Chapter 11.

TABLE 3.12
Ammonia-consumption—1967
(Thousands of short tons)
By major industry groups

Domestic fertilizer	7,790	(72%)
Fiber, plastic intermediates	610	(5.6%)
Explosives	700	(6.5%)
Livestock feeds	250	(2.3%)
Others	1,468	(13%)
Total	10,818	(100%)

Source: Chemical Economics Handbook; Stanford Research Institute; Menlo Park, Calif.

In fertilizer applications, Table 3.13, the trend is to more direct application of anhydrous ammonia to the soil. Ammonium nitrate consumes ammonia directly as well as indirectly through the nitric acid required which was made from ammonia. Urea is gaining wider acceptance as a solid, nitrogen-containing slow-acting fertilizer.

TABLE 3.13
Ammonia consumed—fertilizer
(Thousands of short tons)
1968

Direct anhydrous ammonia	3,150
Aqueous ammonia	225
Ammonium nitrate	2,164
Urea	1,285
Ammonium sulfate	751
Ammonium phosphates	1,425
Other nitric acid	136
Grand total (includes all others and exports)	9,696

Source: Chemical Economics Handbook; Stanford Research Institute; Menlo Park, Calif.

Major fiber and plastic intermediates are shown in Table 3.14. Acrylonitrile is used mainly for acrylic fibers with other uses in acrylonitrile-butadiene-styrene resins (ABS) for molding compounds and in nitrile rubber. The general class of nylon polymers use caprolactam for nylon 6, and hexamethylene diamine and adipic acid go into nylon 66. Isocyanates are one of the main ingredients for polyurethane foams and elastomers. Melamine and urea are widely used in resins made with formaldehyde for molding compounds, laminates, and adhesives.

TABLE 3.14
Ammonia consumed
Major fiber and plastic intermediates
(Thousands of short tons)
1968

Acrylonitrile	284
Caprolactam	31
Hexamethylene diamine	93
Nitric acid	
Adipic acid	161
Isocyanates	41
Urea-formaldehyde	93
Melamine	15
Total	718

Source: Chemical Economics Handbook; Stanford Research Institute; Menlo Park, Calif.

Other non-fertilizer uses of ammonia are given in Table 3.15. The amines and nitriles are comprised of ethanolamines, fatty nitriles and amides, hexamethylenetetramine, and methyl amines for use in a variety of industrial applications. Ammonium nitrate is used nearly 100% for explosives while at least one-half of the remaining nitric acid goes into explosives. The urea in this category is used in animal feeds.

TABLE 3.15
Ammonia consumed
Other nonfertilizer ammonia distribution
(Thousands of short tons)
1968

Amines & nitriles	81
Ammonium nitrate	515
Other nitric acid	542
Urea	200
Total	1,338

Source: Chemical Economics Handbook; Stanford Research Institute; Menlo Park, Calif.

Since many of the derivatives of ammonia are used by all the industry groups, Table 3.16 is presented to show the total consumption of each intermediate.

TABLE 3.16
Ammonia-consumption by major chemical derivatives
(Thousands of short tons)

	1960	*1968*
Total apparent consumption	4,885	12,750
Ammonia—direct application	837	3,375
Nitrogen solutions and mixed fertilizers	432	560
Ammonium nitrate (from NH_3 and HNO_3)	1,484	2,679
Nitric acid total	1,209	2,288
Urea	441	1,593
Ammonium phosphates	228	1,425
Ammonium sulfate	388	750
Acrylonitrile	48	284
Amines and nitriles	33	80
Hexamethylene diamine	30	93
Caprolactam	—	31

Source: Chemical Economics Handbook; Stanford Research Institute; Menlo Park, Calif.

METHANOL

Introduction

Methanol or methyl alcohol is a colorless liquid with molecular formula, CH_3OH. It is extremely poisonous and will cause blindness. It is one of the largest-volume organic chemicals produced synthetically. Its major use is as a raw material for formaldehyde, but large quantities are used as antifreeze, solvent, and chemical intermediate. Until 1923, methanol was produced by the destructive distillation of wood, from which it obtained its common name of wood alcohol. Today nearly all methanol is made using natural gas as a feedstock. Physical properties of methanol are:

Molecular formula	CH_3OH
Molecular weight	32.04
Boiling point	64.5°C.
Freezing point	−97.8°C.
Specific gravity, 20/20°C.	0.792
Lb/gal, 20°C.	6.59

Manufacturing processes

All of the processes for synthetic methanol involve the basic steps of steam reforming of natural gas plus addition of CO_2 to adjust the C–H ratio, compression, syntheses in a catalytic converter, and distillation for purification.

Synthesis gas

The synthesis gas is usually a mixture of CO, CO_2, and H_2, with conversion to methanol according to these equations:

$$CO + 2\ H_2 \rightarrow CH_3OH$$
$$CO_2 + 3\ H_2 \rightarrow CH_3OH + H_2O$$

Since the optimum ratio for C–H is 1–2 for the hydrocarbon feed and natural gas is mostly CH_4 with a 1–4 ratio, CO_2 is normally added to take care of the surplus of hydrogen.

Alternate feedstocks have been used, but their availability and costs have limited their use. One excellent off-gas is obtained from the manufacture of acetylene from natural gas, since it contains approximately the correct proportions of CO, CO_2, and H_2. A second attractive off-gas is produced in a steel-making basic-oxygen furnace; this contains mostly CO which can be converted with steam in a shift reactor to produce CO_2 and H_2.

Methanol process

Once the correct composition of synthesis gas is obtained, the traditional conversion to methanol is obtained under the high pressure of 300 atmospheres, in the presence of a chromium oxide-zinc oxide catalyst at about 660°F. This is shown in detail in Fig. 3.3.

The single most important development in methanol synthesis was ICI's new methanol process using a new active copper catalyst. First plant came on stream in 1966. The process operates at about 50 atmospheres and 500°F. The lower pressures allow the use of centrifugal compressors, rather than reciprocating compressors. It is also possible to operate on hydrogen-rich synthesis gas without having to add CO_2 to obtain the stoichiometric C–H ratio. The process is shown in Fig. 3.4.

Economics of methanol

As more operating experience is obtained on ICI's new low-pressure process, it is becoming apparent that this process will be used

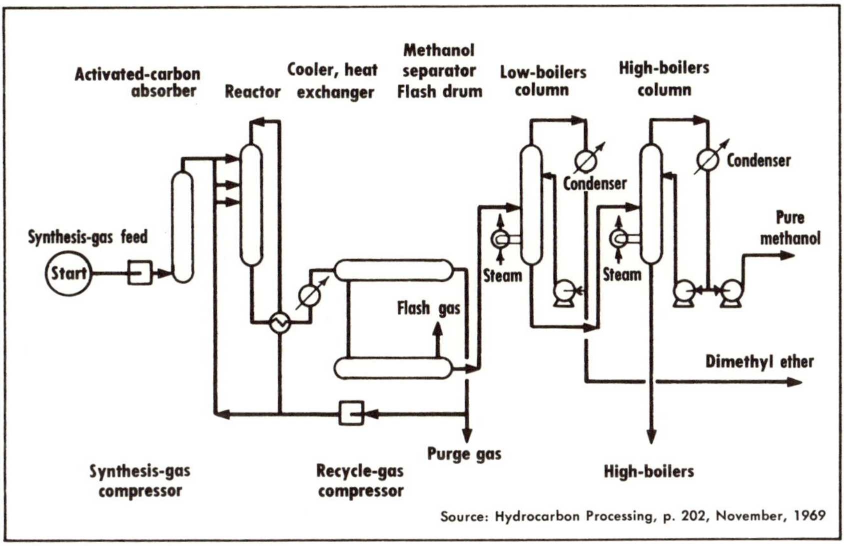

FIG. 3.3 Lurgi high-pressure methanol process

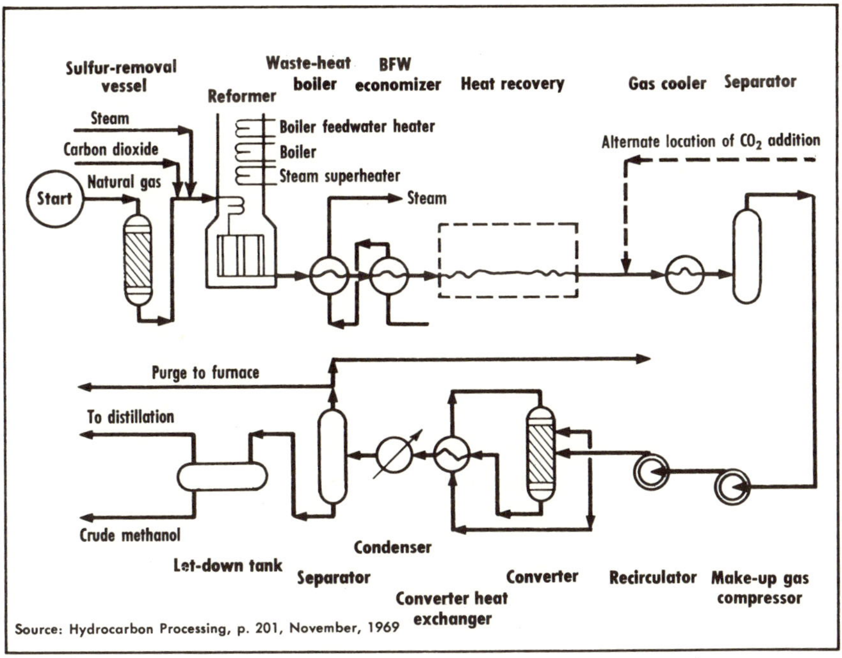

FIG. 3.4 ICI low-pressure methanol process

more and more. The main area of saving is the lower capital and operating costs obtained with the centrifugal compressors. An excellent recent article, "Optimize ICI Methanol Process," by Mehta and Ross of Chemical Construction Corp., appeared in *Hydrocarbon Processing*, November, 1970.

A second major advantage of the low-pressure process is that a small-capacity plant can compete with older high-pressure plants of larger capacity. Now, a large consumer of methanol may justify his own small plant, rather than purchase methanol from one of the very large producers.

Comparative manufacturing costs are given in Table 3.17. This shows the much lower variable costs for the 15 MM-gal/yr low-pressure plant vs. the 30 MM-gal/yr high-pressure plant. Even with very large plants of 80 MM-gal/yr, the low-pressure process is significantly lower in manufacturing costs and capital investment.

TABLE 3.17
Methanol manufacturing costs
(Costs in ¢/gal)

Capacity, tons/day	150	300	800	800
Output, MM gal/yr	15	30	80	80
Process	lp	hp	lp	hp
Compressor	cent.	rec.	cent.	cent.
Capital cost, $ million	$5.4	$8.8	$14.5	$15.2
Variable costs, ¢/gal*	2.65	4.18	2.65	2.87
Labor, maintenance, supr.	2.50	1.73	0.89	0.93
Fixed costs (plant, depr.)	4.31	3.32	1.94	2.05
Cost to manufacture	9.46	9.23	5.48	5.85
S, G & A	0.5	0.5	0.5	0.5
20% return, BFIT	7.20	5.87	3.62	3.80
Sales value (FOB)	17.2	15.6	9.6	10.2
Sales value (FOB) (same basis, naphtha @ 6.5¢/gal)	21.0	19.9	13.4	14.7

*Natural gas at 20¢/1000 cu ft.

Source: B. Hedley, W. Povers and R. B. Stobaugh; Hydrocarbon Processing, Sept. 1970, p. 277.

Producers and capacities

It has been very difficult to keep tabs on the existing methanol plants and the number of new ones being constructed. At the beginning of 1971, there were 14 plants with 900 MM gal capacity, and no less than seven new plants planned for an additional capacity of 800 MM gal. Nine months later, several small plants had been closed and two new 100 MM-gal ICI low-pressure plants had been started by Georgia Pacific and Monsanto. The latest capacity tabulation is shown in Table 3.18.

TABLE 3.18
Methanol—plant capacities—1972
(Millions of gallons)

Allied, South Point, Ohio	25
Borden, Geismar, La.	160
Celanese, Bishop, Tx.	100
Clear Lake, Tx.	200
Commercial Solvents, Sterlington, La.	50
DuPont, Orange, Tx.	130
Beaumont, Tx.	200
Huron, Ohio	30
Escambia, Pensacola, Fla.	50
Georgia-Pacific, Plaquemine, La.	100
Hercules, Plaquemine, La.	80
Monsanto, Texas City, Tx.	100
Rohm & Haas, Deer Park, Tx.	22
Tenneco, Pasadena, Texas	60
Union Carbide, Texas City, Tx.	42
Total	1,349*

*8.9 billion lb/yr.

Source: Oil, Paint & Drug Reporter, Chemical Profile, September 27, 1971.

Plans for a 200 MM-gal plant by Methchem have been dropped. Borden announced a 100 MM-gal plant for the northwest, but nothing new has been heard on this. The latest plant to come on stream is a 200 MM-gal/yr ICI low-pressure plant for Celanese at Clear Lake, Texas (November, 1971). It is anticipated that the smaller plants will close as the larger ones reach full capacity in the next one to two years.

Production, exports and sales are shown in Table 3.19.

Consumption

The end uses of methanol again emphasize the far-reaching markets served by this synthesis gas derivative. An estimate has been made in Table 3.20 for the 1970 pattern of consumption and exports. Market information on the important derivatives is given in Tables 3.21 through 3.26.

TABLE 3.19
Synthetic methanol
(Millions of pounds)

	Production	*Exports*	*Sales*
1960	1,966	198	1,102
1961	2,040	187	1,156
1962	2,228	143	1,123
1963	2,334	103	1,059
1964	2,632	372	1,497
1965	2,869	196	1,395
1966	3,268	52	1,373
1967	3,542	70	1,394
1968	3,852	74	1,490
1969	4,206	103	1,650E
1970	4,945	290	1,950E
1971	4,911	650*	na

Source: US Tariff Commission; and U.S. Census Bureau.

*Based on 6 months Census Bureau data.

TABLE 3.20
Methanol consumption and exports
(Millions of gallons)

	1970 estimate	*% of total*
Formaldehyde	295	40
Dimethyl terephthalate	85	11
Solvents	75	10
Methyl amines	30	4
Methyl methacrylate	30	4
Methyl halides	25	3
Exports	45	6
All others	160	22
Total	745	100

It is easily seen that formaldehyde is the largest outlet for methanol. The three largest methanol producers, Borden, Celanese, and DuPont, are also the three largest formaldehyde producers. It is anticipated that formaldehyde will continue to use over 40% of methanol production.

Dimethyl terephthalate and methyl methacrylate are two of the fastest-growing uses for methanol. It is used directly as a solvent for many applications, as are several of its derivatives, such as glycol methyl ethers. About 3–4% of methanol produced is used as an inhibitor in formaldehyde. However, methanol is decreasing in antifreeze uses and as an antidetonate injection fluid. Exports are expected to increase in the next several years as the available capacity increases. A new technical development is the production of acetic acid from methanol which could become a large consumer of methanol.

Derivatives of methanol—technology and markets

Formaldehyde

Formaldehyde is made by passing methanol and air at 250°C. over heated silver or copper catalyst:

$$CH_3OH + \tfrac{1}{2}O_2 \rightarrow CH_2O + H_2O$$

Formaldehyde is a colorless gas with mp of −92°C. and bp of −21°C. It is soluble in water, alcohol, and ether, and polymerizes

TABLE 3.21
Formaldehyde—producers and capacity—1969
(Millions of pounds)

Producers	*Capacity—37% solution*
Allied	310
Borden	925
Celanese	1,000
Commercial Solvents	125
Cosden	75
DuPont	890
GAF	100
Georgia-Pacific	120
Gulf	45
Hercules	275
Hooker	135
Monsanto	400
Reichhold	230
Rohm & Haas	50
Tenneco	335
Union Carbide	150
Wright Chemical	150
Total	5,315

Source: Oil, Paint & Drug Reporter, Chemical Profile, July 1, 1969.

easily. It is normally handled as a 37% aqueous solution, with or without methanol which acts as an inhibitor.

There are approximately 50 individual plants in the US that produce formaldehyde. Much of the production (about 60% of total output) is used captively to avoid the cost of shipping and handling 37% aqueous formaldehyde. Consequently, a number of relatively small plants are in operation today. Table 3.21 lists producers and their total capacities in 1969.

The end use pattern for formaldehyde is shown in Table 3.22 for 1969. Most of the formaldehyde is combined with phenol, urea, or melamine to form thermosetting resins and adhesives. Its use in hexamethylene tetramine, which is used in explosives, is expected to decline. Polyacetal resins and pentaerythritol should continue growing.

TABLE 3.22

Formaldehyde—end uses—1969

Phenolic resins	24%
Urea-formaldehyde resins	24%
Hexamethylenetetramine	11%
Melamine resins	9%
Polyacetal resins	7%
Pentaerythritol	7%
Fertilizers	4%
Ethylene glycol	4%
Acetylenics	2%
Exports and miscellaneous	8%

Source: Oil, Paint & Drug Reporter, Chemical Profile, July 1, 1969.

Production of 37% formaldehyde solution was 4.31 billion lb (1.59 billion lb on 100% basis) in 1970 (US Tariff Commission).

Dimethyl terephthalate

Methanol is used to esterify terephthalic acid to form dimethyl terephthalate (DMT) of very high purity:

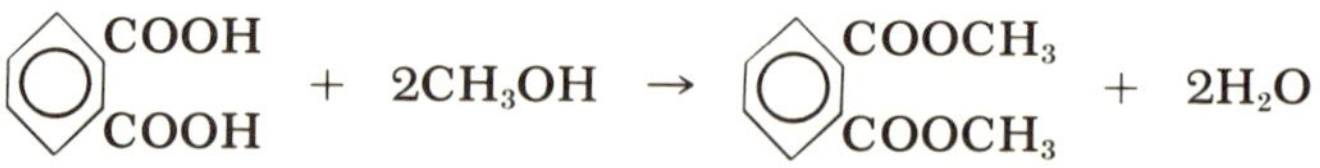

The DMT is then reacted with ethylene glycol to form polyester fiber or film. Up until several years ago, DMT was used exclusively to

make polyesters, but recently, high-purity terephthalic acid (TA) has been used for direct esterification without using methanol. However, the rapid growth of polyesters is expected to be greater than the growth in use of TA, so that methanol use should still grow in this application.

Production of DMT and TA is estimated at 1.8 billion lb in 1970. Growth has been about 30%/yr since 1963. Industry estimates it should continue growing at around 15%/yr over the next five years.

Practically all DMT and TA are used in polyester resins. Fibers take from 85–90%, films take 5–10%, and export and other uses account for about 5%.

Methyl amines

Methyl amines are made by reacting methanol and ammonia in a continuous flow system in the presence of a dehydration catalyst at 450°C., followed by the separation of unchanged raw materials from the three main products: monomethyl amine, dimethyl amine, and trimethyl amine:

$$CH_3OH + NH_3 \rightarrow CH_3NH_2 + H_2O$$
$$CH_3OH + CH_3NH_2 \rightarrow (CH_3)_2NH + H_2O$$
$$CH_3OH + (CH_3)_2NH \rightarrow (CH_3)_3N + H_2O$$

The ratio of 3 amines can be varied by changing reaction conditions.

Current producers and capacities are given in Table 3.23. Rohm and Haas is reported to have 21 *MM* lb standby capacity at Bristol and Philadelphia, Pennsylvania.

TABLE 3.23
Methylamines—producers and capacities
(Millions of pounds)

Producer	*1970 capacity*
Commercial Solvents, Terre Haute, Ind.	18
DuPont, Belle, W.Va.	75
Strang, Texas	26
Escambia, Pace, Fla.	50
GAF, Calvert City, Ky.	10
Pennwalt, Wyandotte, Mich.	10
Total	189

Source: Oil, Paint & Drug Reporter, Chemical Profile, July 1, 1970.

US production of all three methyl amines was reported as 117 million lb in 1969 by the US Tariff Commission. A breakdown by end uses is given in Table 3.24. The distribution between mono, di, and trimethyl amines is, respectively, 32%, 53%, and 15%. These products have grown at 12%/yr, but are expected to grow at a lesser rate in the future.

TABLE 3.24
Methylamines—end uses—1969
(Millions of pounds)

Monomethyl amine	
Carbamyl insecticide	19
Surfactants	10
Other	8
Dimethyl amine	
Dimethyl acetamide and formamide	31
Lauryl dimethylamine oxide	10
Rubber chemicals	9
Others	12
Trimethyl amine	
Choline chloride	17
Other	9
Total production	117

Methyl methacrylate

Polymethyl methacrylate is one of the fastest growing plastics today and is expected to continue growing at 12%/yr. The monomer is prepared according to these reactions:

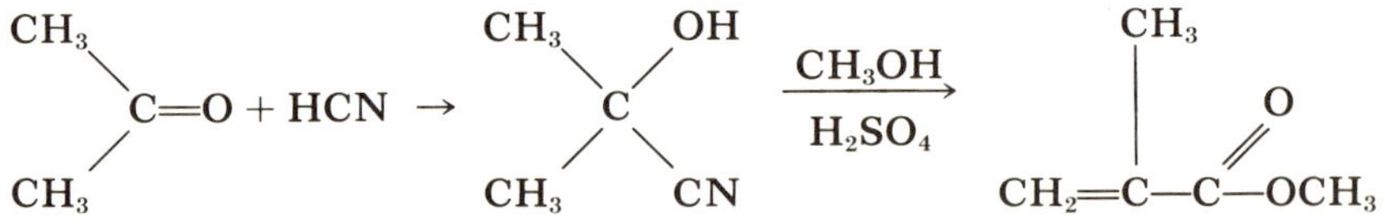

Current producers and capacities are listed in Table 3.25. DuPont is also building a 120-million-lb/yr plant in Memphis, Tennessee. In the US, the acetone-HCN process is used exclusively, but in Europe and Japan there are plans to oxidize isobutylene to methacrylic acid, followed by esterification to the ester.

Nearly all the methyl methacrylate monomer is polymerized to polymethyl methacrylate for use in cast sheet, surface coatings, and molding powders, as shown in Table 3.26. Smaller amounts go into oil

TABLE 3.25
Methylmethacrylate
(Millions of pounds)

Producers	*1970 capacity*
American Cyanamide	
Fortier, La.	80
DuPont, Belle, W.Va.	120
Rohm & Haas, Bristol, Pa.	60
Houston, Tx.	400
Louisville, Ky.	60
Total	720

Source: Oil, Paint & Drug Reporter, Chemical Profile, October 1, 1970.

additives, acrylic syrups, and unsaturated polyester resins. Total production of 430 million lb was reported in 1969 by the US Tariff Commission.

TABLE 3.26
Methylmethacrylate—end uses
1970

Cash sheet	41%
Surface coatings	23%
Molding and extrusion powders	21%
Others	15%

Source: Oil, Paint & Drug Reporter, Chemical Profile, October 1, 1970.

Methyl halides

Methyl chloride and methyl bromide are both made commercially from methanol. A mixture of gaseous CH_3OH and HCl are passed through a converter at 350°C. and atmospheric pressure.

$$CH_3OH + HCl \rightarrow CH_3Cl + H_2O$$

Methyl chloride production was 390 million lb in 1969, with an end use distribution of silicones—38%; tetramethyl lead—38%; butyl rubber—5%; and others—10%.

Methyl bromide production was 21 million lb in 1969, with an end use distribution of soil fumigants—49%; space fumigant—17%

and exports (mainly soil fumigant) – 30%. The methyl halides have been growing at about 18%/yr in the 1960's. Methyl chloride should continue to grow at 8–10%/yr.

Methyl chloride is produced by Allied, Ancon, Dow, DuPont, Ethyl, General Electric, Stauffer, Union Carbide, and Vulcan, with a total capacity of about 500 million lb in 1971. US production was 432 million lb in 1970. Silicones have grown well, but the most dramatic increase has been for use in tetramethyl lead for gasoline. However, its continued growth will depend upon the future of lead in gasoline. Since the major silicone-resin manufactures and tetramethyl lead producers make their own methyl chloride, the merchant market is very competitive.

Methyl bromide is made by American Potash, Dow, Great Lakes Chemical, and Michigan Chemical, with a total capacity of around 30 million lb. Sales are expected to continue to grow at 6–7%/yr with nearly 25-million-lb sales in 1970.

OXO ALCOHOLS

Introduction

The oxo reaction is used to produce alcohols by passing olefins, carbon monoxide, and hydrogen over a cobalt catalyst. The aldehyde is then reduced with hydrogen to yield an alcohol. In some cases, especially with n-butyraldehyde, longer-chain alcohols are produced by an aldol condensation, followed by reduction of the eight-carbon aldehyde to 2-ethylhexyl alcohol.

General oxo chemistry

$$RCH{=}CH_2 + CO + H_2 \xrightarrow[\substack{170^\circ C. \\ 200\ atm.}]{Cobalt} \begin{array}{c} RCH_2CH_2CHO \\ + \\ \quad CHO \\ RCH{-}CH_3 \end{array}$$

$$RCH_2CH_2CHO + H_2 \longrightarrow RCH_2CH_2CH_2OH$$

Specific reactions to make n-butanol and 2-ethylhexyl alcohol

$$CH_3CH{=}CH_2 + CO + H_2 \longrightarrow \begin{array}{c} CH_3CH_2CH_2CHO \\ + \\ \quad CHO \\ CH_3{-}CH{-}CH_3 \end{array}$$

$$CH_3CH_2CH_2CHO + H_2 \longrightarrow \underset{\text{n-Butanol}}{CH_3CH_2CH_2CH_2OH}$$

$$2CH_3CH_2CH_2CHO \xrightarrow{NaOH} CH_3CH_2CH_2CH_2\overset{\overset{C_2H_5}{|}}{C}HCHO + H_2O$$

$$\downarrow H_2$$

$$\text{2 Ethylhexyl Alcohol}$$

The oxo reaction was discovered in Germany in 1938 and was first introduced in the United States by Enjay in 1948. Many alcohols are and can be produced with the oxo process, but most of the production has been of alcohols in the C_8 to C_{10} range, mostly isooctyl and 2-ethylhexyl alcohols.

Synthesis gas is the most convenient method to obtain the proper mixture of CO and H_2, which is required in a 1-to-1 ratio. In the case of oxo synthesis, the olefin feedstock is more of a controlling factor than the supply of synthesis gas. Therefore, the source of synthesis gas will depend upon what is available at the plant.

Manufacturing process

The olefin feed is put into the stainless-steel reactor, along with CO, H_2, and a cobalt catalyst. A temperature around 170°C. and a pressure of 1,500 to 4000 psi is maintained in the reactor. The raw aldehyde product is treated in a decobalter tower to remove cobalt catalyst. The aldehyde is then hydrogenated at 150°C. and 1,500 psi in the presence of a nickel or copper chromite catalyst. If n-butyraldehyde is desired, it is removed prior to the hydrogenation step and sent to a reactor to convert to 2-ethyl hexanol by aldol condensation. A flowsheet for the oxo process is shown in Fig. 3.5.

Producers capacities and production

There are a number of oxo plants in the United States, and each one has quite a bit of flexibility built in, so that several different alcohols may be made at any one plant. Therefore only the producing plants will be listed as shown in Table 3.27.

Data on production and sales of oxo chemicals are quite sketchy. The US Tariff Commission reports on production of butyl alcohol, octyl alcohols, and decyl alcohols, as well as data on several of the specific plasticizers made with oxo alcohols. The picture is also complicated because different routes are used to make the same alcohol, and no distinction is made in what data is reported.

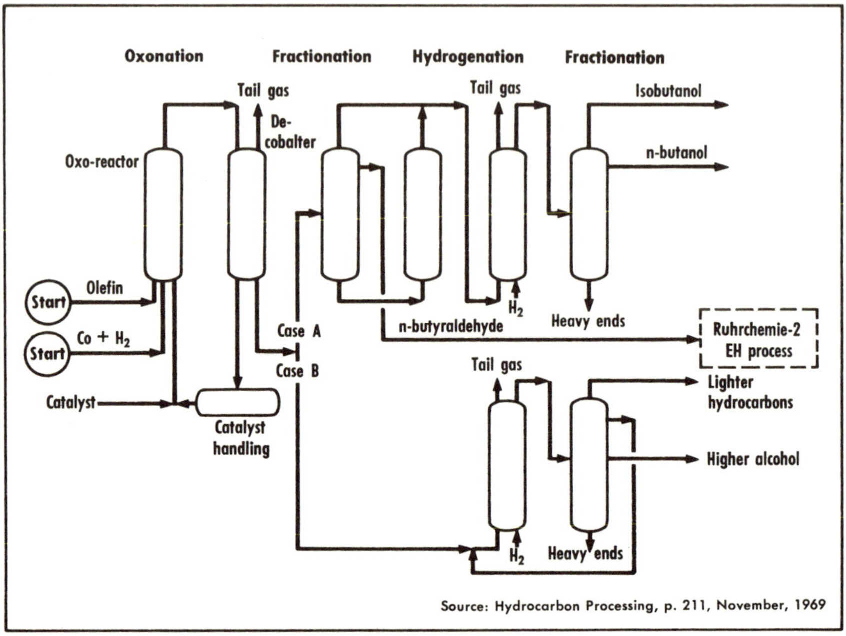

FIG. 3.5 Ruhrchemie A.G. oxo process

TABLE 3.27
Oxo chemicals—producers & capacities—1970
(*Millions of pounds*)

Producer	*Capacity*
Dow Badische, Freeport, Tx.	200
Eastman Longview, Tx.	275
Enjay, Baton Rouge, La.	200
Getty-Air Products, Delaware City, Del.	40
Oxochem, Penuelas, P.R.	250
Shell, Geismar, la.	150
Shell, Houston, Tx.	200
Union Carbide Ponce, P.R.	140
Union Carbide, Seadrift, Tx.	120
Union Carbide, Texas City, Tx.	200
USS Chemicals, Haverhill, Ohio	70
Total	1,845

Source: Oil, Paint and Drug Reporter, Chemical Profile, April 1, 1971.

For example, one of the main chemicals made by the oxo process is n-butyraldehyde, which is converted to n-butanol or dimerized by the aldol condensation to 2-ethylhexenal, prior to conversion to 2-ethylhexanol. n-Butanol is also obtained from fermentation or from n-butyraldehyde derived from acetaldehyde. Other octyl and decyl alcohols are also reported in production data.

Due to the flexibility of oxo producing plants and changing market conditions and feedstock availability, it is most difficult to obtain an exact overall picture.

TABLE 3.28
Oxo chemicals—production by type
(Millions of pounds)

	1966	*1970*
C-4's n-and isobutyraldehyde (mainly for n-butanol and 2-ethylhexanol)	70% (595)	50% (700)
C-8, 10 and 13's isooctanol plus some isodecanol and esotridecanol	27% (230)	28% (390)
C-12–14's (linear)	na	5% (70)
C-3's	na	4% (60)
Others	3% (25)	13% (180)
Total	100% (850)	100% (1,400)

Source: Oil, Paint and Drug Reporter, Chemical Profile, April 1, 1971.

TABLE 3.29
*Alcohols produced by oxo process**
(Millions of pounds)

	1966		*1969*	
	Production sales		*Production sales*	
n-butyl alcohol	398	234	405	256
Isobutyl alcohol	93	74	101	90
2-ethylhexanol	319	149	405	201
Isooctyl alcohols	100	99	109	83
Decyl alcohols	122	72	162	58

*Figures contain production from all processes, not just the oxo process.

Source: US Tariff Commission.

Therefore, a breakdown of production by carbon length is presented in Table 3.28, as well as individual statistics reported by the US Tariff Commission in Table 3.29.

Consumption

The majority of oxo alcohols are used to esterify acids to make plasticizers for use in polyvinyl chloride resins. It is estimated that approximately 70% of production goes into vinyl plasticizers. In all these products there is stiff competition from other alcohols.

Linear alcohols in the C-12 to C-14 range are used in surfactants. In this application, very strong competition comes from other alcohols, particularly those derived from ethylene-based alpha olefins. This use is not expected to grow rapidly.

TABLE 3.30
n-Butanol—end uses—1970

Coatings solvent	21%	Exports	5%
Glycol ether	14%	Amine resins	10%
n-Butyl acetate	14%	Butyl acrylate	7%
Plasticizers	12%	Others	17%

Source: Oil, Paint and Drug Reporter, Chemical Profile, July 1, 1970.

TABLE 3.31
Plasticizers based on oxo alcohols
(Millions of pounds)

	1970	
	Production	*Sales*
Total—all phthalate	855	815
Dibutyl phthalate	23	25
Diisodecyl phthalate	123	121
Di(2-ethylhexyl) phthalate	350	344
Di(isooctyl) phthalate	85	71
Di(tridecyl) phthalate	15	16
Butyl, acetyl phthalates	12	11
Sub total	608	588
Total—all adipate	54	49
Di(2-ethylhexyl) adipate	35	31
Total—all plasticizers	1,336	1,239

Source: US Tariff Commission.

n-Butyl alcohol accounts for most of the other 25% of production. In recent years, oxo-based n-butanol has been increasing its share of production at the expense of other processes. A breakdown of its end uses is as follows in Table 3.30.

In the plasticizer area, a large number of specific materials are made from a wide choice of acids and alcohols. The list of the more important individual plasticizers, Table 3.31, will give an idea of their variety. In fact, over 70% of all phthalate plasticizers and 65% of all adipate plasticizers are made with oxo alcohols. Approximately one-half of all types of plasticizers are made from oxo alcohols.

4

ELLIOT M. MAGER

Aromatics production

AROMATICS for the US petrochemical industry are primarily derived from petroleum refining. No more than 12% of the benzene, toluene, and xylene (BTX) produced in the US still comes from coal tar. The production of aromatics for the chemical market is now a vital part of the US petroleum industry, and one that is receiving increasing attention from petroleum refiners.

Benzene from petroleum sources has grown from less than 100 million gal in 1950 to over 1,000 million gal/yr in 1970. This represents a steady growth rate of 10%/yr.

Xylenes had been growing at the even more astonishing annual rate of 15%, until export market cutbacks in 1969 and 1970 slowed the growth rate down to below 10%.

The petroleum refiner has taken advantage of the most recent technology to press his advantage as the primary supplier of aromatics. But there are competing economic factors which arise within the petroleum industry for use of the aromatics barrel. The most important current factor is the need for the industry to change over to an unleaded, or low-lead gasoline economy. The rate at which this change will occur has not yet been firmly established, but it will have a profound effect on both chemical and petroleum industry economics.

CRUDE OIL AS A SOURCE OF AROMATICS

Aromatics and the cycloparaffins, or naphthenes, which are readily converted to aromatics in all modern petroleum refineries via the catalytic reforming process, are available in crude oil in prodigious quantities.

Of the 41,000,000 bbl/day of crude oil processed in the free world (11,100,000 bbl/day of crude oil refined in the US), almost 1% is either native benzene or those six-carbon-number cycloparaffins, cyclohexane and methylcyclopentane, which are readily convertible to benzene by dehydrogenation in the catalytic reformer. Two to 2.5% of the crude barrel is native toluene or its precursors. The various C_8 alkylbenzenes and the eight-carbon-number cycloparaffins which are convertible to an equilibrium ratio of meta, para, and orthoxylene, plus ethylbenzene, occur in about the same general proportion as do toluene and the toluene formers.

An extension of these figures, based on the conversation efficiencies usually obtained in reformers, indicates that the supplies of both toluene and xylenes are virtually inexhaustible compared to current and projected future chemical requirements, provided, of course, the

TABLE 4.1
Occurence of aromatics and aromatic-forming cycloparaffins in various crude oils

	Louisiana Gulf	*West Texas*	*Venezuelan*	*Libyan*	*Nigerian*	*Iranian*
Crude-oil gravity, °API	31	41	27	40	35	34
1b/gal	7.25	6.83	7.43	6.87	*7.08*	7.12
Yield of 150–300°F. naphtha	13	18	10	17	17	15
Total C_6-C_8 aromatics,						
% on crude	1.1	1.79	1.85	1.0	2.50	1.80
% on naphtha	8.5	11	18.5	5.8	20.7	12.0
Benzene, % on crude	0.15	0.18	0.15	0.07	0.11	0.19
Toluene, % on crude	0.45	0.51	0.60	0.37	0.92	0.56
C_8 aromatics, % on crude	0.50	1.10	1.10	0.56	1.47	1.05
Total C_6-C_8 naphthenes,						
% on crude	3.87	6.37	3.4	2.50	7.2	2.92
Benzene formers, % on crude	0.67	0.97	0.50	0.55	1.2	0.65
Toluene formers, % on crude	1.3	2.0	1.6	1.05	3.5	1.19
C_8-aromatics formers,						
% on crude	1.9	3.4	1.3	0.90	2.5	1.08

The basic aromatics and their companion cycloparaffins occur in the 150° to 300°F. boiling fraction known as intermediate naphtha, or straight-run gasoline.

economics favor recovering them. Benzene supplies will probably have to be supplemented, as is already occuring to some extent, by converting toluene or going after other sources, such as ethylene-plant byproduct pyrolysis gasoline.

Table 4.1 shows the occurrence of both native aromatics and the available aromatic-forming cycloparaffins in some of the world's standard, large-volume crude-oil supplies.

Table 4.2 provides some extrapolations of the figures on aromatics potential in crude oil against both the worldwide and US chemical markets for aromatics.

TABLE 4.2
Aromatics availability and current chemical usage

	Free world			USA		
	%	*Bbl/day*	*MM gal/yr*	*%*	*Bbl/day*	*MM gal/yr*
Crude oil processed	100	41,000,000	614,000	100	11,100,000	168,000
Available benzene	0.75	307,500	4,716	0.90	100,000	1,440
Chemical benzene market	0.40	166,000	2,550[1]	0.68	84,000	1,200
Available toluene	2.0	820,000	12,507	2.3	282,000	4,300
Chemical markets				0.43	52,000	800
Available xylenes	2.0	820,000	12,507	2.5	308,000	4,500
Chemical markets				0.33	41,000	600

[1]47% USA, 32% Europe, 15% Japan.

Except for a 15 to 20% volumetric shrinkage due to the density difference between the cycloparaffin and the resultant aromatic, the conversion efficiency in producing both toluene and xylenes is essentially 100%. Not so for benzene, since the 5-carbon-number methylcyclopentane ring (MCP) does not readily convert to benzene even in the most efficient catalytic reformer. The cyclohexane, which does convert essentially 100%, generally occurs in crude oils in smaller concentrations than MCP. The overall conversion efficiency for benzene is therefore in the 60 to 70% range.

The extrapolated figures on "available toluene" and "available xylenes" are actually quite conservative because they consider only the conversion of seven and eight-carbon-number cycloparaffins, via dehydrogenation to toluene and xylenes. There is even a greater pool of aromatics to draw upon, typically 50 to 100% more toluene and xylenes. This is all based on level of reforming severity, the ability of the reformer to cyclize straight-chain hydrocarbons, the hydrocracking of heavier alkylbenzenes, etc.

CATALYTIC REFORMING—PRIMARY SOURCE OF AROMATICS

During the early and mid-1950's the platinum-reforming process for crude naphthas became commercialized.

There are presently over 2.9 million bbl/day of installed reforming capacity in the US. This represents some 22% of the total US crude-oil-processing capacity. Total free world capacity is some 6.1 million bbl/day, or 15% of the 41,000,000 bbl/day of crude oil processed. In 1953 there was probably less than 500,000 bbl of reforming capacity in the US.

Reforming answered the refiners need for high-octane gasolines to meet the compression and horsepower races of the mid 1950's and at the same time permitted the automobile people to indulge in those races. Without reforming there simply could not have been sufficient high octane fuel available to permit the industry the luxury of engaging in that race.

Today catalytic reformate represents about 40% of the US gasoline pool, and at $300–$500 of investment/bbl of installed capacity, is still both the most direct and least expensive route to high-quality motor fuels meeting today's specifications.

The effect of catalytic reforming on the gasoline marketing picture can be seen in Fig. 4.1, which shows how the research octane ratings of US regular and premium has increased in recent years. The average research octane rating on premium increased from 92 to 99+ in the seven years between 1953 and 1960. During this same period regular-grade octane increased from 85 to 92. Today's regular is equivalent in octane to 1954's premium.

A typical catalytic reformer (for motor-fuel production) takes a low-octane heavy gasoline or naphtha and upgrades its antiknock quality substantially. This is not done, however, for just the cost of fuel and other necessary utilities. Condensation of saturates into the denser aromatics, plus other side reactions, such as hydrocracking of heavy paraffins, results in a yield of high-octane reformate amounting to 70 to 80 vol % of the naphtha charged. The 20 to 30% of "lost" gaso-

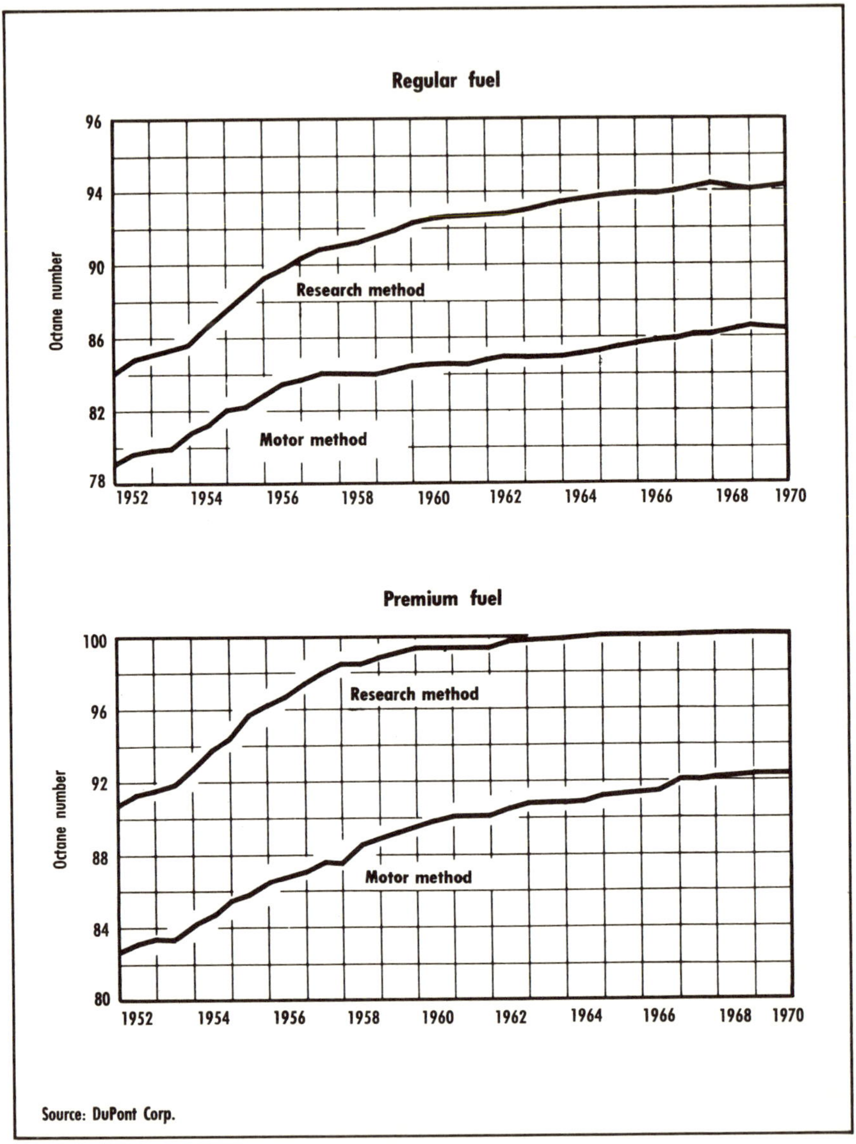

FIG. 4.1 Trends in antiknock quality of US gasoline

line volume shows up as light-ends yields. Substantial volumes of hydrogen and saturated light hydrocarbons: methane, ethane, propane, and butanes are produced.

Fig. 4.2 shows typical yields of stabilized (C_5+) reformate and

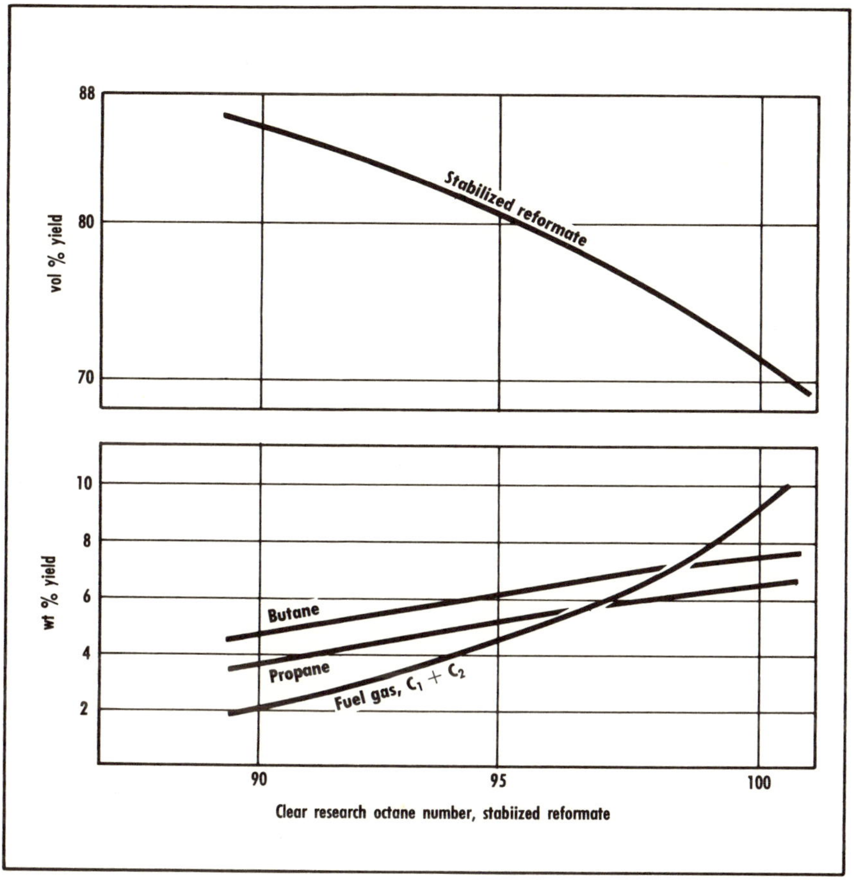

FIG. 4.2 Typical yields, stabilized reformate

light ends vs. product clear research octane number (a common measure of reforming severity). This is for a naphtha of average reformability.

Catalytic reforming is carried out over a platinum-based catalyst at temperatures up to 1,000°F., pressures between 150 and 500 psig, and in a hydrogen atmosphere, where hydrogen produced by the dehydrogenation of naphthenes and dehydrocylization of paraffins is recycled to suppress coking. Usually, a mol ratio of hydrogen to fresh charge of at least 4 to 1 is maintained. Fig. 4.3 is a flow chart for a typical reforming process.

Naphtha charge is generally passed through a hydrodesulfurization and denitrification step for essentially complete removal of trace

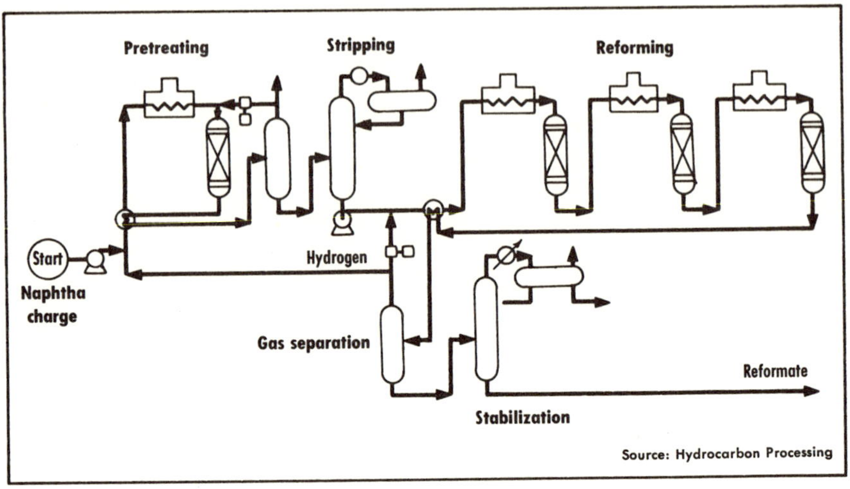

FIG. 4.3 Typical catalytic-reforming scheme

contaminants. The hydrogen produced by the reformer is used as treat gas in the desulfurization reaction which takes place over a cobalt or nickel molybdenum catalyst at elevated temperatures and pressures (300 to 700 psig, 600° to 800°F.).

Desulfurized feed is generally reformed in a three, four, or even five-reactor system, with interheaters to make up heat lost in the highly endothermic dehydrogenation reaction.

The effects of various processing variables on yields, catalyst life, and investment are summarized in Table 4.3.

TABLE 4.3
Effect of process variables in catalytic reforming

Process variable	*Yield*	*Catalyst life*	*Invest-ment*	*Variable operating costs*
1. Increase hydrogen/hydrocarbon feed ratio	no effect	increases	increases	increases
2. Increase system pressure	reduces	increases	increases	decreases
3. Require greater product aromaticity, octane number	reduces	reduces	increases	increases
4. Charge more naphthenic feed stock	increases	increases	reduces	no effect
5. Increase charge stock end point	increases	reduces	no effect	no effect

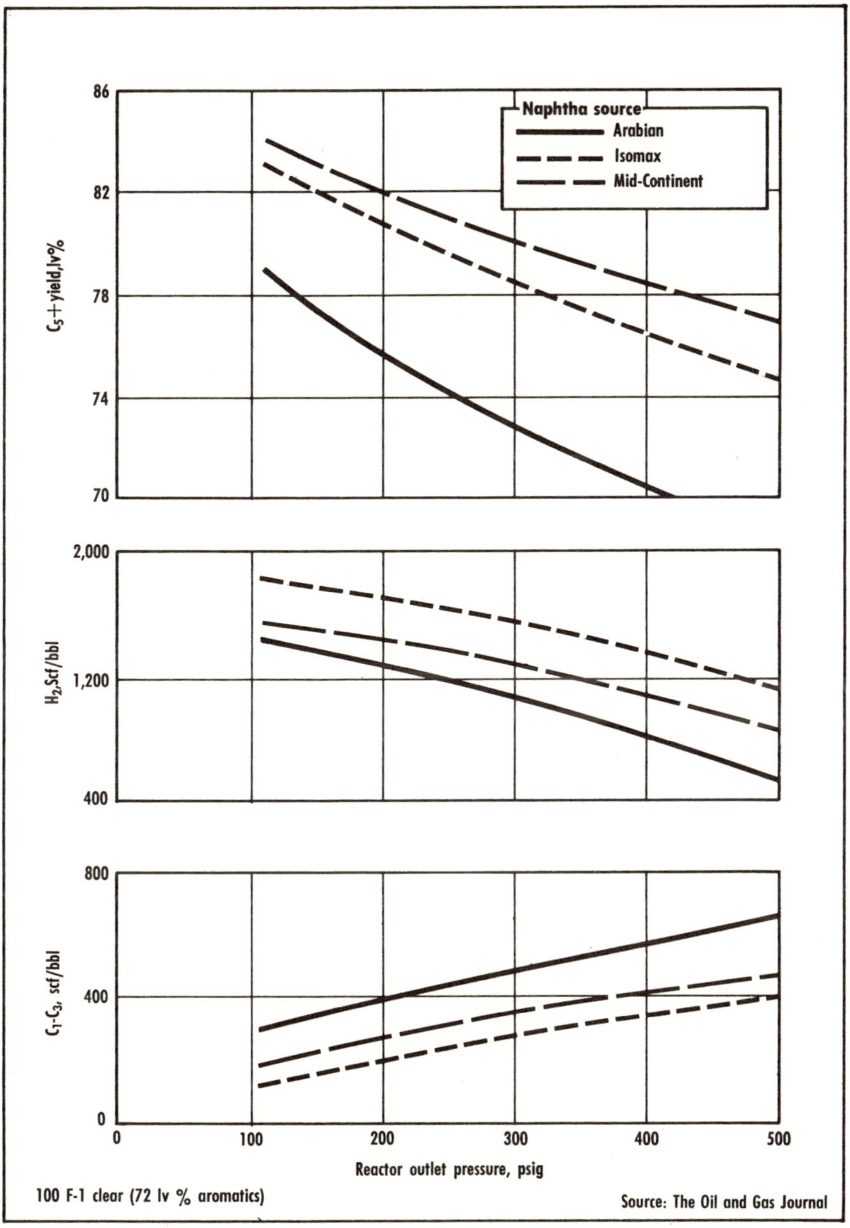

FIG. 4.4 Effect of pressure on reforming yield

The effect of reactor outlet pressure, a major process variable, on the yields of C_5+ stabilized reformate, hydrogen, and other light ends for some typical feeds are shown in Fig. 4.4. The incentive to go to lower pressures and develop catalysts resistant to deactivation due to the faster coking rates at these lower pressures is seen to be very great.

Modern reforming catalysts consist basically of platinum dispersed on an alumina base. Bimetallic catalysts which utilize rhenium and possibly palladium in addition to platinum, represent the latest advancements in catalytic-reforming technology. These new catalysts help the trend toward lower pressure, where the yield losses are significantly reduced if a suitable catalyst life can be maintained.

The ideal reforming catalyst first dehydrogenates all available naphthenes in the charge and then makes additional aromatics via dehydrocyclization of paraffins. At the same time the catalyst system should suppress the less desirable hydrocracking of heavier naphthenes and paraffins to form light ends which typically have only fuel-gas value.

Typical reforming catalysts contain between 0.3 and 1.0 wt % metal. The newer bimetallic catalysts are supposed to maintain their selectivity toward the desirable dehydrogenation of naphthenes and cyclization of paraffins to naphthenes as reaction temperature increases and as coke lays down on the catalyst.

A small concentration of "acid centers" on the catalyst in the form of chloride or fluoride, activated by small quantities of moisture, is necessary to promote the all important naphthene-dehydrogenation activity of the platinum, as well as controlled acid-catalyzed hydrocracking. In practice, a delicate balance must be maintained, since excessive moisture hydrolyzes the acid centers and in effect washes them away. Total halide content on catalyst generally runs between 0.5 and 1.0 wt %.

Fig. 4.5 summarizes the three major reactions which occur in the reforming process:

1. Naphthene dehydrogenation.
2. Paraffin dehydrocyclization.
3. Hydrocracking of naphthenes and paraffins.

Due to the highly endothermic nature of the naphthene dehydrogenation and paraffin dehydrocyclization reactions, the net reaction is endothermic and is usually carried out in a series of three to five reactors with fired reheat between reactors.

Well over 40% of the total installed catalytic-reforming capacity

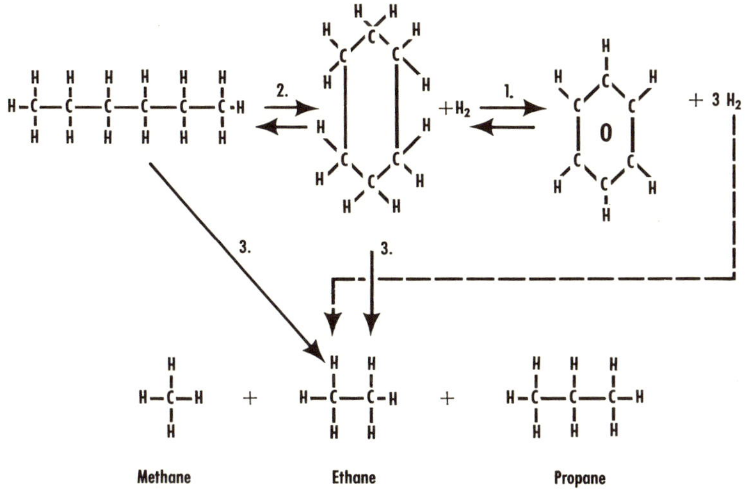

Source: The Oil and Gas Journal

FIG. 4.5 Reforming reactions

in the US is licensed from UOP under the commercial name of Platforming. The reformate from the UOP process is called Platformate. The other 50-odd percent of the domestic capacity is shared among these processes: Englehard, Standard of Indiana (Ultraforming), Mobil Oil, Esso (Powerforming), Houdry (Houdriforming), and Atlantic Richfield (Catforming).

CONCENTRATIONS OF PURE AROMATICS IN REFORMATE

The aromatics contained in high-octane reformates are in concentrations ranging from 45 to 70 vol %. The remaining constituents are the normal and iso paraffins which go through the reforming processing largely unconverted, and the small quantities of naphthenes which are not converted. Reformate typically contains less than 1% olefins.

A typical 93 to 95-octane reformate (unleaded) produced from a 200° to 375°F.-boiling-range naphtha charge stock will contain the approximate quantities of aromatics shown in Table 4.4. These are expressed as percentages on both stabilized reformate and fresh feed.

TABLE 4.4
Typical aromatics content in reformate

	Vol % of reformate	*Vol % of fresh feed*
Benzene	5	4
Toluene	24	19.2
Ethylbenzene	4	3.2
Paraxylene	4	3.2
Metaxylene	9	7.2
Orthoxylene	5	4
C_9 and C_{10} aromatics	4	3.2
	55	44.0

The total aromatics content can be brought up to 65 to 70%, as reformate octane is increased to 100. However, the net increase in aromatics based on naphtha charge is only 3 to 4%, due to the fact that the higher octanes are achieved by hydrocracking of low-octane paraffins. This produces less valuable light ends and results in considerably lower reformate yield.

The value differential between aromatics and paraffins is not sufficient in the normal refinery situation to support the marginal increase in aromatics yield gained by higher severities. This is illustrated in Table 4.5. A 3.5% increase in total aromatics yield results in a 11.5% decrease in raffinate available for gasoline or jet fuel, and an 8% increase in less valuable light ends. The current trend towards lead reduction in gasolines may force severity up despite the increased light-ends yields.

TABLE 4.5
Aromatics yield as function of severity

Reformer severities RON, clear	*C_5+ reformate, vol % feed*	*Aromatics, vol % of reformate*	*Aromatics, vol % feed*	*Paraffin raffinate vol % feed*	*Fuel gas and light ends, vol % feed*
89–91	85	48	41	44	15
93–95	80	55	44	36	20
99–101	72	66	47.5	24.5	28

SEPARATION OF AROMATICS

Aromatics are normally separated from the remaining paraffins by solvent extraction. The high-purity aromatics concentrate from extraction is then fractionated to produce the individual aromatic products.

Since the converted paraffins and naphthenes cover the whole spectrum of the naphtha boiling range and would interfere with any attempt to separate high-purity aromatics by conventional fractionation, it is not normally attempted. Instead, various solvent extraction processes, the most common of which are Udex (di and triethylene glycol) and more recently Sulfolane, are employed to obtain virtually a perfect separation between aromatics and non-aromatics.

A typical aromatics-extraction plant consists of a trayed or rotating disk contactor, at the top of which lean solvent in ratios greater than 3 to 1 on feed, is introduced. Feed enters at an intermediate point. Aromatic rich solvent taken off the bottom of the extractor is charged to an extract stripper which separates the aromatics from the solvent

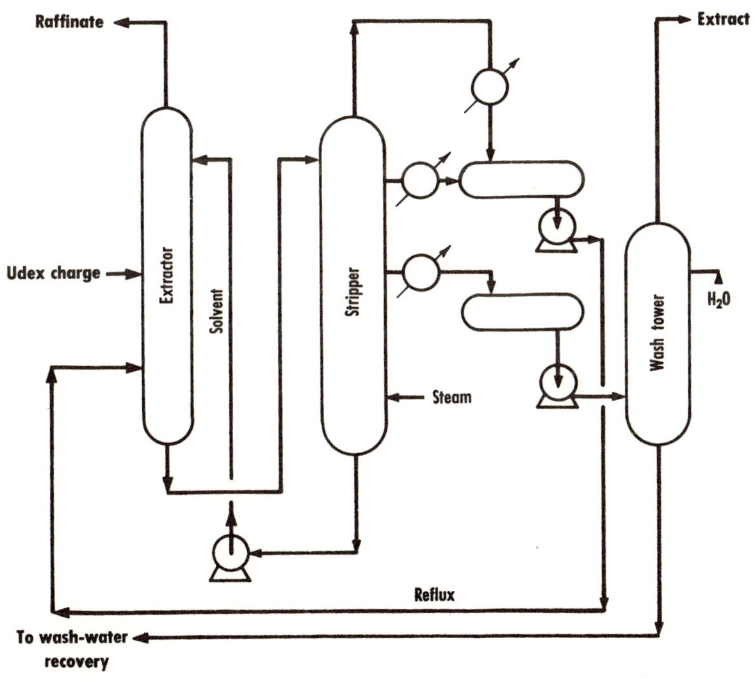

Source: The Oil and Gas Journal

FIG. 4.6 Aromatics-extraction flow scheme, Udex process

by steam stripping, which also drives light residual paraffins overhead. The dissolved aromatics are vaporized and stripped from the Sulfolane as it flows down the tower. Stripped aromatic vapors are condensed into the extract receiver. Intermediate stages of separation are often employed to collect impure aromatics for return to the extractor.

The extractor overhead, or raffinate product, is water-washed to remove any residual solvent.

A typical extraction flow scheme for the Udex Process is given in Fig. 4.6.

Recoveries of over 99% are normally obtained on benzene. Toluene recoveries typically run between 98 and 99%. Over 95% of the xylenes contained in the charge are also normally recovered. Lesser quantities of Cq and heavier aromatics are recovered into the extract.

The processing costs associated with the various solvent-extraction processes are directly related to the required solvent-to-feed ratio. Utility costs for pumping solvent/extract mixtures and steam for heating the rich solvent to drive the aromatics off overhead are by far the largest operating cost components for aromatics extraction. The more recently developed Sulfolane Process, originally developed by Shell and licensed through UOP uses a highly polar compound, tetrahydrothiophene dioxide (sulfolane), which permits reduced solvent-to-feed ratios and at the same time increases total recovery of xylenes and heavy aromatics. The chemical structure of sulfolane is:

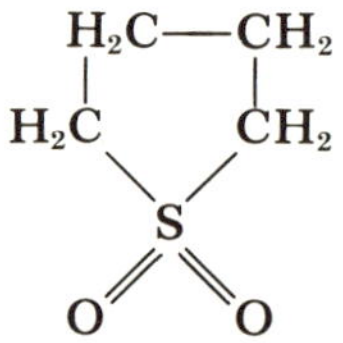

As reformate severity goes up, or if reformate to extraction is supplemented by the more concentrated aromatics streams available in pyrolysis gasolines, the solvent-to-feed ratios will go even higher. The logical next step in aromatics recovery would be a process to selectively extract the lower volumes of non-aromatics from a 60 to 70% aromatics-rich feed instead of the other way around. No process has yet been commercialized, but the idea may have great potential.

After clay treating for trace olefin and sulfur removal, the aromatic-rich extract from the solvent stripper is normally separated into

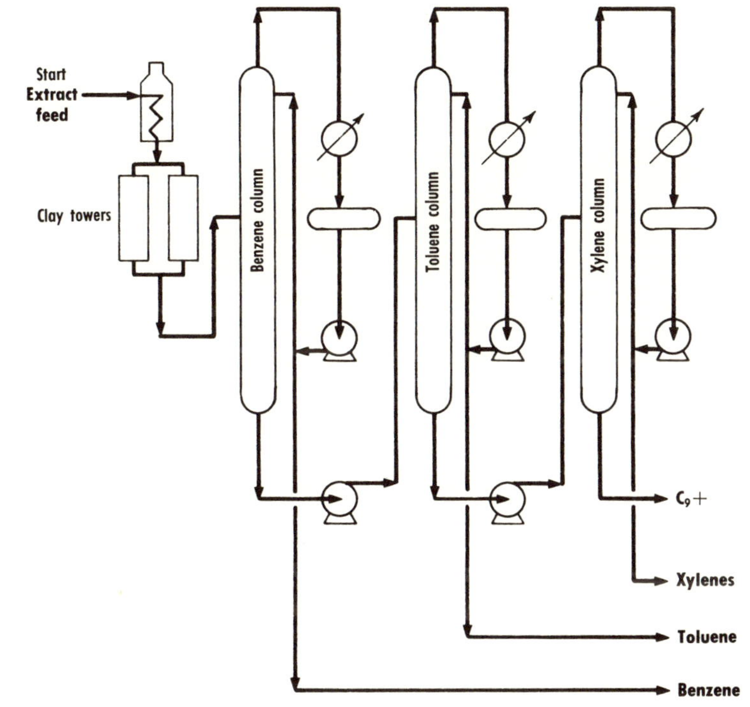

Source: The Oil and Gas Journal

FIG. 4.7 Fractionation of high-purity aromatics

TABLE 4.6
Product qualities

	Benzene	*Toluene*	*Mixed xylenes*
Benzene content	*99.9%*	300 ppm	nil
Toluene content	300 ppm	*99.8%*	700 ppm
C_8 aromatics content	nil	900 ppm	*99.0%*
ASTM distillation range, °F.	1	1.2	4–5
Specific gravity @ 15.5°C.	0.884	0.870	0.872
Lb/gal	7.36	7.25	7.26
Non-aromatics content, ppm	150	250	750

a 99.9+% pure benzene, 99.8% pure toluene and a 99% pure mixed xylene stream in a series of conventional overhead refluxed fractionation columns, as shown in Fig. 4.7.

Typical product qualities produced in modern extraction and aromatics fractionation facilities are given in Table 4.6.

ACTUAL PRODUCTION OF BTX

Expressed in terms of actual aromatics recovered for chemical and solvent use from petroleum vs. availability in reformate, only benzene recovery approaches a level which is significant. About 50% of the potential benzene in US reformate is recovered.[1] The comparable figures for toluene and xylene are both about 10%. If we ignore toluene going back into gasoline, toluene recovery for chemical use is closer to 8%. About 9% of the available xylenes are recovered.

Rather than tieing back to crude oil and "native BTX," as was done in Table 4.2, 1970 US reforming capacity is extrapolated in Table 4.7 in terms of aromatics potential based on 89 to 91-RON reformate compositions, this being typical of current severity levels.

TABLE 4.7
Aromatics contained in average US reformate (89 to 91 octane)

1970 US reforming capacity, Bbl/day	*Aromatics yields from reforming*			
	As % of charge	*Bbl/day*	*MM gal/yr*	*Mm lb/yr*
2,700,000	3.7% benzene	100,000	1,470	10,800
	17.8% toluene	480,000	7,050	51,000
	16.3% xylene	440,000	6,500	47,000

Of the 1,470 million gal of potential reformate benzene available, no more than 740 MM gal, or 50%, is recovered. This is based on the estimated overall benzene sources in Table 4.8.

Realistically speaking, the actual recovery figures for benzene may be more like 60% of that produced, since the refiner (who is typically short of reforming capacity) will cut his naphtha to a higher initial boiling point when not going after maximum benzene recovery. This will permit him to fill his reformer with his lowest-octane heavy

[1]This figure must allow for an important assumption as to exactly how much of the 500 MM gal/yr of toluene-dealkylation capacity was actually operated in 1970. The indications are that 50% to 60% was operating.

TABLE 4.8
Estimated benzene sources

	MM gal/yr
Coal tar	120
Pyrolysis gasoline	90
Toluene dealkylation	250
Reformate	740
	1,200

naphtha. This practice rejects native benzene and the benzene-forming C_6 naphthenes to the lighter straight run gasolines still used in blending leaded regulars.

Fig. 4.8 is a historical benzene-production chart. Projections of 1980 benzene requirements for chemical use generally run between 2.5 and 3.0 billion gal.

US toluene production reached 800 MM gal in 1970, with 40 to 45% of the recovered toluene going back into benzene as dealkylation feed. Solvent and chemical use accounted for about 20% of the

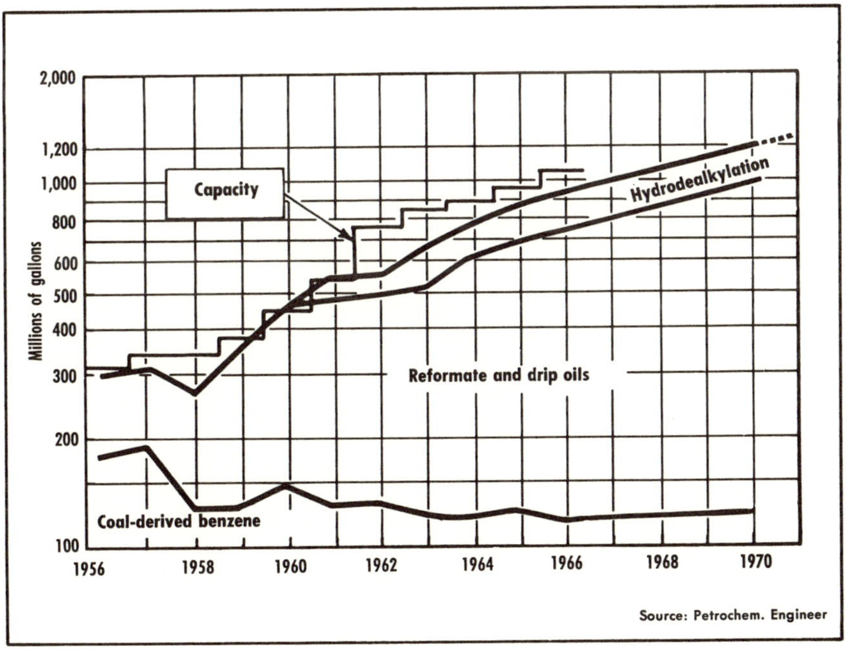

FIG. 4.8 US benzene production

toluene, and at least 200 MM gal went back into gasoline as blending stock.

The historical figures for toluene are presented in Fig. 4.9.

Current US demand for recovered C_8 aromatics is about 600 MM gal/yr, with 92 to 93% coming directly from reformate. Slightly more than half of this is represented by the para and ortho-xylene markets, most of the rest going to solvents or even back into gasoline as mixed xylenes for blending.

ANOTHER MAJOR SOURCE OF AROMATICS

The steam cracking of ethane and propane to produce ethylene and propylene yields small quantities of an aromatics concentrate which is sometimes called drip oil, or pyrolysis gasoline, Table 4.8. However, the cracking of heavier feeds such as naphtha and gas oil for olefin production produces substantial volumes of pyrolysis gasoline. A billion-lb/yr ethylene plant charging gas oil will produce 8,000 to 10,000 bbl/day of pyrolysis gasoline.

The pyrolysis materials are very rich in aromatics, but at the same time are extremely unstable due to high diolefin and mono-olefin

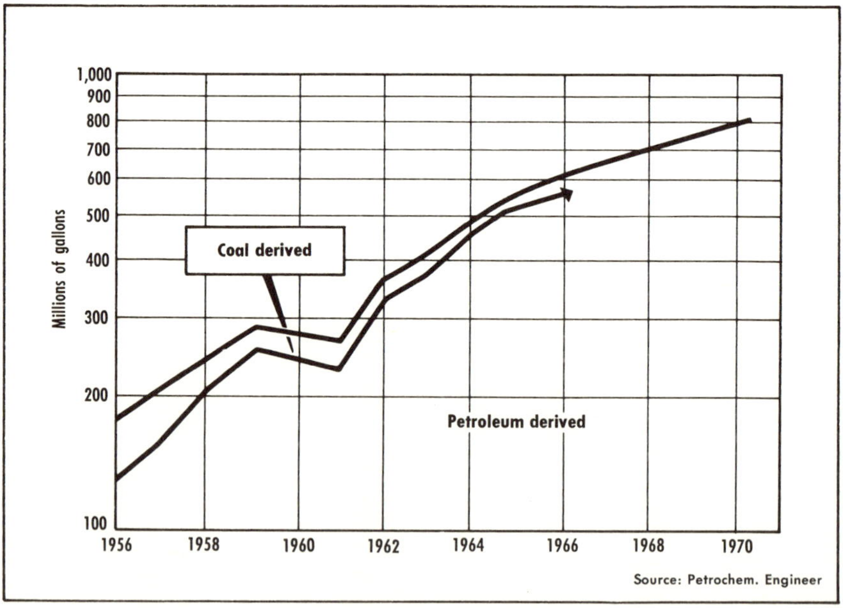

FIG. 4.9 US toluene production

contents. Also, most of the residual sulfur in the olefin-plant feed is concentrated in the drip oil.

A typical pyrolysis gasoline may contain over 70 wt % aromatics, about half of which is benzene.

TABLE 4.8
Typical hydrocarbon analysis, pyrolysis gasoline

Diolefins	15
Mono-olefins	8
Saturates	7
Benzene	32
Toluene	14
C_8 aromatics	11
C_9+ aromatics	13
	100

The billion-lb/yr ethylene plant cracking gas oil could thus produce 3,000 bbl/day of benzene, or over 300 million lb/yr (40 million gal/yr) of benzene.

Probably no more than half the presently available pyrolysis gasoline is currently processed for aromatics recovery. However, with the expected trend toward large naphtha and gas-oil cracking complexes on the Gulf Coast and in Puerto Rico, it is virtually certain that this source of aromatics will grow faster than any other in the next five to ten years.

Marketable aromatics can be produced from pyrolysis gasoline

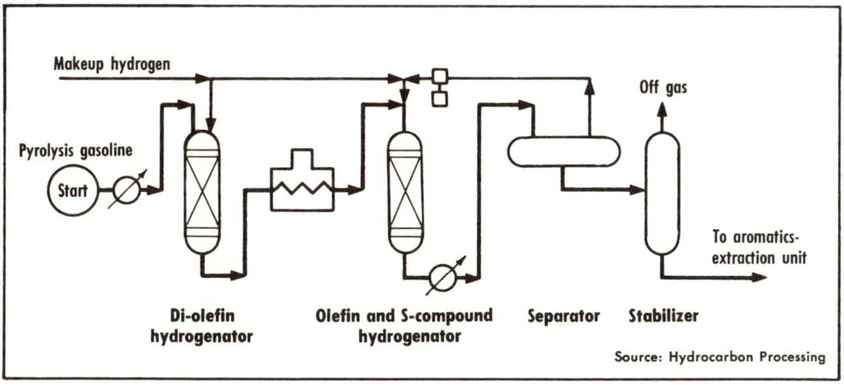

FIG. 4.10 Two-stage hydrogenation of pyrolysis gasoline

after hydrogenation and virtual complete desulfurization. This is typically a two-stage process, using a low-temperature catalyst for diolefin saturation and a more conventional high-temperature hydrogenation stage for mono-olefin saturation and sulfur removal. A typical pyrolysis gasoline treating facility is shown in Fig. 4.10.

BENZENE FROM TOLUENE BY HYDRODEALKYLATION

Both potential and actual yields of toluene from naphtha reforming are two to three times that of benzene. On the other hand, the demand for benzene has always been, and for the foreseeable future will remain, substantially greater than that for toluene, which has not as yet really found a large-scale chemical use.

Both catalytic and thermal processes for dealkylation of toluene to benzene have been available since about 1959. Pure benzene results from the quantitative demethylation of the toluene to yield benzene and methane.

$$\begin{array}{ccccc} C_7H_8 + & H_2 \rightarrow & C_6H_6 + & CH_4 \\ 92 \text{ lb} + & 2 \text{ lb} & 78 \text{ lb} + & 16 \text{ lb} \end{array}$$

To avoid coking, the dehydrogenation reaction is carried out at elevated pressures, in the presence of excess hydrogen. Pressures between 600 and 900 psig are generally used. Temperatures as high as 1,400°F. are employed for thermal dealkylation. The catalytic processes typically operate at 200 to 300°F. lower temperatures, but are more sensitive to coke laydown.

Since the dealkylation reaction is highly exothermic, cold hydrogen is used as quench in the reactor system. Mol ratios of up to 8 to 1 hydrogen-to-toluene are used. Toluene conversion/pass is generally in the 70 to 80% range. One thermal process claims 85%. A typical toluene dealkylation plant flow scheme is given in Fig. 4.11.

A 100% efficient operation with 100% ring retention would yield 84.7 wt % of benzene, or just about 83 vol % benzene on toluene charge. Most licensors claim yields in excess of 97% of theoretical, or about 80 wt % of benzene from toluene.

In the US we currently have about 500,000,000 gal/yr of hydrodealkylation capacity available. It is split about evenly between the catalytic processes (UOP's Hydeal and Houdry's Detol) and thermal processes (Atlantic Richfield and HRI's HDA process and Gulf's

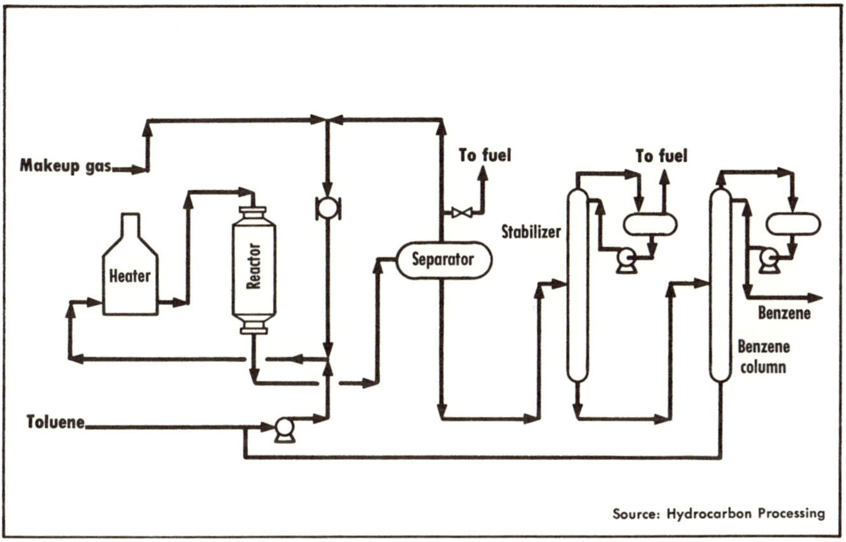

FIG. 4.11 Toluene hydrodealkylation plant

THD). The companies which offer catalytic processes are the ones which also happen to be catalyst manufacturers.

Even if toluene-dealkylation processing were free, which it certainly is not, benzene would have to command a 4 to 5¢/gal premium over toluene before it would pay to dealkylate. This is based on shrinkage alone. To support the capital and operating costs for a new facility, the refiner normally requires a guaranteed 8 to 9¢/gal price differen-

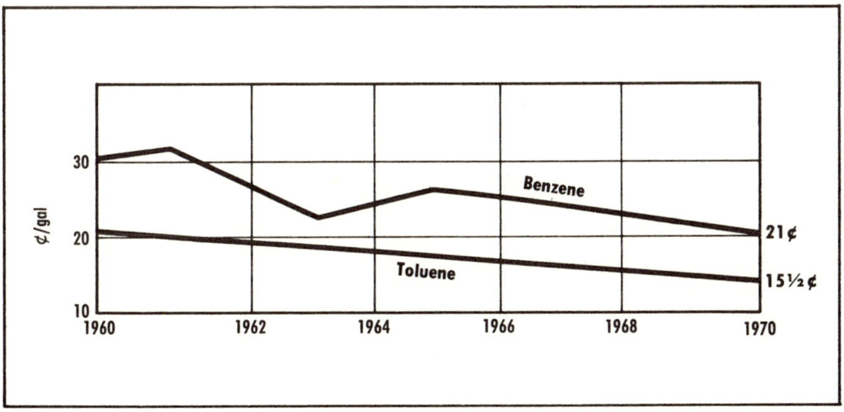

FIG. 4.12 Recent benzene-toluene price history

tial for benzene over toluene. The facts of life in the recent aromatics markets have been more like a 6 to 7¢/gal spread. Toluene has realistically been selling for 15 to 16¢/gal, while benzene has been at 21 to 22¢/gal, with recent sales reported below 20¢ (see Fig. 4.12).

The price spreads of the early 1960's resulted in a great deal of dealkylation capacity being installed, but many dealkylation plants were shut down beginning in 1963. As a general rule of thumb, it is pretty certain that toluene-dealkylation plants shut down when benzene values drop below 22 to 23¢/gal. Toluene is almost always worth a minimum of 15¢/gal in gasoline. Some estimates of 1970 and 1971 benzene sources indicate that only about 250 million of the 1,200 million gal produced came from dealkylation of toluene. This would

TABLE 4.9
Toluene dealkylation economics

	$/day	*$M/350-day yr*
Raw material		
3,000 bbl/day toluene at 15.5¢/gal	19,550	6,850
17 MMscfd hydrogen at 25¢/Mscf	4,280	1,500
	23,830	8,350
Operating costs		
Fuel, power, steam, cooling water	715	0,250
Catalysts and chemicals	30	0,011
Labor and supervision	285	0,100
Maintenance, 3% of erected cost ex royalty	220	0,078
Insurance and taxes, 2% of onsites	150	0,052
Interest on capital, 7%/yr	600	0,210
Depreciation, 15-yr straight line	490	0,170
Total operating costs	2,490	0,871
Products		
2,400 bbl/day benzene at 24¢/gal	24,200	8,460
17 MMscfd vent gas at 15¢/Mscf (to fuel)	2,550	890
Total product value	26,750	9,350
Profitability		
Total product value	26,750	9,350
Raw material cost	(23,830)	(8,350)
Operating cost	(2,490)	(0,871)
Pretax profit	430	0,150
After-tax profit—tax @ 50%	215	0,075
Depreciation	490	0,170
Net cash flow	705	0,245
Annual cash flow, % of investment	—	8.2%

indicate that no more than 60% of available dealkylation capacity was operating.

The economics of a 3,000-bbl/day dealkylation unit, costing a total of $3,000,000 to erect (including $400,000 to obtain a paid up royalty) is extremely sensitive to raw material and product values. An economic calculation shows a generally unacceptable profit situation with a 15.5¢/gal alternate disposal value for toluene and 24¢ for benzene. See Table 4.9.

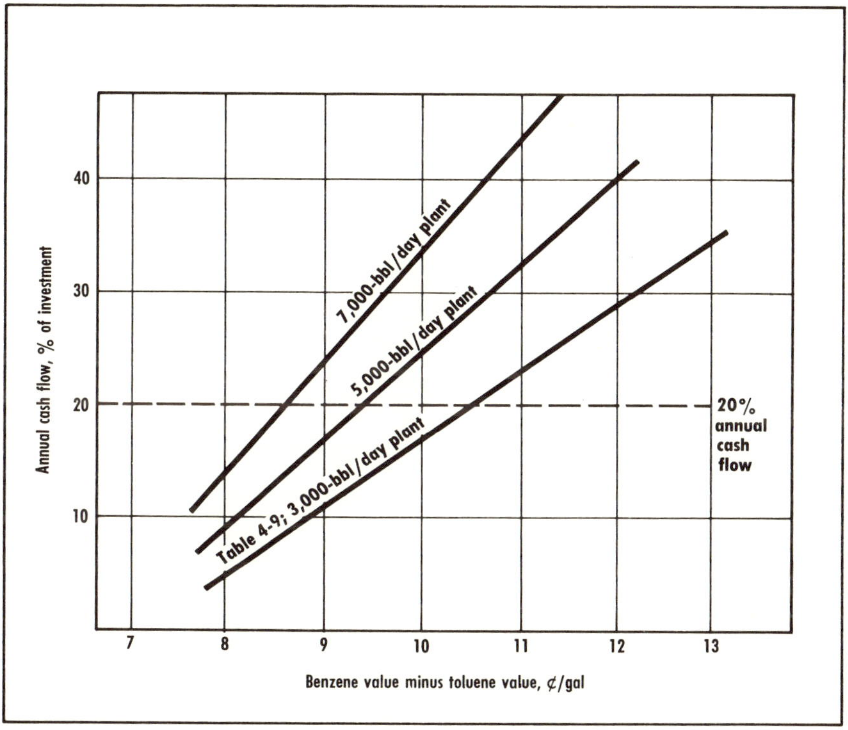

FIG. 4.13 Toluene-dealkylation cash flow

The general relationship between the benzene/toluene value differential and the annual cash flow is plotted in Fig. 4.13, with plant size as parameters. The value difference would have to be 10.5¢/gal to show an annual cash flow of 20% on investment for the plant detailed in Table 4.9. A 7,000-bbl/day plant might achieve this same percentage return at an 8.5¢/gal price spread.

XYLENE PRODUCTION

Over 98% of the C_8 aromatics used in chemical processing are derived from petroleum sources. Of the petroleum-derived xylenes, 95% come from reformate and about 5% from pyrolysis gasoline.

The total yield of C_8 aromatics available from refinery catalytic reforming is about the same as that of toluene when the refiner maximizes BTX reformer feed end point. C_8 aromatics are therefore about 22% on reformate, or 18% on naphtha charge at a 93 to 95-octane severity level.

The C_8 aromatics found in catalytic reformate generally occur in the ratio shown in Table 4.10, with some variation based on crude source and processing severities.

TABLE 4.10
C_8 aromatics, typical breakdown as found in catalytic reformate

	Typical %	*Range* %
Ortho-xylene	23	19–26
Meta-xylene	40	35–40
Para-xylene	17	16–20
Ethylbenzene	20	17–21

Some significant physical properties of the C_8 aromatics are given in Table 4.11.

TABLE 4.11
C_8 aromatics physical properties

	Ortho-xylene	*Meta-xylene*	*Para-xylene*	*Ethylbenzene*
Boiling point, °C.	144.4	139.1	138.3	136.2
Freezing point, °C.	−25	−48	+13	−95
Density, lb/gal	7.37	7.23	7.21	7.26
Structure	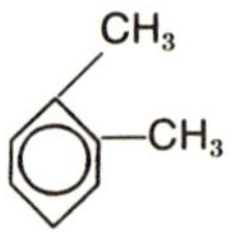	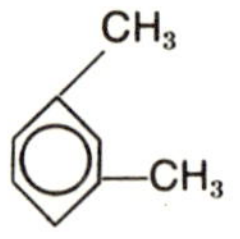		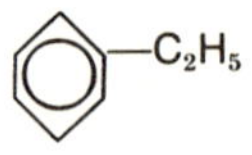

These properties indicate that conventional fractionation to separate the four isomers is virtually impossible. However, 95+% pure ortho-xylene can be taken as a bottoms product on a 100-tray fractionation tower (150 to 200 ft tall) with reasonable reflux ratios. Some refiners do fractionate out ethylbenzene but this takes over 300 trays and fantastic reflux ratios. Most of the ethylbenzene used in styrene manufacture is produced by alkylating benzene with ethylene. The utilities involved in the ethylbenzene superfractionation operation are usually prohibitively expensive, except possibly on the Gulf Coast. The much higher freeze point of p-xylene versus the other isomers has led to successful separation via the crystallization route.

Domestic para-xylene recovery has thus far been based on various crystallization processes. These generally use two stages of crystallization to produce a 99+ % pure product. In the first, or coldest stage, crude para-xylene is frozen out at −80° or −90°F. via indirect refrigeration. The para-xylene is separated from the filtrate (mother liquor) either by a filter-scraping device or by centrifugation. The relatively impure p-xylene from the first stage is concentrated, either by successive stages of partial remelting and recrystallation or a second recrystallation step at about −25°F. Filtrate from the second stage is recycled back to the first stage crystallizer.

Recently a modified crystallization process licensed by Phillips Petroleum has gained attention. The Phillips process employs a final "crystal-purification column," in which several zones for filtration, crystal purification, and contact with high-purity reflux are controlled by a pulsating-energy source.

In all crystallization processes, the formation of eutectics with ortho and meta-xylene normally limit the effective recovery of para-xylene to about 60% of that contained in the reformate, or not much more than 10% of the total C_8 aromatics in the stream.

Para-xylene recovery complexes often include an isomerization step which is used to shift methyl groups, converting meta and ortho-xylene to additional para. Some of these processes claim conversion of the ethylbenzene to xylenes.

A xylene complex can be designed to afford flexible yields of para and ortho-xylene in order to take maximum advantage of market shifts, etc.

A flow chart for a flexible C_8 aromatics processing plant is shown in Fig. 4.14. The ortho-xylene recovery step can be placed upstream of the isomerization unit if a greater emphasis is placed on this product.

Xylene isomerization processes typically are of the fixed-bed type. UOP and Englehard offer noble-metal catalysts (platinum on silica-

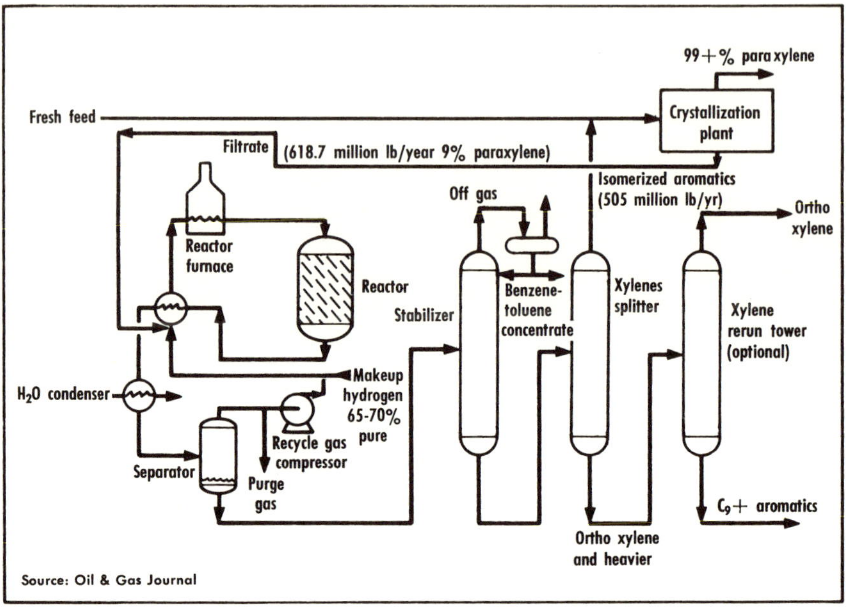

FIG. 4.14 Xylenes processing

alumina) which will both isomerize the ethylbenzene and disproportionate some of it to toluene and benzene. Fresh feed would normally consist of crystallization–unit effluent which is mixed with recycled hydrogen (up to 10-to-1 mol ratio hydrogen to feed) and heated to reaction temperatures (700° to 900°F.). The reaction products are cooled and fractionated.

The C_8 aromatics in the reactor effluent are an equilibrium mixture consisting of the same relative proportions of isomers as in the starting reformate, since the isomerization process conditions are similar.

Isomerization reactor effluent is recycled directly back to the crystallizer, or possibly to an ortho-xylene recovery column first, but only after fractionation to remove both the lighter products (mostly benzene and toluene) and the C_9+ aromatics produced in the reaction.

The licensors of the non-noble metal catalyst processes (Esso, Atlantic, ICI) claim less production of both light and heavy byproduct streams, but significant quantities of unconverted ethylbenzene is withdrawn in a mixed-xylene byproduct stream. The relative economics of the two isomerization routes are generally determined by the alternate disposal value to the processor of C_8 aromatics other than paraxylene, and total feed availability and quality vs. para-xylene requirements.

The most recent advances in C_8-aromatics processing technology have been aimed at getting around the costly crystallization step for p-xylene recovery.

UOP has come out with the "Parex" process which separates p-xylene with a selective adsorbent, and for which over 90% para-xylene recovery is claimed. This in itself is potentially a very valuable step, since it will have the overall effect of both reducing the amount of charge required to xylene isomerization and at the same time improve the isomerization reaction equilibria, since the charge will be freer of p-xylene.

In the Parex process, a liquid-phase mixture is passed over a bed of solid adsorbent which retains the "more slender" para-xylene molecule. A hydrocarbon wash is used to desorb the para-xylene. The hydrocarbon used for the wash is selected so that it can either easily be separated from the para-xylene by distillation. The first Parex unit came onstream late in 1971 in Europe.

The Parex process simulates a batchwise adsorbtion process by making use of a complicated multiport value which permits moving the feed and desorbent inlets and product withdrawal points around the bed. This permits various parts of the bed to act in adsorbent and desorbent services in continuous sequence.

The Japan Gas-Chemical Co. has announced a process which selectively extracts the meta-xylene from the other isomers and then isomerizes them to an equilibrium mixture of xylenes at conditions favorable to the maximization of para-xylene in the isomerizate.

The meta-xylene forms a complex with HF and boron trifluoride (BF_3) and is extracted at ambient temperature. By applying heat to the extract, the complex is broken down and the HF-BF_3 solvent mixture is recycled to the extractor. The 99+% purity meta-xylene is sent to low temperature isomerization which uses HF-BF_3 as a catalyst. Low-temperature isomerization favors the yield of para-xylene vs. the 800°F.+ temperatures used with noble-metal catalysts.

CONVERSION OF TOLUENE AND HEAVY AROMATICS TO BENZENE AND XYLENE

The marginal economics associated with toluene dealkylation has led to the commercialization of processes for "disproportionating" two molecules of toluene into one molecule of benzene and one of xylene:

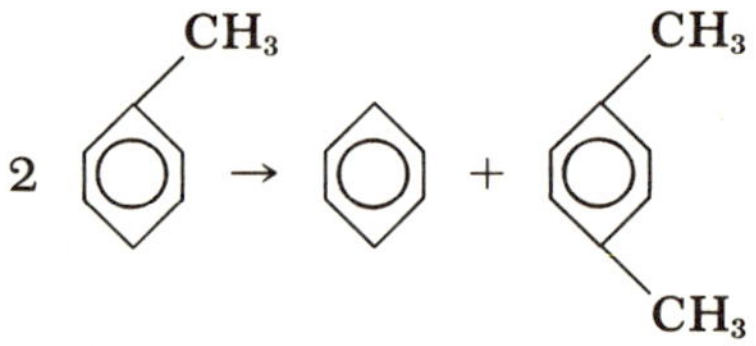

One process is licensed by Atlantic-Richfield and is called "Xylenes-Plus," since it can also be used to "transalkylate" some of the heavy C_9 aromatics into additional xylenes by reacting them with toluene.

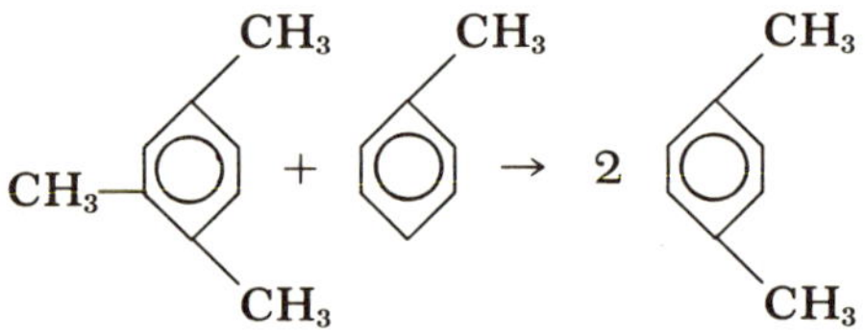

UOP licenses the other process in conjunction with Toyo Rayon, and it is called "Tatoray." Both processes are vapor-phase catalytic, using non-precious metals.

Atlantic-Richfield claims a great deal of flexibility to vary the ratios of xylene to benzene between 1.5 and 10, depending on the amount of C_9 charged. The process can therefore be operated in accordance with changing market demand patterns and the relative price strengths of benzene and xylene. C_9 aromatics are not particularly valuable to most refiners.

The xylenes produced by disproportionation have the additional advantage to most processors of containing essentially no ethylbenzene. Literature data published by Atlantic indicate that ortho and para-xylenes are each about 25% of the C_8 product:

	Reformate xylenes	*Disproportionation/ transalkylation xylenes*
Ortho-xylene	23	25
Meta-xylene	40	50
Para-xylene	17	25
Ethylbenzene	20	—

Although the theoretical disproportionation yield for a 100% toluene feed is 50/50 benzene/xylene, in reality only about 37% of benzene and 55% of xylene is obtained. Some C_9 is produced, and there is some cracking to light paraffins and olefins. However, on the whole 95 to 97% of the feed is retained as valuable aromatics vs. hydro-dealkylation's 80% benzene yield.

Disproportionation and transalkylation also have the advantage of not consuming any outside hydrogen, although the UOP process uses a hydrogen atmosphere to minimize coke laydown on the catalyst.

Table 4.12 compares typical yields for hydro-dealkylation and disproportionation/transalkylation, showing how the xylene-to-benzene ratio can be varied as C_9's are added.

TABLE 4.12
Various product ratios possible with disproportionation/transalkylation

		Products		
	Charge	*Benzene*	*Xylenes*	*Xylene/benzene ratio*
Hydrodealkylation	2,000 toluene	1,600	—	—
Disproportionation	2,000 toluene	740	1,100	1.5
Transalkylation	2,000 toluene 1,000 C_9A	400	2,470	6.2
Transalkylation	2,000 toluene 2,000 C_9A	350	3,490	10

In general, the overall economics favor toluene dealkylation only when incremental benzene, and not xylenes, is the requirement, or when xylene does not command a price differential over toluene.

Disproportionation is favored when both incremental benzene and xylenes are required. Transalkylation would only be attractive when xylenes are really the desired end product, since benzene yield is low in this situation.

UNLEADED GASOLINE

The need to move positively in the direction of removing the lead from 6,100,000 bbl/day, of motor gasoline produced and sold in the US will have some interesting, if not startling, effects on the aromatics business.

In recent years the aromatics content in premium and regular gasolines in the US has leveled out at 24% in premium and about 20% in regular. With premium sales now at about 40% of the total, this means that 21.5% of the total gasoline sold, or 1,300,000 bbl of aromatics find their way into gasoline daily. This is equivalent to 20,000 MM gal/yr, or eight times the total chemical markets for aromatics. Not all of the aromatics in gasoline comes from reforming, since catalytically cracked gasoline, which makes up 40% of the gasoline pool, also contains about 25% aromatics. This is about half the total contained in gasoline; however, a large proportion of the aromatics in cat-cracked gasoline are C_9's and C_{10}'s.

Trends in gasoline hydrocarbon type composition are plotted in Figure 4.15.

Lead normally adds 8 research octane numbers to the gasoline pool. Today the unleaded pool averages 89 octane. If we assume that some time within the next five to eight years it is highly likely that all the lead will have to be out of gasoline, and further that the unleaded regular will average 3 octanes below today's 94 and the premium 4 octanes below today's 100, the entire unleaded pool will have to be 4 to 5 octane numbers higher than today's pool.

Without the need for lead removal, the aromatics used in gasoline would continue to grow by the same 3–4% as total gasoline sales, or 600 to 700 MM gal annually. But, in order to satisfy this set of assumptions regarding the future no-lead posture, the aromatic content of the gasoline is likely to go up from today's 21.5% to at least 30%. This

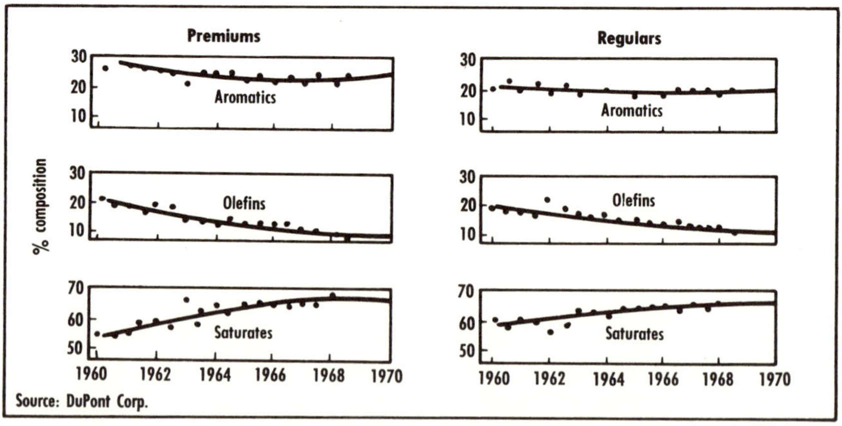

FIG. 4.15 Hydrocarbon types in US gasoline

will require not only increased reforming severities, but extraction of aromatics concentrates from reformate for use in gasoline only. This will create strong competition for the future aromatics barrel from the gasoline pool.

A distillation curve for a typical gasoline is given in Figure 4.16, with some typical individual hydrocarbons, including the important aromatics, spotted on the curve.

An examination of the leaded and unleaded octanes of the major refinery gasoline-blending components shows that most refinery streams are suitable for blending into current leaded gasolines. But when the refiner begins to look for high-octane unleaded components

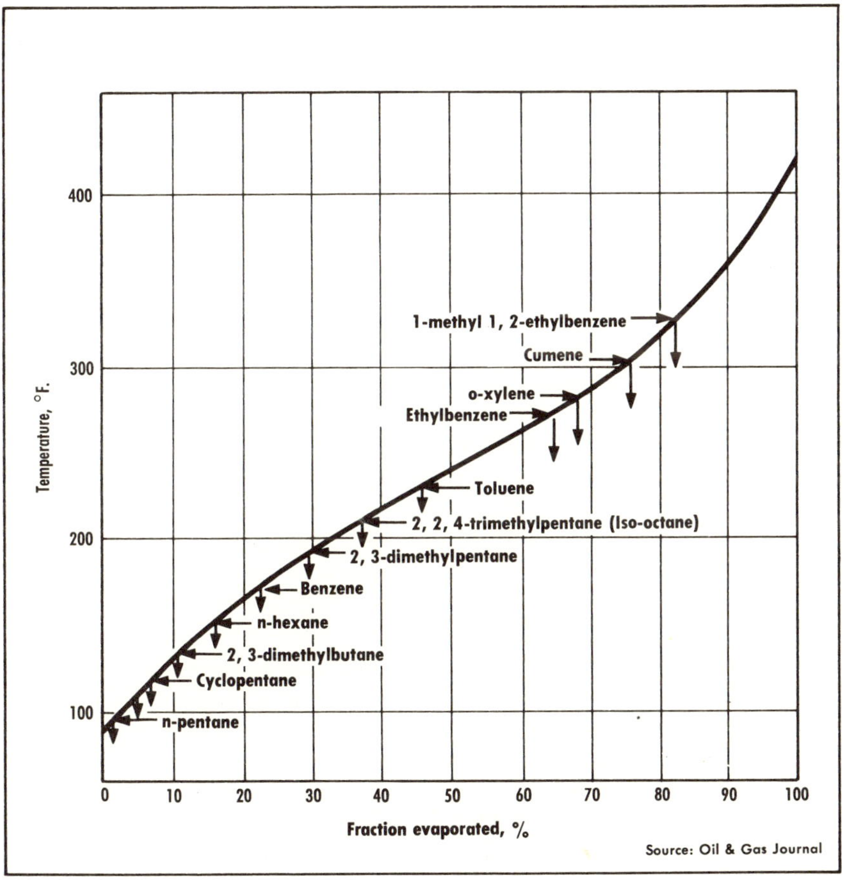

FIG. 4.16 Distillation curve for typical gasoline

his attention will be (and already has) focused on aromatics as the highest-octane replacement components readily available.

A C_6-C_9 aromatic concentrate from extraction will blend into unleaded gasoline with an apparent or blending octane number of 111 to 113. Removal of the more valuable benzene for chemical use will marginally improve the blending characteristics. Removal of ortho-xylene for chemical use will noticeably improve the blending value of the remaining aromatics. Table 4.13 gives typical blending octane numbers for aromatics.

TABLE 4.13
Aromatics clear & leaded research octane blending values

	0 gm lead	*3 gm lead*
Benzene	108	112
Toluene	112	115
Ethylbenzene	113	116
P-xylene	114	115
M-xylene	114	116
O-xylene	100	96
Typical C_9A mix	103	107

Descriptions of all the major refinery gasoline-blending components, their source, major hydrocarbon constituents, and octane ratings, clear and leaded, are given in Table 4.14.

Much engineering and computer time has been, and will continue to be spent by the refining industry on evaluating alternate routes to producing unleaded gasoline. These economics invariably consider the alternate disposal value for potential gasoline components into chemical outlets, other refinery feedstock, or product blending use. The aromatics will always be examined in light of their value in the chemical market. Low-octane raffinate will be looked at as olefin feed, both for sales and for internal use, to make iso-butylene for alkylation. Or it will be considered as a constituent of jet fuel. The cost of isomerizing light straight run will be compared to selling it as cracking feed, etc.

All of this gets further complicated by its effects on the entire refining operation—the amount of crude to be run, the supplying of raw materials to the refinery, and the availability of plant capacity. Components which are sold off as no longer being suitable for gasoline blending must in the end be replaced by more crude run to stills,

TABLE 4.14
Major refinery gasoline blending components

Gasoline-blending component	*Source*	*Major hydrocarbon constituent*	*Research unleaded*	*Octane rating, 3 gm Pb/gal*
Light straight run naphtha-LSR	Crude-oil distillation	C_5 & C_6 paraffins & cyclo-paraffins	60–72	82–89
Cat naphtha	Catalytic cracking	C_5 to C_{10} paraffins, olefins, some aromatics	91	98
Alkylate	Alkylation	Primarily C_7-C_8 branched paraffins	94	105
Isomerate	Isomerization	C_5 & C_6 branched paraffins	82–92	92–101
Reformate	Catalytic reforming	C_6 to C_9 aromatics, paraffins	90–95	97–101
Hydrocrackate	Hydrocracking	C_5 to C_7 paraffins	85	97
Extract	Solvent extraction of reformate	C_6 to C_9 aromatics	111	114
Raffinate	Solvent extraction of reformate	C_6 to C_6 paraffins	65	85

more catalytic reforming, more alkylation, etc. Investment programs must be time-phased into a lead-removal program which as yet has no firm timetable.

No matter how complex the unleaded gasoline planning picture, there is one simplifying factor. It is quite safe to assume that on a short-term basis that the individual refiner looks at his economic evaluations and makes many of his decisions in terms of his primary product which, of course, still is gasoline.

What does the refiner find when he stacks up aromatics as gasoline-blending components vs. their alternate value as chemicals? The answer right now is one which encourages him to keep the aromatics in the gasoline.

Leaded 100-octane premium gasoline sells for about 13¢/gal on the US Gulf Coast in cargo quantities. Seasonal variations and other supply/demand factors generally keep the price in the 12.5 to 13.5¢/gal range. Regular sells for 1.75 to 2.0¢/gal less than premium.* Most refiners actually consider regular to cost about 1.5¢/gal less to produce than premium.

On a marginal basis, the value of a mixed toluene/xylene stream in upgrading regular to premium can be calculated. These economics include a "volatility bonus" for aromatics, which permits additional low-cost butane to be blended into the gasoline for constant vapor pressure. These figures are presented for a hypothetical 1,000 bbl of regular gasoline in Table 4.15.

TABLE 4.15
Regular gasoline upgraded to premium

	Bbl	*%*	*¢/gal*	*$/bbl*	*Leaded RON*	*Vapor pressure*
Regular gasoline	1,000	70.0	11.5	4.83	94.5	10.5
Mixed toluene-xylene	380	26.5	17.9	7.50	113.0	2.0
Butane	50	3.5	6.0	2.52	101.0	60.0
Premium	1,430	100.0	13.0	5.46	100.0	10.5

By back calculating the mixed toluene-xylene value from the known premium and regular prices, the cost of available butane, the cost of lead, etc., the refiner comes up for a marginal aromatics value of $7.50/bbl in premium gasoline. Backing out the cost of lead at 0.65¢/gal (27.5¢/bbl) for 3 gm/gal, the aromatics are worth a net 17.25¢/gal in the gasoline pool.

Value at 3 gm lead	17.90¢/gal
Less cost of lead	0.65
Net value in pool	17.25¢/gal, or $7.23/bbl

With both chemical-quality toluene and xylenes selling for 16¢/gal, or less, it is obvious that substantial quantities of aromatics are going into gasolines right now, even in advance of lead-removal legislation.

Obviously, there is not an unlimited market for premium, and every refiner could not follow through with this exercise to any great extent without violently upsetting the market. Nevertheless, the aro-

*East Coast prices are generally 1.0¢ per gal above the Gulf Coast to make up for additional transportation costs—either as crude to East Coast Refineries, or as finished gasoline to bulk terminals.

matics-blending route to meet incremental premium demand is a tried-and-true fact of refinery economics.

During the first stages of the transition to unleaded gasoline, which may well be a 91 to 92-octane unleaded regular, with a continuation of 100-octane leaded premium until most of today's high compression ratio engines are off the road a few years hence, the refiner will be forced to blend another 3 to 5% of aromatics to regular, or about 130,000 bbl/day (1,900 MM gal of aromatics/yr). This is likely to put a strain on chemical supply which has traditionally been growing at a much faster annual rate than gasoline.

5

ELLIOT M. MAGER

Olefins production

STEAM PYROLYSIS OF HYDROCARBONS, THE PRIMARY SOURCE OF OLEFINS

ANALOGOUS to catalytic reforming as the primary source of aromatic petrochemicals, steam cracking, or pyrolysis, of various hydrocarbon feedstocks is the major source of oelfins throughout the world. In the US, due mainly to feedstock cost considerations, the pyrolysis has concentrated on the lighter hydrocarbon feedstocks for production of predominantly ethylene. Due to feed stock availability and pricing differences, European and Japanese plants produce a full range of olefins and other products from the cracking of naphtha and other heavier feed stocks.

Fig. 5.1 presents a block flow diagram of a typical olefins plant. The same scheme is used, with very minor modifications, for feeds ranging from ethane to heavy gas oils.

The hydrocarbon feed is cracked at high temperatures and low pressures in tubular furnace coils with steam present. Cracked gaseous products, the lightest being hydrogen and the heaviest equivalent to heavy fuel oil, are quenched as quickly as possible in a quench tower, after suitable heat recovery. The heavier fuel oil and pyrolysis-gasoline products are separated in the quench tower. The remaining gases are charged to a series of multi-stage centrifugal compressors.

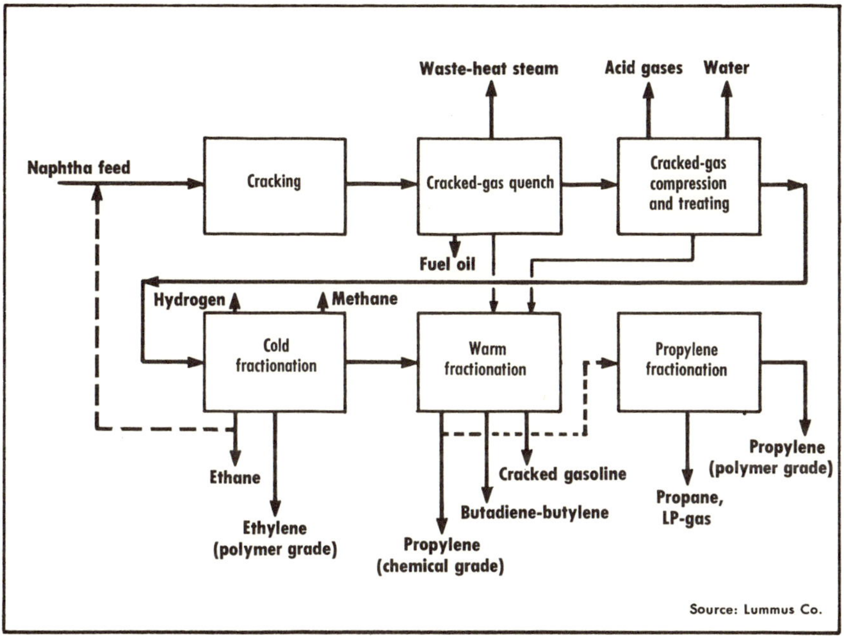

FIG. 5.1 Typical ethylene plant

The great amount of heat released in the quenching step is used to produce steam to drive the compressor turbines and to reboil the downstream fractionation towers.

Before the final compression stage, acid gases such as H_2S, CO_2, and often the small amount of acetylene which is formed, are removed in order to allow further high pressure cold processing and to meet product quality requirements. Water is also removed after compression.

The fractionation train is generally divided into a cold train where the lightest gases, hydrogen and methane, are separated from ethylene and heavier. An ethylene-ethane superfractionator is required to produce high-purity polymer-grade ethylene. Ethane is usually recycled to the cracker, particularly if maximum ethylene yield is desired.

The cold fractionation stage uses both propylene and ethylene-refrigeration steps. The centrifugal compressors are driven by the quench-generated steam.

More conventional, or warm fractionation, is used to separate propylene from propane and a cracked C_4 stream rich in butylenes and butadiene, from the cracked pentane and heavier. The pyrolysis gasoline is a prime source of aromatics or of high-octane gasoline after suitable hydrogenation. Propane is usually recycled to the cracking coil, or sold as LP-gas.

Many new developments and improvements in cracker technology have taken place during the past 15 or 20 years and have been applied to olefin-plant technology. Cracking furnaces have been increased both in thermal efficiency and size. Improvements in furnace-tube metallurgy have permitted more severe cracking, with improved yields and lower utilities costs, as recycle requirements for unconverted products are reduced. More efficient compressor systems have reduced total horsepower requirement. And the cracking furnace heat is now recovered more efficiently than before by means of transfer-line heat exchangers placed upstream of the quench tower, higher-pressure steam generation, etc.

The design of ethylene plants has been forced to cope with progressively more stringent product specification requirements. Contaminant levels on acetylene, sulfur, and moisture which would have been acceptable 15 years ago are no longer tolerated. Total ethylene content has gone from 97–98% to 99.9+%.

The increasing demand for polymer-grade ethylene and the development of direct-oxidation processes for ethylene oxide created the need for higher and higher product purities.

Table 5.1 gives typical current ethylene marketing specifications.

TABLE 5.1
Typical ethylene product specifications

Ethylene, % mol	99.9–99.95
Methane, % mol	0.01–0.1
Ethane, % mol	0.05–0.1
Propylene and heavier, ppm, wt	below 10
Acetylene, ppm, wt	5–10
Total sulfur, ppm, wt	1–10
H_2S, ppm, wt	1 max
H_2, ppm, wt	1 max
CO, ppm, wt	1–5
CO_2, ppm, wt	5–10
Oxygen, ppm, wt	2–5
Water, ppm, wt	5–10

NATURAL-GAS LIQUIDS—SOURCES AND PRODUCTION LEVELS

The natural-gas liquids which feed the domestic ethylene industry are mainly derived from gas-processing plants. The processing plants produce ethane, propane, and butane, plus a light natural-gasoline stream often called "plant condensate." Some of the natural gasoline is recovered right at the gas well, in which case it is usually referred to as "field condensate." The natural gasoline is basically a C_5/C_8 hydrocarbon, usually containing about 50% of iso and normal pentane. Table 5.2 shows US gas-liquids production for 1969.

TABLE 5.2
US gas liquids production—1969

	Gas plants		*Refineries*		*Total*		
Source	*MMbbl/ yr*	*M bbl/ day*	*MMbbl/ yr*	*M bbl/ day*	*MMbbl/ yr*	*M bbl/ day*	*% to petro-chemicals*
Ethane	63	172	9	25	72	197	100%
Propane	195	535	77	211	272	746	30%
Butane (Iso & n)	120	330	38	104	158	434	45%
Nat gasoline	158	433	—	—	158	433	—
	536	1,470	124	340	660	1,810	

US Bureau of Mines.

The refinery ethane figures in Table 5.2 really represent the original ethylene source, which was refinery cracker gas. The contained ethylene was separated, and ethane was cracked to ethylene in the earliest steam pyrolysis plants.

Additional ethane is recoverable at the gas plants, but recovery is effectively limited by current pipeline-gas BTU standards which require some ethane content.

The refinery propane figure represents total refinery production. However, the refinery butane figure represents only the 16% of the total 655,000 bbl/day production which gets into the market. The rest was either blended directly, or further processed into gasoline at the refinery.

All 197,000 bbl/day of the gas plant and refinery ethane finds its way into petrochemicals. About 220,000 bbl/day of propane goes to chemicals, the remainder goes into the LP-gas energy market. In addition to the 746,000 bbl/day shown, about 20,000 bbl/day of propane was imported in 1969, mainly from Canada.

In the US today more than 85% of the olefin plant capacity charges natural-gas liquids. Less than 5% of the capacity charges naphtha or gas oil. About 10% of the olefins still come directly from olefin-rich refinery gas streams.

With the weakening domestic supply position in natural gas, imports and refineries are likely to become a considerably more important source of light hydrocarbon liquids to those olefin producers whose plants are designed to handle light hydrocarbons only. Alternative feeds to ethane and propane will, of course, be seriously considered for future olefins plants.

Based on 1980 projections of anywhere between 35 and 50 billion lb/yr of ethylene, it has been estimated that by 1980 only 40 to 50% will come from natural-gas liquids, 20 to 25% will come from naphtha, and 30 to 40% from gas-oil cracking.

FEEDSTOCK, MAJOR PROCESS VARIABLES, AND YIELDS

Feedstock quality

In general, the higher the molecular weight of the feedstock, the easier it cracks. For example, typical pyrolysis-furnace coil-outlet temperatures for ethane are 1,600°F., whereas a 1,200° to 1,300°F. coil outlet is more typical for naphthas and gas oil.

The lighter feedstocks have higher hydrogen-to-carbon ratios, expressed as wt % hydrogen. This H/C ratio is probably the most important measure of a feedstock's ethylene potential.

As cracking severity increases, ethylene yield is favored, but the production of hydrogen and methane also increases with severity. Since hydrogen and methane have the highest H/C ratios of all cracker products, the point where carbonaceous coke and tar begins to form and foul both the heater, transfer-line exchanger, and quench facilities occurs sooner on heavier, low H/C ratio feeds. Generally a H/C ratio between 7 and 8 in the C_5+ portion of the reaction products defines limiting severity, where coking rate in the transfer-line exchange and quench becomes intolerable. Typical hydrogen-to-carbon ratios for various feeds are:

	H/C ratio, wt
Ethane	20.2
Propane	18.2
Butane	17.3
Typical naphtha	15
Light gas oil	13.5

In addition to feed hydrogen content and molecular weight, the basic hydrocarbon type composition of the stock has an important affect on its crackability and anticipated yield patterns.

Paraffins are the easiest of the basic hydrocarbon types to crack.

Straight-chain paraffins such as normal butane give very high ethylene yields, whereas branched paraffins such as isobutane and isopentane give higher propylene yields.

The cycloparaffins, or naphthenes are harder to crack since the ring must be opened first. At low cracking temperatures there is some tendency to dehydrogenate into aromatics. Side chains do split off the naphthenes to form light olefins, however. Higher cracking temperatures tend to open the ring, forming olefins and diolefins.

Aromatic rings in the feedstock tend to pass through the pyrolysis furnace unchanged. However, side chains do break off via hydrodealkylation, and this accounts for the relatively high benzene concentration in the pyrolysis gasoline, at the expense of toluene and xylenes. Coke and tar-forming tendencies, due to ring condensation, are also evidenced with high aromatic content feeds.

Major process variables

The major process variables over which the plant operator has a certain degree of control are:

1. Temperature.
2. Cracking coil residence time.
3. Total pressure and hydrocarbon partial pressure (steam to hydrocarbon ratio).

The major reactions which occur in ethane cracking can be represented as:

$$C_2H_6 \rightarrow C_2H_4 + H_2 \qquad \text{Primary cracking reaction}$$

$$\left.\begin{aligned} C_2H_4 + 2H_2 &\rightarrow 2CH_4 \\ C_2H_4 &\rightarrow C_3,\ C_4,\ C_5 + H_2 \end{aligned}\right\} \text{Secondary reactions}$$

Primary reactions are favorable. In these the feed cracks to olefins, hydrogen, and methane.

Secondary reactions are less favorable. The olefins themselves are involved in further cracking, in hydrogenation back into lighter saturates, and in condensation and polymerization reactions which form aromatics and coke.

With propane the reactions are:

$$\left.\begin{aligned} C_3H_8 &\rightarrow C_3H_6 + H_2 \\ C_3H_8 &\rightarrow C_2H_4 + CH_4 \\ 2C_3H_8 &\rightarrow C_4H_8 + 2CH_4 \\ 2C_3H_8 &\rightarrow C_2H_6 + C_3H_6 + CH_4 \end{aligned}\right\} \text{Primary reactions}$$

$$\left.\begin{aligned} C_3H_6 + 3H_2 &\rightarrow 3CH_4 \\ C_3H_6 &\rightarrow C_4,\ C_5 \text{ and heavier, } H_2 \end{aligned}\right\} \text{Secondary reactions}$$

In general, low residence time favors the desirable primary cracking reactions and minimizes the less desirable secondary reactions. Ethylene yield increases substantially as residence time decreases.

Typical residence times at temperatures about 1,200°F. in crackers designed for 1,500°–1,600°F. coil-outlet conditions is often 0.5 second, or less.

The steam/hydrocarbon ratio, usually expressed in terms of pounds of steam to pounds of hydrocarbon feed, is an important process variable.

Low hydrocarbon partial pressure favors the production of olefins. Steam is used to lower the hydrocarbon partial pressure and acts to encourage the primary olefin-forming reactions. At the same time the presence of steam reduces the partial pressure of the high-boiling aromatic secondary-reaction products and lessens the coke-forming tendency on the tube walls and in the quench heat exchangers.

Steam also has another important side effect. It diminishes the adverse catalytic effects of iron and nickel in the tube walls which promote carbon formation.

The most desirable combination of pressure and steam dilution ratio is the one which strikes a balance between the uneconomically high steam rates which would be required at high total pressure and the uneconomically high compression power which would be required for the cracked gases if total pressure were very low. Normally coil outlet pressures are in the 10 to 20 psig range.

Typical steam/HC ratios for various feedstocks are:

Feed	*lb steam/lb hydrocarbon*
Ethane	0.25–0.40
Propane	0.30–0.50
Naphtha	0.50–0.80
Gas oil	0.80–1.00

Yields

Ethane is normally cracked at about 1,600°F. coil outlet temperature. This achieves about a 60% conversion/pass. Normally, a total ethylene yield of 80 wt % is obtained. Yield of propylene and heavier products is minimal.

Feed

Ethane–98% purity, 1% each of methane and propane.

Ultimate yields	*wt %*
Methane + hydrogen	14
Ethylene	80
Propylene	1.6
Propane	0.8
C_4's (60% butadiene)	2.0
C_5+	1.6
	100.0

Propane is normally cracked at about 90% once-through conversion, also with coil outlet temperatures in the 1,575°–1,625°F. range. Normally, a total ethylene yield of 45% is obtained. Propylene yield amounts to 15%. Ethylene yield can be increased to 50% if propylene is not desired and is included in the recycle.

Feed

Propane – 93% purity; 6% ethane, 1% butane

Ultimate yields	*With propylene recovery wt %*	*Without propylene recovery wt %*
Methane + hydrogen	28	31
Ethylene	45.5	50
Propylene	15	–
C_4's (60% butadiene)	2.5	3.5
C_5+	9.0	15.5
	100.0	100.0

Normal butane is cracked at about 95% conversion/pass and yields up to 40% ethylene and 20% propylene. Somewhat lower coil-outlet temperatures are used (1,475°–1,500°F.) since normal butane would tend to overcrack at conditions optimum for ethane and propane. Overcracking would result in excessive coke laydown. Butane is therefore not normally cracked in mixture with ethane and propane. Typical yields for a feed containing 97% normal and 3% isobutane are:

	wt %
Methane + hydrogen	24
Ethylene	37
Propylene	17
Propane	1
Butadiene	1.6
Other C_4	6.4
C_5+	13
	100.0

The effect of cracking severity on the yield structure when cracking naphtha and heavier feeds is much more pronounced than for light hydrocarbons. The propylene/ethylene ratio can be adjusted within a considerable range. Total conversions of up to 75% C_3 and lighter are obtained. A mild type of operation might yield 25% ethylene and 16% propylene. A more severe operation would yield 35% ethylene and only 10% propylene, plus more aromatics. A great deal of flexibility exists.

Gas oils are usually cracked to yield 25 to 30% ethylene and 10 to 15% propylene. Butadiene yields of 4 to 5% are obtained on these heavier feed stocks.

PRODUCT-PURIFICATION SECTIONS OF THE OLEFINS PLANT

Although the design variables and operating flexibility incorporated into the cracking system have the most bearing on ultimate yields, the overall plant economics and performance depend to a great degree on the downstream part of the complex.

The cracking furnace area represents only about 25% of the total investment involved in an olefins plant. The remaining investment is tied into the effluent quench, gas compression, treating, dehydration, product recovery, and refrigeration sections.

The cracking section is generally considered to consist of the furnace itself. The modern furnace has three main sections:

1. A convection heating section where feed is preheated and diluent steam is superheated.
2. The radiant reaction section itself.
3. An economizer to recovery heat from the flue gas. The heat going either into boiler feedwater preheat or steam generation.

The function of the quenching system is to cool the effluent gases as rapidly as possible and at the same time recover maximum heat. Rapid cooling is required in order to minimize olefin loss due to the more undesirable secondary reactions.

The first quench step in all modern plants is a quench heat exchanger where the preheated boiler feed water is turned into the steam used in the cracking coil.

The final cooling, or quenching of the reaction, usually occurs by direct mixing with cooling water or hydrocarbon distillate in the quench, or scrubber tower. The quench water flows from the tower to a settling drum where heavy tar is taken out of the bottom. The lighter distillates floating on the quench water are skimmed off and sent to storage or further processing. The oil-saturated quench water is recirculated to the tower after further cooling. The closed quench-water system avoids loss of valuable hydrocarbons and prevents contamination of effluent water.

The increased fouling tendencies associated with heavier feeds and the increase in heavy product yield have dictated use of a direct oil quench. Fig. 5.2 presents a schematic flow diagram of the cracking and quench systems.

Except for the small quantity of distillate and heavy gasoline condensed in the quench system, almost all of the reaction products, including C_5 and C_6, end up in the gaseous phase with the low-pressure systems used through the plant's quench stage. The quench tower overhead gas is suction to the first stage of compression.

Due to the need to drop out liquid products along the way and the limitations on compressor discharge temperatures required to avoid rapid olefin and diolefin polymerization, it is usually necessary to have up to five stages of compression.

Cracked gas compression is a very "dirty" service, particularly in the first and second stages. Rapid polymer formation and subsequent fouling can occur. Sometimes a stable gas oil is atomized into the compressor suction to act as a flushing oil for washing down and dissolving resins.

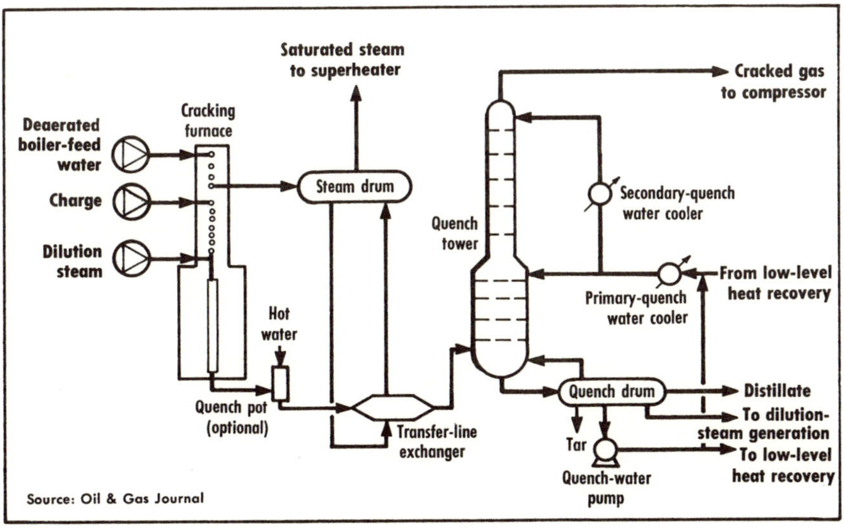

FIG. 5.2 Cracking and quench system

Single-train centrifugal compressors have all but replaced the reciprocating machines used 10 to 15 years ago. This also goes along with the general increase in plant size.

Acetylene and acid gases are usually removed between the third and fourth compression stages at about 200 psig. Acetylene is catalytically hydrogenated in order to meet the increasingly stringent acetylene specifications on ethylene product. Contained hydrogen is utilized in the hydrogenation step. Due to the high diolefin content of the cracked gas, the catalyst bed tends to coke rapidly, and a swing reactor system with offline regeneration is used.

Conventionally, a caustic scrubbing system is used for CO_2 and H_2S removal. The amount of H_2S formed will depend on feed sulfur content, and a regenerative type of acid-gas-removal process such as amine scrubbing may be used upstream of quantitative removal by caustic.

The cracked gas is dried by use of solid absorbants, either aluminum oxide materials or molecular sieves.

Only after compression, acetylene and acid gas removal, drying, etc., can the actual recovery of the specific olefins be undertaken. A series of six or seven fractionation towers, the first three of which make up the cold train, are utilized to separate the primary olefin products.

The typical fractionation train starts with the low temperature demethanizer which operates at about 450 psig and takes −140°F. methane and hydrogen as the overhead product. The tower bottoms (+50°F.) goes to a 375-psig deethanizer which takes an overhead ethylene/ethane stream at about 275 psig with about a 40°F. overhead temperature. It usually has over 100 trays.

The deethanizer bottoms go to a depropanizer which operates at about 220 psig and contains 30 to 40 trays. The hardest split of all is in the propylene-propane fractionation which may require up to 200 trays and operates at about 260 psig.

THE US ETHYLENE ECONOMY AND THE SEARCH FOR FEEDSTOCKS

As opposed to aromatics, where the petroleum industry dominates, the demand for the major olefin product, ethylene, has grown so rapidly that the chemical industry developed its own primary source of ethylene, the large supply of inexpensive natural-gas liquids available mainly on the Gulf Coast. Well over half the ethylene is produced by chemical companies (the figure was 66% in 1960), as opposed to only 15% for aromatics.

Using these natural-gas byproduct liquids, the generally accepted "cost," including 20% after tax return on investment, of producing ethylene on the US Gulf Coast is in the 2.3 to 2.5¢/lb range.

Naphtha pyrolysis processes worldwide have been able to produce ethylene at a comparable "cost" from naphtha valued at between 4.5 and 5.5¢/gal. (However, co-product values are also an important factor in determining the actual "cost" to produce ethylene from naphtha.) Naphtha has historically been surplus in the rest of the world, with many European refineries actually using it as fuel in the face of limited gasoline markets.

Yet naphtha in the US is worth up to 10¢/gal in our gasoline-oriented economy as reformer charge stock. The calculation given in Table 5.3 illustrates how a naphtha stream, which the refiner can reform into 100-octane leaded gasoline (94 octane clear) is worth essentially twice its world market value.

Assuming 100 octane leaded gasoline is worth 13.0¢/gal, then 94-clear reformate (which can be upgraded to 100 octane with 3 gm of lead costing about 0.22¢/gm, or 0.66¢/gal) is worth 13.0−0.66, or 12.34¢/gal. We can calculate the feed naphtha's value using 12.3¢ for the C_5 + reformate yield and the appropriate typical values for the other products yielded.

TABLE 5.3
Value of domestic naphtha as reformer feed

	Yield, %	*Value, ¢/gal*	*¢/lb*	*¢/gal feed*
C_5+ reformate	85 vol	12.34		10.49
Butane	6 vol	6.0		0.36
Propane	5 wt	—	1.1	0.34
Fuel gas	4 wt	—	0.75	0.19
Hydrogen	1 wt	—	1.5	0.08
Gross feed (6.3 lb/gal)				11.46
Less operating cost				1.50
Net feed value				9.96

This situation is not unique to naphtha.

Table 5.4 summarizes the differentials which exist between various potential feedstocks as 1¢/lb olefin feed and their typical refinery alternate values.

TABLE 5.4
US feedstocks, refinery values vs olefin feed

	Value as olefin feed stock		Refinery alternate value ¢/gal
	¢/lb	¢/gal	
Ethane	1.0	3.11	none
Propane	1.0	4.23	none
Butane	1.0	4.86	5–6 in gasoline
Naphtha	1.0	6.3	9.96 as reformer feed
Light gas oil	1.0	7.0	9.0 as distillate fuel oil
Heavy gas oil	1.0	7.5	9.5 as cat cracker feed

The feedstock requirements for the US ethylene industry have been increasing at a faster rate of late than the natural-gas liquids supply. In fact the shortage of natural gas in general is a major contributing factor in the current US "energy crisis."

The chemical industry has been looking for alternate feedstocks. Logically, these are naphthas and gas oils, which are locked up and controlled by the refining industry and further tied into the entire Oil Imports Administrative structure as "unfinished oils" which essentially require a crude-oil import ticket.

Cracking of natural-gas liquids still accounts for over 80% of the US ethylene plant feed, but a battle royal has been brewing between the oil and the chemical people of late, with the Government caught in the middle, over "free access" to feedstocks.

Stopping short of attempting to resolve the problems of "free access" and all the other complications of the oil imports program, it should be noted that other important factors are currently acting on the historical "built-in" domestic feed-stock cost differentials between natural gas liquids and heavier feeds such as naphtha and gas oils. These factors may very well tip the balance in favor of the heavier feed stocks in the not-too-distant future:

1. The shortage of natural gas has created a tight market situation for propane for the LP-gas markets, at times driving the spot market price up to well over 6¢/gal (1.4¢/lb). This cuts into the availability of economical quantities of the traditional feed stock for the domestic ethylene industry.
2. The pressure toward low lead or unleaded gasolines will increase the value of the byproduct aromatics yielded from pyrolysis gasoline, derived mainly from naphtha and other heavier ethylene feed stocks. Butylenes, which are yielded in greater

quantities with heavier feedstocks, will also increase in value as alkylation feed.

3. At the same time that aromatics are increasing in value in the gasoline pool, the low-octane raffinate streams which the refiners produce from their own aromatics recovery operations will decrease substantially in value. The highly paraffinic raffinate, having a better hydrogen/carbon ratio as olefin feed will probably be the first domestic "naphtha" type stream to find its way economically into the pyrolysis unit for olefin production. Low octane natural gasoline and field condensates will also be dropped out of the gasoline pool.

A recent study by the Lummus Co., presented at the spring 1970 joint ACS and Chemical Institute of Canada meeting in Toronto, showed the combined affects of increasing the cost of natural gas liquids and aromatics byproducts values on the value of naphtha feed stock to an olefins plant. The study showed that 1.5¢/lb naphtha (9.5¢/gal) is breakeven with 1.4¢/lb ethane and 1.3¢/lb propane at current aromatics values, Fig. 5.3. If aromatics go up 5¢/gal, due to lead

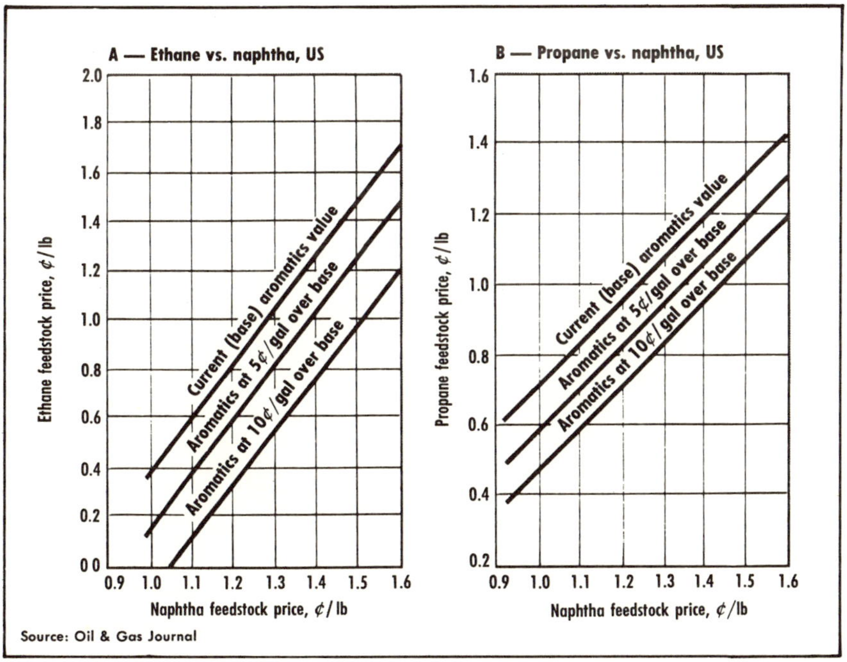

FIG. 5.3 Break-even prices for ethane and propane vs. naphtha

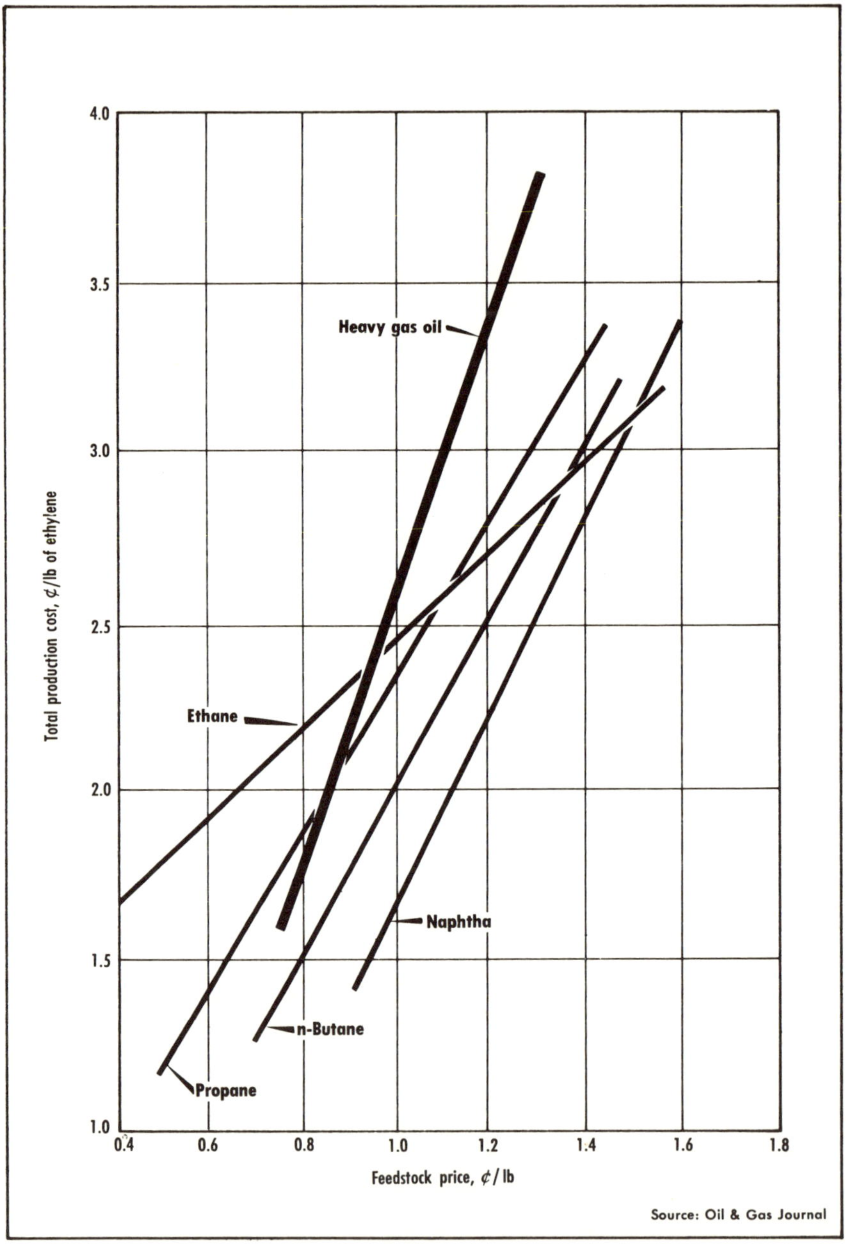

FIG. 5.4 Effect of feedstock price on US ethylene costs

removal pressures, the 1.5¢/lb naphtha breaks even with 1.25¢ ethane and 1.2¢ propane. The study was based on a plant sized to produce 1 billion lb/yr of ethylene. Such a plant would cost about $45,000,000 on a naphtha/light gas oil charge and about $35,000,000 if designed to handle ethane and propane only (battery limits only).

The total effect of raising the feed stock costs from current 1.0¢/lb levels to the 1.3 to 1.4¢/lb range for ethane and propane and 1.5¢/lb range for naphtha is to increase the production cost of ethylene from the 2.3 to 2.5¢/lb, to about 3.0¢. This effect is shown in Fig. 5.4.

As domestic refinery streams such as raffinate and naphtha begin to find their way into the ethylene production picture (based on the need for new sources of feedstock, increased byproduct aromatics values, and decreased value of low octane components) it is likely that the refiners will play an increasingly important role in the ethylene industry.

THE REFINING-PETROCHEMICAL INTERFACE

All of the interdependencies and interactions between the refining and petrochemical industries touched on in the production of olefins and aromatics – the feed stock situation, the dominance of the refining industry in aromatics, etc. – are basically the result of both the forward integration of the petroleum industry and backward integration of the chemical industry.

The integrated petroleum companies used their cash flow to create further opportunities to utilize crude oil by getting into faster growing petrochemicals. The chemical companies integrated back towards raw materials to assure themselves of reliable raw material sources at reasonably stable prices. And they took advantage of the availability of low-cost natural-gas liquids to control the major portion of the domestic ethylene industry.

There has been a great deal of discussion of late about the so-called refining-petrochemicals Interface, or demarcation line between the petroleum and petrochemical segments of the total hydrocarbon-processing industry.

Since virtually the whole spectrum of petrochemical raw materials can be produced either as a refinery gasoline byproduct or by steam pyrolysis of hydrocarbon feed stocks, the location of the refining-petrochemicals Interface really depends more on the production yields and costs via the various routes, rather than any theoretical ability to do it all one way or the other. Serious consideration will have to be given to these points before any simple demarcation line can be established.

1. The growth rate of chemical vs. energy requirements? The long-term trend of faster chemical growth has been abruptly reversed by the recent economic situation which affected the chemical industry much more seriously than the domestic petroleum industry. Petroleum product demand, particularly for heavy fuels has been continuing to grow, putting a strain on crude-oil-production capacity.
2. The capital availability for investment? The chemical companies have been progressively showing lower profits over the past few years and are, therefore, generating less of the capital which they require for further backward integration. On the other hand, the oil industry is soon to embark on one of the most expensive privately funded ventures of all time—bringing the Alaskan crude oil to market. This will probably significantly tighten up the capital-availability situation, as far as petrochemical expansion is concerned.
3. Feed-stock availability and cost as affected by low lead gasoline or other anti-pollution requirements?
4. Feed-stock availability and cost as affected by future oil import policies and/or special access for petrochemical producers?
5. The development of new technology? For example, the production of more olefins via catalytic-cracking technology or propane dehydrogenation, or the cracking of whole crude for ethylene production, could have significant effects on the refining and petrochemical industries.

Many computer studies, using highly sophisticated LP models, have been made by the petroleum and the chemical industries and by many outside consultants on how best to set up a highly integrated, "chemical refinery." Such a refinery would be designed to bridge the interface and result in a maximum profit situation.

It is impossible to generalize, since the optimum selection of processes, the ratio of chemical to fuels, etc., are influenced by a great many factors, starting from crude oil type and going on through geographical location, marketing estimates, etc. But the optimum "chemical refinery" generally turns out to produce 30 to 50% petrochemicals. This situation would tend to favor the petroleum companies who have the raw materials and the required gasoline outlets to tackle projects of this magnitude.

SPECIAL PROPYLENE CONSIDERATIONS

A fundamental difference exists in the factors controlling production and availability of ethylene and propylene in the US. As opposed to ethylene, the emphasis which must be made on propylene is that the petroleum refining industry controls its availability. Since propylene has other refinery uses, its price is set by "alternate use" economics.

Over 85% of the propylene available for chemicals comes from refining operations (mainly catalytic, some thermal cracking). The chemical industry accounts for production of only 15% of its own propylene requirement.

The US ethylene industry is based on the cracking of natural-gas liquids along the Gulf Coast, which yields minimum quantities of byproduct propylene. The major source of propylene is petroleum refining, really the catalytic cracker within the refinery complex.

As a refinery stream, propylene is also a byproduct. And chemical outlets have to compete within the usual complex set of refinery variables because propylene is also feed to an important refinery gasoline process, alkylation. The alky unit reacts butylenes and propylene with isobutane to form highly branched C_7 and C_8 paraffins for high-octane gasoline.

CATALYTIC CRACKING IN THE REFINERY—THE PRIMARY SOURCE OF PROPYLENE

Catalytic cracking has been one of the mainstays in the US refiner's arsenal of processing weapons since the 1940's.

The first fixed-bed Houdry unit was put into operation by Socony in 1936. Oil vapor was passed through a fixed bed of natural clay pills or pebbles. Each reactor was on reaction cycle for about 10 minutes and was then regenerated off-line for about 20 minutes.

The fluid-bed process was developed jointly by Esso, UOP, Kellogg, and others during World War II. The first unit went into operation in 1942, and 32 were built before 1945, mainly to speed the production of aviation gasoline for the war effort. While Socony and Houdry developed their fixed-bed technology first into moving bed and later into air-lift TCC Units (which use high velocity gas rather than buckets to move catalyst), the fluid process has become the major present-day cat-cracking process.

Conversion in a cat cracking unit is generally defined as

100% − net yield of material boiling above a standard gasoline end point (usually 430°F.).

Thus, if total yield of products boiling above 430°F. is 28 vol % of feed, the conversion would be 72%. Since the individual products recovered may be hard to measure, it is more convenient to define conversion via the unconverted heavy products.

The original cracking catalysts were acid-treated natural clays which contained 85% S_iO_2 and 15% alumina. Through the 1950's a general changeover to synthetic high-alumina catalysts containing 25 to 30% alumina occurred.

Changeover to molecular-sieve, or zeolite, catalysts, which are copolymers of alumina and silica, has been almost universal since 1964. These catalysts permit higher gasoline yields and lower coke yields at constant conversion. Since most cat units are effectively limited in conversion and throughput by ability to burn coke off the catalyst, the refiners were able to further increase gasoline yield by raising conversion with the newer, more active catalysts.

Typical comparative cat-cracking yields on a 25° API feed stock run at high conversion for constant coke yields are given on Table 5.5.

TABLE 5.5
Cat cracking yields

	Silica-alumina	*Zeolite*
Coke yield, wt %	7.3	7.4
Conversion	76.0	83.5
Dry gas,* wt %	3.0	2.9
Propylene, vol %	6.2	5.5
Propane, vol %	2.2	2.2
Butylenes, vol %	6.7	7.8
Butanes, vol %	5.9	8.1
C_5+ gasoline, vol %	50.4	55.4
Cycle oils, vol %	24.0	16.5

*Hydrogen, methane, ethane, ethylene.

Ethylene yield from cat cracking is very low, no more than 10% as much as propylene, and is seldom recovered in conventional cat cracker light-ends-recovery plants.

Zeolite catalysts have a lower selectivity to light olefin production, but this can be made up for by running to the higher conversions attainable, with the lower coke yields characteristic of the zeolite catalysts. For the feed stock in Table 5.5, the net propylene yield is, however, reduced by about 11%.

FACTORS AFFECTING AVAILABILITY OF REFINERY PROPYLENE

The future of cat cracking and its growth rate have recently become clouded by hydrocracking as an alternate means of producing light distillates from heavier crude-oil components. It is generally acknowledged that high-pressure hydrocracking, requiring an outside hydrogen source, is more expensive than cat cracking, but it has feed flexibility over cat cracking on extra-heavy stocks.

Although total US hydrocracking capacity is still less than 700,000 bbl/day, it is expected to grow at a faster rate than cat cracking over the next few years. This trend was to be accelerated by the need to destroy heavy high-sulfur residual oils for anti-pollution purposes. Actually most US hydrocracking is installed on the West Coast where the first restrictions were placed on high-sulfur residual fuel oils.

Hydrocracking produces no olefins whatsoever. The total C_3 yield is propane. However, hydrocracking often produces 20% or more butane, most (75%) of which is iso-butane. Refinery alkylation units have historically been short of isobutane, with sufficient butylenes and propylene available. However, hydrocracking and the new zeolite catalysts are shifting the refineries from being short to being long on isobutane. Therefore, the refiners have more incentive to utilize their propylene as alkylation feed as the value of IC_4 drops.

The preferred olefins for alkylation have always been the butylenes which yield a 2 to 3 higher-octane-number alkylate than mixed propylene-butylene, and a 5 to 6 higher-octane-number alkylate than propylene alone. However, when excess isobutane is available, the refiner naturally turns to propylene to supplement alkylation feed.

Motor alkylate has generally been worth 14 to 15¢/gal as a refinery gasoline component. With isobutane worth about 5¢/gal, propylene in alkylate can usually be shown to be worth about 13¢/gal, or 3¢/lb. As the second-highest octane component available for the unleaded, or low-lead gasoline pool, and a very clean burning one at that, it is quite likely that alkylate value will go up 2 to 3¢/gal in the gasoline pool. This will increase the value of propylene to the refinery to 3½¢/lb.

OTHER PROCESSES FOR PRODUCING PROPYLENE

Isobutane is not a particularly good pyrolysis feed stock for ethylene manufacture because it yields essentially all propylene and isobutylene. Yet with the advent of hydrocracking and higher-severity gasoline reforming operations, this material has been turning up in excess. Pure isobutane can be subjected to conventional steam pyrolysis to yield 80 to 85% of propylene plus butylene, when the unconverted isobutane is recycled to the cracker.

Processing severity can be adjusted to produce virtually 100% of a balanced alkylation feed or can be shifted in favor of excess propylene at higher pyrolysis temperatures. Refiners are beginning to use this route for conversion of surplus isobutane. For example, 10,000 bbl/day of isobutane can be converted into 6,300 bbl/day of alkylate (1,500 bbl/day propylene, 2,000 bbl/day butylene, and 4,400 bbl/day isobutane required to form 6,300 bbl/day of alkylate) with no net propylene yield. Higher-severity cracking could be used to recover some propylene from the system while alkylating only the remaining butylenes. Typical isobutane-pyrolysis yields for balanced alkylation case are provided in Table 5.6.

TABLE 5.6
Typical isobutane pyrolysis yields

	Wt %
Hydrogen	0.9
Methane	9.5
Ethylene	2.0
Ethane	0.5
Propylene	16.0
Propane	0.2
Butylene	22.2
Isobutane	48.4
Normal butane	0.2
Pyrolysis gasoline	0.1

Cracking severity can be increased by recycling iso-butane and alkylating only butylenes, with the net propylene stream yielded for chemical feed stock. The normal volume ratio of isobutane to olefin in alkylation is 1.25 volumes isobutane plus 1.0 volume olefin, yielding 2.0 volumes of alkylate.

The feasibility of catalytic dehydrogenation of propane to propylene has also been demonstrated. The Houdry Catadiene process which was originally developed to produce butadiene from normal butane has been adopted to the production of propylene from propane. Due to high coke-deposition rates, this is a cyclic operation, utilizing five or more catalyst beds, each of which has an onstream period of 10 to 15 minutes. Catalyst is regenerated by burning off coke with preheated air. An alumina-chrome oxide catalyst is utilized.

It has been suggested that fluid cat-cracking techniques could be adopted to light naphtha feeds, or even to propane to produce required refinery olefins. Gulf Research and Development has cat-cracked naphtha in pilot-plant studies. Conversions of 35 to 50% to C_5 and lighter were obtained, with a total yield (selectivity) of C_2 to C_5 olefins ranging up to 56%. Of the olefins produced, propylene and butylenes account for 40 and 30% respectively.

Propane may also be eventually dehydrogenated to upwards of 70 wt % propylene by means of fluid-bed dehydrogenation using fluid cat-cracking techniques. Studies in this field have indicated attractive payouts if propylene is worth 3.5¢/lb, and propane could be obtained for 4.5 to 5¢/gal. To the author's knowledge units have been proposed, but none as yet built.

6

ARTHUR M. BROWNSTEIN

Toluene and benzene

TOLUENE and benzene are colorless, flammable liquids which boil at 110°C. and 80°C. respectively. Together they constitute the principal building blocks on which most aromatic chemical intermediates are based. Toluene and benzene are intimately related. Not only are they produced from the same coke oven and petroleum sources, but the principal chemical use for toluene is the manufacture of benzene.

TOLUENE

As discussed in Chapter 4, petroleum sources, notably catalytic reforming, are the principal supply of toluene and benzene. While toluene production grew from 2.0 billion to 5.7 billion lb/yr from 1960 to 1971, contribution from coke oven distillates declined steadily (Table 6.1). In 1971, toluene production from coke ovens was at an all time low of 15 MM gal, and constituted less than 2% of US production. A decade ago, coke oven distillates accounted for more than 12% of production. No reversal of this trend in the foreseeable future appears likely.

Until 1969, hydrodealkylation of toluene to benzene constituted more than 60% of its chemical usage. This end use dropped precipitously in 1969 and declined even lower in 1970 and 1971 (Table 6.2).

TABLE 6.1
Toluene production[1]
(MM gal)

Year	*Coke oven*	*Petroleum*	*Total*	*MM lb*[2]
1960	34	242	275	2,000
1961	28	231	259	1,890
1962	27	333	360	2,620
1963	26	380	406	2,960
1964	26	470	496	3,600
1965	25	524	549	3,990
1966	23	561	584	4,250
1967	19	624	643	4,670
1968	20	675	695	5,050
1969	20	740	760	5,530
1970	17	678	695	5,050
1971[3]	15	770	785	5,730

[1] US Tariff Commission.
[2] 7.27 lb/gal.
[3] Based on 7 months' data.

TABLE 6.2
Toluene end use pattern
(MM lbs)

Year	*Benzene*[1]	*TDI*[2]	*Solvents*[3]	*Misc.*[4]	*Total*
1960	1,370	26	200	404	2,000
1961	1,410	42	123	315	1,890
1962	1,650	57	262	651	2,620
1963	1,980	80	296	604	2,960
1964	2,250	70	360	920	3,600
1965	2,540	90	399	961	3,990
1966	2,920	106	425	799	4,250
1967	2,980	122	467	1,101	4,670
1968	3,070	135	505	1,340	5,050
1969	1,730	158	553	3,325	5,530
1970	1,180	183	505	3,071	5,050
1971	1,020	220	571	3,919	5,730

[1] 1.2 lb toluene/lb of benzene.
[2] 0.61 lb/lb tolylene diisocyanate.
[3] Estimated at 10% of total.
[4] Mostly gasoline; also includes TNT, phenol, benzoic acid, benzyl chloride, and exports.

This was a consequence of benzene oversupply due to a depressed economy. Benzene prices had dropped to such low levels that most producers who employed hydrodealkylation of toluene could not afford to continue production (Chapter 4). Consequently, benzene derived from less expensive reformate satisfied most of market demand.

That amount of toluene which had been already isolated from reformate was largely consigned to the gasoline pool, and is largely represented in the miscellaneous column of Table 6.2. During 1971, it has been estimated that benzene production accounted for less than 20% of total chemical demand for toluene.

There is no dearth of toluene capacity in the US. Chemical capacity is estimated at 6.9 billion lb, substantially in excess of demand in recent years. The list of producers in Table 6.3 relates to these com-

TABLE 6.3
Major toluene producers (1972)

Company		*MM lbs*
Ashland	(Catlettsburg, Ky.)	150
Arco	(Wilmington, Calif.)	160
	(Houston, Texas)	275
Coastal States	(Corpus Christi, Tx.)	150
Chevron	(El Segundo, Calif.)	175
Corco	(Guayanilla Bay, P.R.)	150
Conoco	(Ponca City, Okla.)	150
Enjay	(Baton Rouge, La.)	475
	(Baytown, Tx.)	475
Mobil	(Beaumont, Tx.)	330
Monsanto	(Alvin, Tx.)	225
Phillips	(Guayama, P.R.)	650
Shell	(Houston, Tx.)	320
Sun	(Marcus Hook, Pa.)	225
	(Tulsa, Okla.)	150
	(Corpus Christi, Tx.)	175
Texaco	(Port Arthur, Tx.)	150
Union Carbide	(Taft, La.)	150
Others[1]		2,515
	Total	6,900

[1]Includes 225 MM lb from coke-oven producers.

panies who isolate toluene for chemical use, and do not include product not separated from reformate and used for gasoline. Prices have remained relatively uniform in the last decade and are summarized in Table 6.4.

TABLE 6.4
Petroleum toluene prices[1]
(¢/lb)

Year	*USTC*	*OPD (lowest)*
1961	2.7	3.4
1962	2.6	3.4
1963	2.4	2.9
1964	2.3	2.9
1965	2.3	2.9
1966	2.4	2.9
1967	2.7	3.0
1968	2.6	3.2
1969	2.5	3.2
1970	na	3.2
1971	na	3.2

[1]Nitration grade, delivered (tanks).

Tolylene diisocyanate

Other than benzene production, tolylene diisocyanate (TDI) represents the only large, individual chemical use for toluene. Although toluene demand for this application grew nearly nine fold from 1960–1971, it still accounts for only 4% of the total. TDI is principally employed for production of flexible polyurethane foams.

TDI manufacture is a multi-step process comprising nitration, hydrogenation, and phosgenation. Nitration is effected with mixed nitric and sulfuric acids to yield principally an 80/20 mixture of 2,4 to 2,6-dinitrotoluene. The washed and neutralized dinitrotoluenes are catalytically hydrogenated to the corresponding diamines and purified by distillation.

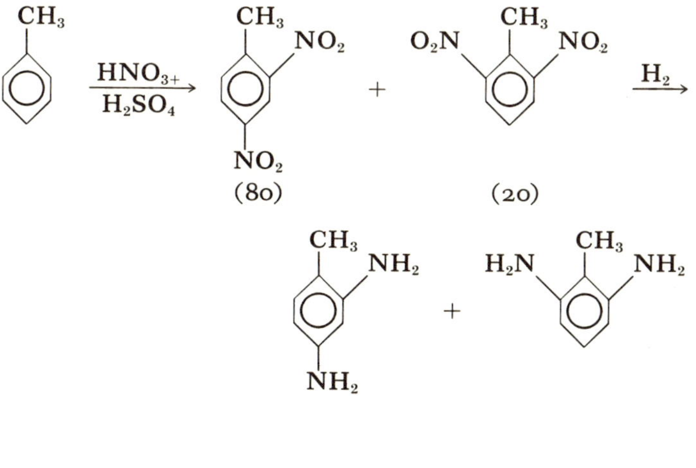

The tolylene diamine mixture is dissolved in o-dichlorobenzene and contact with phosgene in three stages at temperatures ranging up to 200°C.

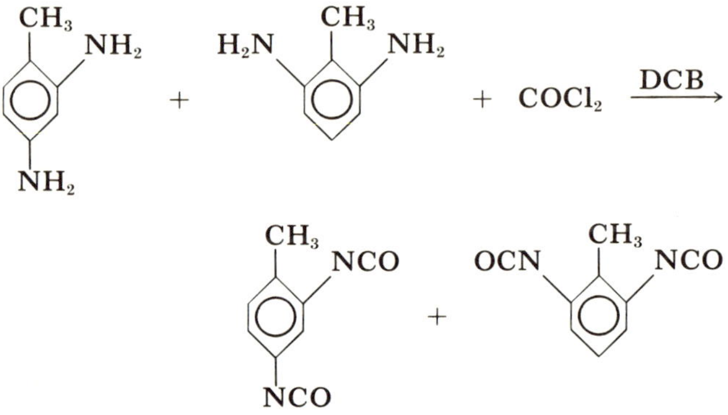

Yields on toluene average about 85%.

Most facilities are integrated encompassing phosgene production as well as toluene nitration and reduction. A flow diagram for the phosgenation step is shown in Fig. 6.1, which is taken from Encyclopedia of Chem. Technol., Kirk-Othmer, 2*nd* Ed., Vol. 12, p. 56 © 1967; John Wiley & Sons.

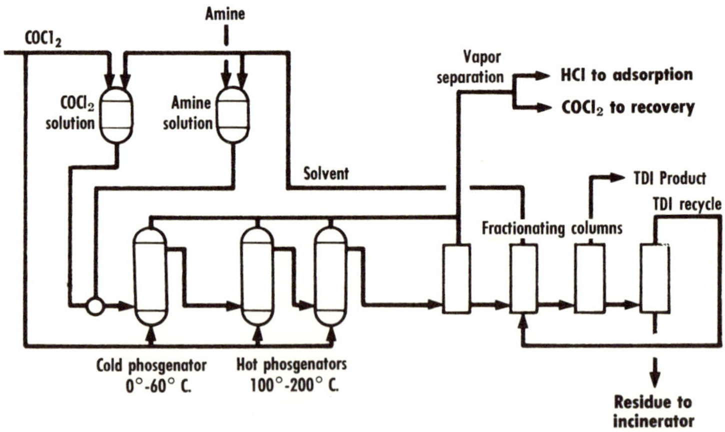

Source: Encyclopedia of Chem. Technol., Kirk-Othmer, 2nd Ed., Vol. 12, p. 56

FIG. 6.1 Production of tolylene diisocyanate

Demand for TDI has grown from 43 to 360 MM lb in the 1960–1971 period as shown in Table 6.5. This represents a compounded annual growth rate greater than 20%. A 65/35 blend of 2,4- to 2,6-TDI is used on a minor scale in the US.

Flexible polyurethane foams are essentially made by reacting TDI or other diisocyanate with a polyether glycol. The number of applications for these foams is extensive. Cushioning materials, such as found in furniture, automobiles, and aircraft are large-volume applications.

TABLE 6.5
Tolylene diisocyanate production[1]

Year	*MM lb*	*Unit value (¢/lb)*
1960	43	71
1961	68	66
1962	94	61
1963	131	57
1964	114	50
1965	147	35
1966	173	33
1967	200	29
1968	221	28
1969	258	30
1970[2]	300	na
1971[3]	360	na

[1] US Tariff Commission. Represents an 80/20 mixture of 2,4- and 2,6-tolylenediisocyanate.
[2] Oil, Paint & Drug Reporter (Nov. 2, 1970), p. 11.
[3] *Ibid,* May 10, 1971; p. 11.

TABLE 6.6
1972 tolylene diisocyanate capacity

Company		*MM lb*
Allied Chemical	(Moundsville, W.Va.)	70
BASF-Wyandotte	(Geismar, La.)	35
DuPont	(Deepwater, N.J.)	105
Mobay	(New Martinsville, W.Va.)	100
Olin	(Ashtabula, O.)	30
Rubicon Chemicals	(Geismar, La.)	25
Union Carbide	(Institute, W.Va.)	55
	Total	420

For years, the TDI market has been in chronic oversupply. This situation precipitated the enormous price drop from 50 to 35¢/lb in 1964–65 (Table 6.5). Installed capacity for 1972 appears to be adequate to accommodate demand. (Table 6.6).

Toluene foreign trade

The export market has long accounted for a sizable fraction of US toluene production. This represented a mean 50 MM gal or about 350 MM lb annually through 1966. Following that year, US exports dipped sharply, and imports, which had been miniscule, rose rapidly. The long term US positive trade balance quickly vanished, and since 1969 the US has been a net importer of toluene in increasingly greater quantities (Table 6.7). Although there are no duties on toluene imports, this surge in imports coincided with initiation of Kennedy Round tariff reductions in 1967. Japan has consistently led the list of exporters to the US as shown in Table 6.8.

TABLE 6.7
Toluene foreign trade[1]
(MM gal)

Year	*Imports*	*Exports*	*Balance*
1960	3	51	−48
1961	2	42	−40
1962	1	56	−55
1963	3	61	−58
1964	5	57	−52
1965	14	47	−33
1966	29	51	−22
1967	31	19	−12
1968	32	39	−7
1969	50	32	+18
1970	93	20	+73
1971[2]	108	14	+94

[1] US Census Bureau.
[2] Based on 5 months' data.

BENZENE

As one would expect, benzene production from petroleum has paralleled that of toluene. Although benzene production grew nearly three-fold in the last decade, contributions from coke-oven sources have steadily dwindled. Whereas coke oven distillates provided 30%

TABLE 6.8
Major toluene imports[1]
(MM gal)

Exporting country	*1970*	*1971*[2]
Canada	6.6	10.3
Italy	21.5	5.1
Japan	26.0	32.2
Netherlands	4.1	–
Netherlands Antilles	17.2	5.5
Rumania	2.1	0.7
Trinidad	9.0	13.5
USSR	2.3	–
Other	4.4	40.7[3]
Total	93.2	108.0

[1] US Census Bureau.
[2] Based on 5 months' data.
[3] Includes Belgium, Spain, Bulgaria et al.

of US benzene production in 1960, they were only 8% in 1971 (Table 6.9). This trend will undoubtedly continue as changes in steel production technology continue to require less coke.

The general downturn in the US economy in 1970 and 1971 also saw a decline in continuous benzene growth. Production in 1970 was

TABLE 6.9
Benzene production[1]
(MM gal)

Year	*Coke oven*	*Petroleum*	*Total*	*MM lb*[3]
1960	136	309	445	3,260
1961	121	413	534	3,920
1962	116	418	534	3,920
1963	115	526	641	4,700
1964	119	611	730	5,350
1965	122	705	827	6,050
1966	114	837	951	6,950
1967	91	879	970	7,100
1968	93	908	1,001	7,320
1969	102	1,060	1,162	8,500
1970	93	1,073	1,166	8,555
1971[2]	85	932	1,017	7,500

[1] US Tariff Commission.
[2] Based on 7 months' data.
[3] 7.38 lb/gal.

substantially at the same level as in 1969, and 1971 exhibited an actual decline. As a consequence, installed capacity is clearly excessive and posted prices declined to an all time low of 2.98¢/lb (Tables 6.10 and 6.11).

Unlike toluene, benzene has a broad based chemical end use pattern. Since 1960, styrene, phenol, and cyclohexane have clearly dominated the markets for benzene (Table 6.12). Maleic anhydride and aniline have vied for a distant fourth and fifth position in this market.

TABLE 6.10
1972 petroleum benzene capacity
Major producers

Company		*MM lbs*
Amoco	(Texas City, Tx.)	440
Ashland	(Catlettsburg, Ky.)	220
Coastal States	(Corpus Christi, Tx.)	510
Commonwealth	(Guyanilla Bay, P.R.)	730
Cosden	(Big Spring, Tx.)	210
Dow Chemical	(Freeport, Tx.)	220
	(Bay City, Mich.)	200
Enjay Chemical	(Baton Rouge, La.)	550
	(Baytown, Tx.)	610
Gulf Oil	(Phila., Pa.)	220
	(Port Arthur, Tx.)	290
Hess	(Corpus Christi, Tx.)	260
Mobil	(Beaumont, Tx.)	370
Monsanto	(Alvin, Tx.)	500
Shell	(Wood River, Del.)	220
Sinclair	(Houston, Tx.)	230
Standard of Calif.	(El Segundo, Calif.)	180
Sun Oil	(Corpus Christi, Tx.)	180
Texaco	(Westville, N.J.)	500
Union Carbide	(Taft, La.)	370
Union Oil	(Lemont, Ill.)	220
Others[1]		4,320
	Total	11,450

[1] Includes coke oven producers at 950 MM lb.

Styrene

Without question, styrene production represents the major use for benzene. Styrene manufacture has consistently accounted for 35 to 40% of US benzene production.

TABLE 6.11
Petroleum benzene sales price[1]
(¢/lb)

Year	*USTC*	*OPD (lowest)*
1961	4.36	4.06
1962	3.42	3.39
1963	3.14	3.39
1964	3.14	3.39
1965	3.28	3.39
1966	3.28	3.52
1967	3.28	3.66
1968	2.87	3.12
1969	2.97	3.39
1970	na	3.12
1971	na	2.98

[1] Nitration grade, tanks (freight equalized).

TABLE 6.12
Benzene end use pattern
(MM lb)

Year	*Styrene*[1]	*Maleic anhydride*[2]	*Phenol*[3]	*Cyclohexane*[4]	*Aniline*[5]	*Others*[6]	*Total*
1960	1,370	109	710	460	103	508	3,260
1961	1,381	95	731	596	117	1,000	3,920
1962	1,520	102	773	675	131	719	3,920
1963	1,685	105	884	841	147	1,038	4,700
1964	2,015	143	1,063	1,069	161	899	5,350
1965	2,245	155	1,173	1,330	186	961	6,050
1966	2,500	204	1,289	1,580	227	1,150	6,950
1967	2,563	203	1,297	1,480	215	1,342	7,100
1968	2,899	220	1,463	1,690	250	798	7,320
1969	3,600	243	1,643	1,850	318	846	8,500
1970	3,400	261	1,709	1,510	354	1,321	8,555
1971	3,180	286	1,625	1,290	316	803	7,500

[1] 0.87 lb benzene/lb styrene—Assumes 90% of styrene is from synthetic ethylbenzene.

[2] 1.21 lb benzene/lb maleic anhydride.

[3] 1.00 lb benzene/lb as average for all processes.

[4] 0.98 lb benzene/lb cyclohexane.

[5] 0.95 lb benzene/lb aniline.

[6] Includes anthraquinone, DDT, detergent alkylate, solvent, and exports.

Styrene is manufactured by catalytic dehydrogenation of an ethylbenzene intermediate. About 10% of the ethylbenzene is obtained from reformate, with the balance (90%) being produced by reaction of benzene and ethylene via $AlCl_3$ catalysis:

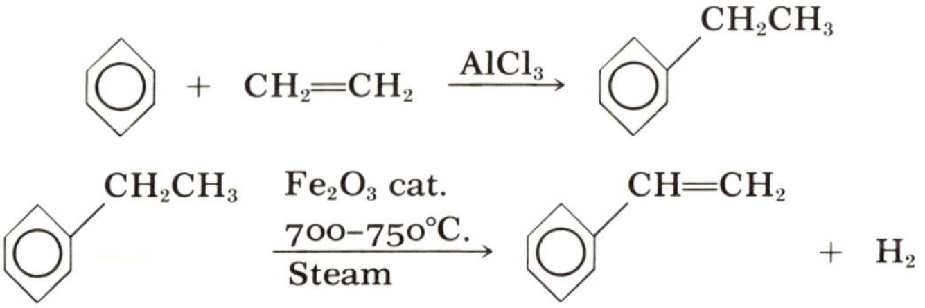

The ethylbenzene is vaporized, mixed with primary steam and is superheated in the convection section of a steam superheater (Fig. 6.2). Selectivity to styrene from ethylbenzene is usually 91%. The estimated process economics for a 500 MM-lb plant are summarized in Table 6.13.

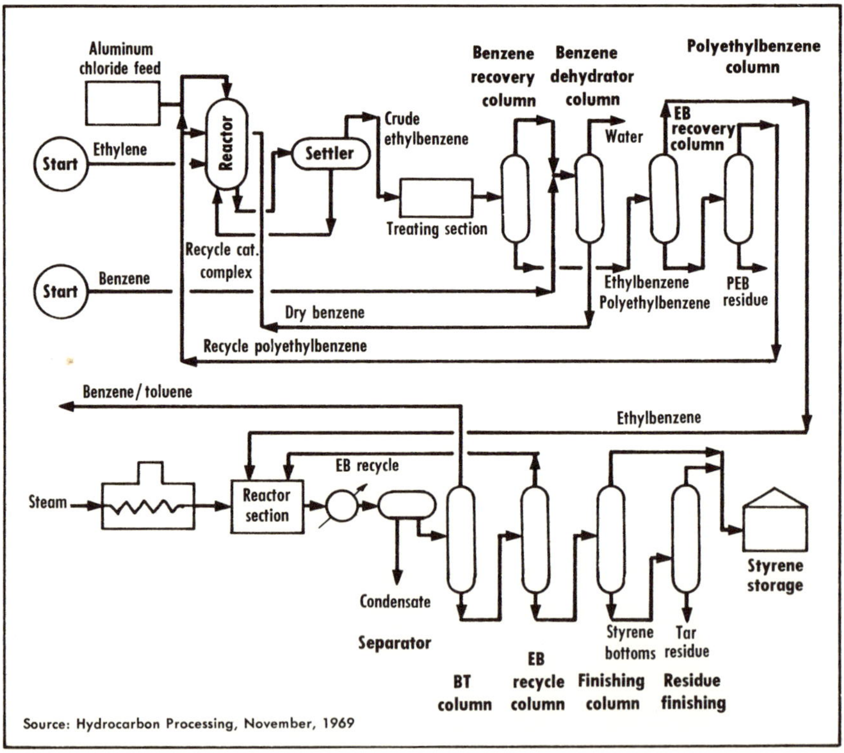

FIG. 6.2 Styrene flow diagram

TABLE 6.13
Estimated styrene economics[1]
(500 MM-lb plant; 1972 construction period)

A. *Total fixed capital* = $35.0 MM
B. *Production costs*

	¢/lb styrene
Raw materials[2]	3.95
Labor	0.13
Utilities	0.91
Maintenance (6% ISBL + 3% OSBL)	0.34
Overhead (45% maint + labor)	0.56
Taxes (1.5% of invest)	0.10
Depreciation (10 yr)	0.70
Total	6.69

[1] Dehydrogenation process.
[2] 1.10 lb ethylbenzene at 3.50¢/lb + catalyst and chemicals.

These data show that the production cost for a 1972 plant is 6.69¢/lb. If one excludes depreciation, this is the same as the probable current sales price for this material (Table 6.15). Evidently, the value of ethylbenzene may be overstated. Although 3.5¢/lb was the price for ethylbenzene in 1969 (Table 6.14), industry sales upon which this value is based only account for 10% of production. The actual captive

TABLE 6.14
Ethylbenzene market[1]
(MM lb)

Year	*Production*	*Sales*	*Unit value (¢/lb)*
1962	2,010	83	4.9
1963	2,451	347	4.0
1964	2,954	473	4.2
1965	3,023	580	4.1
1966	3,245	459	4.0
1967	3,347	470	4.0
1968	4,034	507	3.8
1969	4,907	458	3.5
1970	na	na	na
1971	na	na	na

[1] US Tariff Commission.

TABLE 6.15
Styrene production[1]

Year	*MMM lb*	*Unit value, ¢/lb*
1960	1.745	11
1961	1.762	11
1962	1.942	10
1963	2.154	9
1964	2.571	8
1965	2.864	8
1966	3.192	8
1967	3.278	8
1968	3.698	7
1969	4.594	6
1970	4.353	8[3]
1971[2]	4.050	8[3]

[1] US Tariff Commission.
[2] Based on 7 months' data.
[3] Oil, Paint & Drug Reporter; posted price in tanks (delivered).

TABLE 6.16
Styrene capacity
(MM lb)

Company		*1967*[1]	*1970*[1]	*1972*
Amoco	(Texas City, Tx.)	300	800	800
Cosden	(Big Spring, Tx.)	100	100	100
Cos-Mar	(Carville, La.)	—	500	500
Dow	(Freeport, Tx.)	500	650	650
	(Midland, Mich.)	300	350	350
El Paso	(Odessa, Tx.)	85	120	120
Foster-Grant	(Baton Rouge, La.)	200	250	250
Gulf Oil	(Donaldsville, La.)	—	—	500
Marbon	(Baytown, Tx.)	125	135	shut down
Monsanto	(Texas City, Tx.)	650	800	1,300[2]
Shell	(Torrance, Calif.)	210	240	240
Sinclair-Koppers	(Houston, Tx.)	70	110	110
	(Kobuta, Pa.)	200	430	430
Sun Oil	(Corpus Christi, Tx.)	60	80	80
Union Carbide	(Sea Drift, Tx.)	300	300	300
	(Institute, W.Va.)	110	shut down	shut down
	Total	3,210	4,865	5,730

[1] Oil, Paint & Drug Reporter, July 7, 1969 and earlier profiles.
[2] New plant that replaced 800 MM-lb unit.

value of ethylbenzene is probably below 3.5¢/lb. It is noteworthy that the utilities cost of 0.91¢/lb is quite high and reflects the high cost of steam.

Styrene production grew rapidly in the last decade, from 1.7 to 4.6 billion lb/yr by 1969 (Table 6.15). Coincident with this rapid growth of 12% annually has been a corresponding price decline.

From a value of 11¢/lb at the beginning of the decade, the price of styrene had declined nearly 50% by 1970. Two factors facilitated this decline in price: (1) an apparent decline in ethylbenzene transfer price, and (2) a rapid increase in plant size.

Installed plant capacity nearly doubled from 1967–72 as shown in Table 6.16.

Plant capacity in 1972 appears adequate to accommodate demand which has been in decline since 1970. One factor was the general downturn in the US economy. Another was the installation of sizeable European capacity which crimped soaring US exports of styrene.

Styrene is used for a variety of polystyrene homo- and copolymers. Usage in 1971 was estimated to be:

Use[1]	*MM lb*
Straight polystyrene	1,000
Rubber-modified polystyrene	810
Styrene-butadiene elastomers	690
Acrylonitrile-butadiene-styrene and styrene acrylonitrile resins	365
Exports	238
Others	947
Total	4,050

[1]Based on Oil, Paint & Drug Reporter, May 17, 1971 (p. 11).

TABLE 6.17
Styrene exports[1]

Year	*MM lb*	*Unit value, ¢/lb*
1965	367	8.18
1966	360	7.06
1967	436	6.60
1968	538	6.28
1969	832	6.29
1970	568	6.35
1971[2]	238	6.30

[1]US Census Bureau.
[2]Based on 5 months' data.

Exports only accounted for about 6% styrene production in 1971 as opposed to 13% in 1970 and 18% in 1969. As illustrated in Table 6.17, the 1971 export market was the smallest in recent years. The dramatic shrinkage in exports to Western Europe is apparent in Table 6.18.

TABLE 6.18
Principal purchasers of US styrene exports[1]
(MM lb)

Importing country	*1970*	*1971*[2]
Brazil	46.1	41.5
Belgium	15.8	—
Italy	23.1	11.0
Japan	16.9	0.1
Netherlands	267.4	145.0
Spain	20.0	—
Switzerland	14.5	1.5
Other	163.8	38.9

[1] US Census Bureau.
[2] Based on 5 months' data.

Phenol

Phenol ranks second only to styrene as the greatest use for benzene. It has been produced by a variety of processes, only one of which does not employ benzene. Kalama Chemical recently purchased a small plant from Dow Chemical, which is based on toluene. As shown in Table 6.19, this unit constitutes less than 2% of total capacity, and there appears to be little trend away from benzene-based processes.

In 1968 phenol technology was highly varied, with 55% of domestic capacity based on the cumene route. There has been a rapid acceleration toward cumene as the preferred process, with the result that 83% of active capacity in 1972 was based on this technology.

The cumene process comprises two steps wherein cumene must first be produced via alkylation of benzene with propylene. This is a vapor phase process employing a phosphoric acid catalyst supported on kieselguhr:

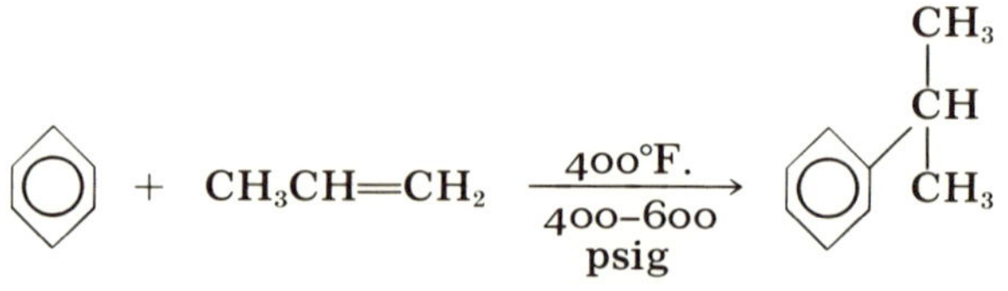

TABLE 6.19
Phenol capacity
(MM lb)

Company		*1968*	*1970*	*1972*	*Process*
Allied Chemical	(Phila., Pa.)	325	400	500	Cumene
Chevron Chemical	(Richmond, Calif.)	65	65	65	Cumene
Clark Oil Co.	(Chicago, Ill.)	60	75	75	Cumene
Dow Chemical	(Midland, Mich.)	235	235	235	Chlorobenzene
	(Freeport, Tx.)	–	–	400	Cumene
Georgia Pacific	(Plaquemine, La.)	–	–	200	Cumene
Hercules, Inc.	(Gibbstown, N.J.)	50	50	–	Cumene
Hooker Chemical	(South Shore, Ky.)	65	65	–	Raschig
	(N. Tonawanda, N.Y.)	65	65	–	Raschig
Kalama Chemical	(Kalama, Wash.)	35	50	50	Toluene
Monsanto	(Alvin, Tx.)	140	415	415	Cumene
	(Monsanto, Ill.)	115	–	–	Sulfonation
Reichhold	(Tuscaloosa, Ala.)	90	135	135	Sulfonation
Shell Chemical	(Houston, Tx.)	50	50	50	Cumene
Shelby Oil	(El Dorado, Kans.)	50	50	50	Cumene
Union Carbide	(Marietta, O.)	110	110	–	Raschig
	(Bound Brook, N.J.)	150	150	150	Cumene
USS Chemicals	(Haverhill, O.)	–	200	200	Cumene
	Total	1,605	2,115	2,525	
	% Cumene-based	55	69	83	

Selectivity is 96% on benzene and 90% on propylene. Much of the cumene is purchased on the merchant market (Tables 6.20 and 6.21). There is a tendency for many producers to employ converted polymer gasoline units for this purpose. Refinery-grade propylene containing up to 45% propane is usable. Propane passes through the reactor unchanged and is sold as LP-gas.

TABLE 6.20
Cumene production and sales[1]
(MM lb)

Year	*Production*	*Sales*	*Unit value (¢/lb)*
1965	663	na	na
1966	895	na	na
1967	1,134	na	na
1968	1,347	na	na
1969	1,687	909	3.7

[1] US Tariff Commission. Data for 1970 and 1971 not available.

TABLE 6.21
US cumene capacity[1]
(*MM lb*)

Company	*1969*	*1970*	*1972*
Amoco (Texas City, Tx.)	50[2]	50[2]	50[2]
Ashland (Catlettsburg, Ky.)	88	88	88
Clark (Blue Island, Ill.)	100	100	100
Coastal States (Corpus Christi, Tx.)	130	130	130
Dow (Midland, Michigan)	10	10	10
Gulf (Phila., Pa.)	400	400	450
(Port Arthur, Tx.)	—	—	450
Hercules (Gibbstown, N.J.)	80	80	80
Marathon (Texas City, Tx.)	210	210	210
Monsanto (Alvin, Tx.)	200	640	640
Shell (Houston, Tx.)	80	80	80
Skelly (El Dorado, Kan.)	80	80	80
Sun (Corpus Christi, Tx.)	120	120	200
Texaco (Westville, N.J.)	120	120	120
Total	1,668	2,108	2,688

[1] Based on Oil, Paint & Drug Reporter.
[2] Primarily α-methylstyrene production.

Essentially, cumene has no other market than for phenol synthesis. The process diagram for cumene is illustrated in Fig. 6.3.

In order to produce phenol, cumene is oxidized with air in an

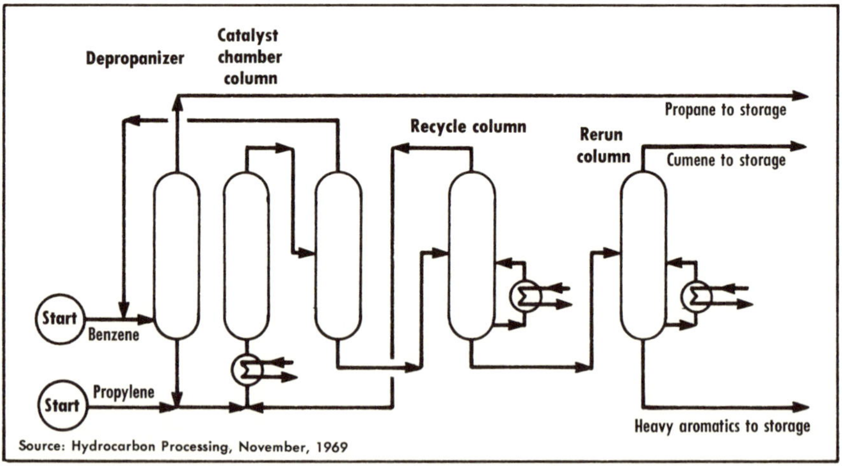

FIG. 6.3 Cumene flow scheme

aqueous alkaline emulsion to yield the hydroperoxide. The reaction proceeds in about 90% selectivity at 100°C. The hydroperoxide so formed is then contacted with 10% H_2SO_4 and is decomposed to phenol and acetone. Again, selectivity is of the order of 90 to 95%.

Coproduction of acetone (0.6 lb/lb phenol) greatly enhances the process economics. The process scheme for phenol production is shown in Figure 6.4.

Until recently the Raschig Process was a competitive phenol route. This comprised catalytic chlorination of benzene in the vapor phase (200°C.) followed by catalytic hydrolysis at 500°C. to yield phenol. The HCl used as the chlorinating agent is recovered and recycled.

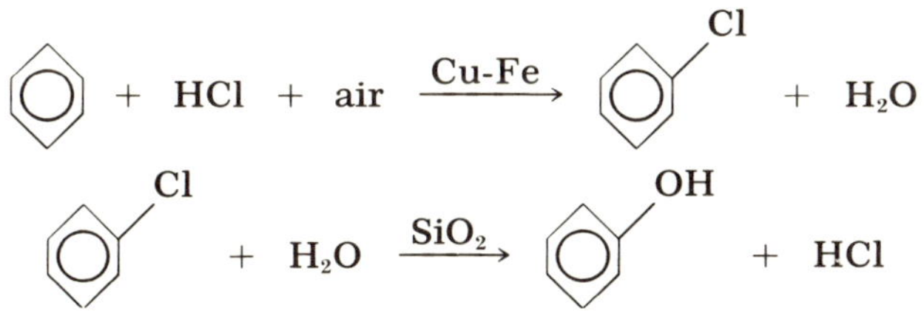

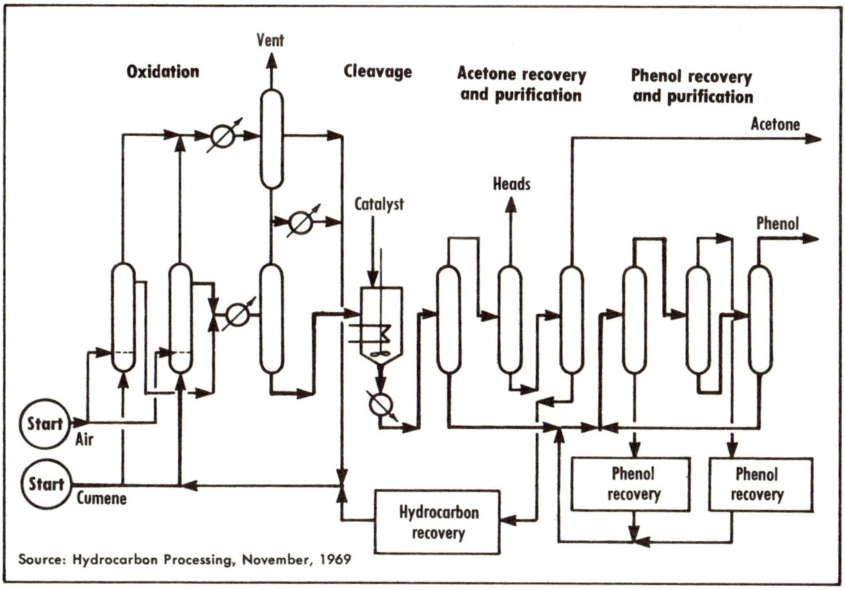

FIG. 6.4 Cumene-based phenol process

Overall selectivity is about 89%. The last active unit in the US was shut down in late 1971.

The toluene process alluded to earlier comprises liquid-phase oxidation of toluene to benzoic acid, followed by copper catalyzed decomposition to phenol and CO_2:

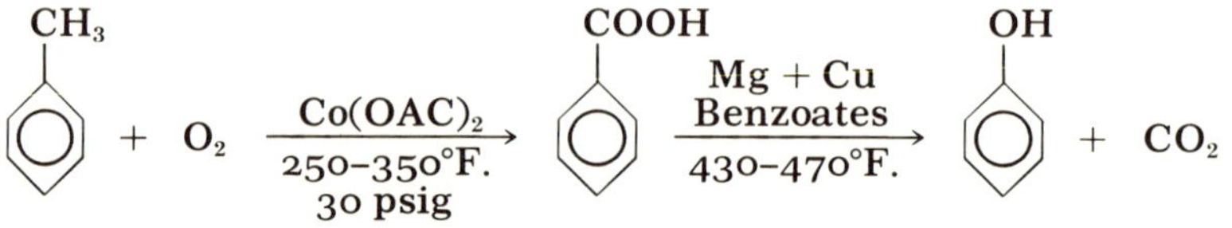

Selectivity to phenol is about 80%. Substantial production of tar supposedly occurs in the process.

TABLE 6.22
Estimated economics for phenol production
(400-MM-lb plant; 1972 construction)

Fixed investment costs

Process	*$ MM*
Cumene	26.6
Toluene	30.0
Raschig	36.1

Production costs, ¢/lb

	Cumene	*Toluene*	*Raschig*
Raw materials	5.81[1]	3.45[3]	3.67[4]
Labor	0.29	0.29	0.29
Utilities	0.92	0.71	0.78
Maintenance (6% ISBL + 3% OSBL)	0.32	0.36	0.43
Overhead (45% maint. & labor)	0.27	0.30	0.32
Taxes & insurance (1.5% of invest.)	0.10	0.11	0.13
Depreciation (10%)	0.67	0.76	0.91
Total	8.38	5.98	6.53
By-product credit[2]	2.74	—	—
Net	5.64	5.98	6.53

[1] 1.45 lb cumene/lb at 3.7¢/lb + catalyst and chemicals.
[2] 0.60 lb acetone/lb phenol at 4.6¢/lb.
[3] Includes 1.3 lb toluene at 2.5¢/lb.
[4] 0.94 lb benzene/lb at 3.4¢/lb + catalyst and chemicals.

Two other processes are still employed for phenol production. These are both based on classical chemistry and consist of chlorination or sulfonation of benzene to either chlorobenzene or benzenesulfonic acid. In a succeeding step, either of these is submitted to caustic fusion to yield a sodium salt of phenol. The latter is isolated via acidification and purified.

By virtue of co-product acetone formation, the cumene process appears to be more economical than its competitors. This is illustrated in Table 6.22. In the foreseeable future, it is likely that expansions in phenol capacity will all be via the cumene route.

US growth in phenol production advanced an average 9% annually in the last decade. Synthetic processes have long accounted for nearly all US production. Naturally occurring phenol from coal tar and petroleum represented little more than 2% of 1971 production, down from 6% in 1961 (Table 6.23).

TABLE 6.23
Phenol production[1]
(MM lb)

Year	*Natural*[3]	*Synthetic*	*Total*	*Unit value, ¢/lb*
1961	48	731	779	14
1962	52	773	825	12
1963	51	884	935	10
1964	50	1,063	1,113	10
1965	56	1,173	1,229	9
1966	57	1,289	1,347	9
1967	59	1,297	1,356	9
1968	50	1,463	1,513	7
1969	54	1,643	1,697	8
1970	46	1,709	1,755	8[4]
1971[2]	39	1,625	1,664	8[4]

[1] US Tariff Commission.
[2] Based on 7-months' data.
[3] Coal tar and petroleum.
[4] Oil, Paint & Drug Reporter posted price in tanks (freight allowed). USTC values not available.

Growth in phenol demand has been paced largely by phenolic resins. These compounds are readily produced by condensation of formaldehyde with phenol. The 1971 end use pattern for phenol can be summarized:

End use	*MM lb*[1]
Phenolic resins	830
Caprolactam	333
Bisphenol-A	166
Adipic acid	80
Surfactants	85
Others	170
Total	1,664

[1] Based on Oil, Paint & Drug Reporter (Aug. 4, 1969).

Of these uses, bisphenol-A has demonstrated the greatest growth potential. This product is readily produced by acid-catalyzed condensation of acetone and phenol, both co-products from the cumene process. Yields of 96% are obtainable with dry HCl as a catalyst.

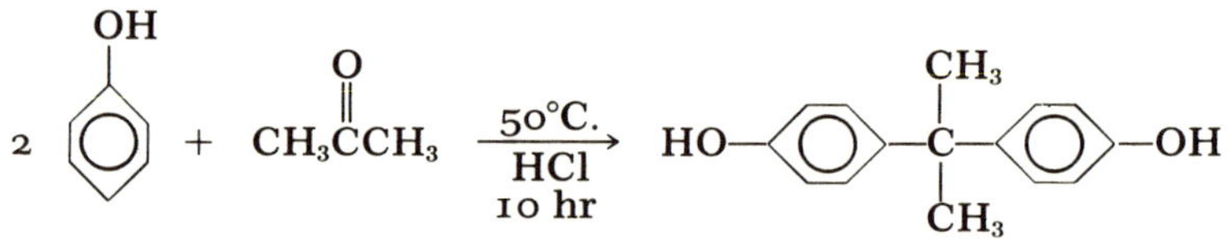

Bisphenol-A is largely used for production of polycarbonate and epoxy resins.

Cyclohexane

Conversion of benzene to this material proceeds in about 95% yield via liquid phase, catalytic hydrogenation. The catalyst is

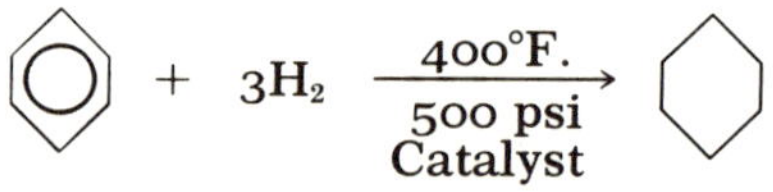

usually platinum-supported on alumina and contains a small amount of lithium salt as a promoter.

Cyclohexane is also directly recoverable from petroleum, and Phillips has two units in Texas that use this technique (Table 6.24). The presence of several hydrocarbons that boil in the same range makes recovery difficult. An 85% concentrate is manufactured commercially, in contrast to the nearly 100% material produced by catalytic hydrogenation of benzene. In preparing an 85% concentrate from petroleum, the alicyclics from natural gasoline undergo careful fractionation.

TABLE 6.24
Cyclohexane capacity[4]
(MM gal)

Company	*1967*[1]	*1972*[2]	*Process*
Ashland (Catlettsburg, Ky.)	20	30	Benzene
Arco (Wilmington, Calif.)	15	15	"
Conoco (Lake Charles, La.)	40	shut down	"
(Ponca City, Okla.)	40	shut down	"
Cosden (Big Spring, Tx.)	8	8	"
Enjay (Baytown, Tx.)	40	40	"
Gulf (Port Arthur, Tx.)	33	33	"
Phillips (Borger, Tx.)	47	47	Petroleum
(Las Mereas, P.R.)	46	46	Benzene
(Sweeney, Tx.)	30	53	Petroleum
Pontiac (Corpus Christi, Tx.)	12	12	Benzene
Shell-Corco (Guayanilla Bay, P.R.)	30	30	"
Texaco (Port Arthur, Tx.)	40	40	"
Union (Nederland, Tx.)	33	33	"
Total	434	387	
Total[3] (MM lb)	2,820	2,520	

[1]82% based on benzene hydrogenation.
[2]74% based on benzene hydrogenation.
[3]6.5 lb/gal.
[4]Based on Oil, Paint & Drug Reporter Profile, Jan. 1, 1969.

Installed cyclohexane capacity declined in the 1968–72 period as a consequence of decreased demand. As shown in Table 6.25, 1971 production was about 600 MM lb lower than 1969. Decline in production has been attributed to a decrease in demand for nylon and to a steady drop in the export market.[1]

In recent years, the export market has claimed a full third of US production, and growing overseas capacity has made its impact. Moreover, cyclohexane possesses a very narrow end-use market, that of nylon-6 and nylon-66. Cyclohexane is used as the starting material for synthesis of adipic acid, hexamethylenediamine, and caprolactam, which are used to prepare these resins. Since 1969, the domestic nylon market has been a victim of the general downturn in the US economy.

[1]Oil, Paint & Drug Reporter, September 13, 1971.

TABLE 6.25
Cyclohexane production[1]

Year	*MM lb*	*Unit value, ¢/lb*
1961	701	5.0
1962	863	4.0
1963	1,079	4.0
1964	1,367	4.0
1965	1,700	4.0
1966	1,901	4.0
1967	1,777	4.0
1968	2,039	3.0
1969	2,231	3.0
1970	1,817	na
1971[2]	1,650	na

[1] US Tariff Commission.
[2] Based on 7 months' data.

Adipic acid

Adipic acid can be prepared either directly from cyclohexane via air oxidation or by passing through cyclohexanone and cyclohexanol as intermediate oxidation steps. Some producers are understood to form the cyclohexanone and cyclohexanol intermediates from phenol via catalytic hydrogenation. In these instances, cyclohexane is not employed at all (Table 6.26).

TABLE 6.26
Adipic acid capacity
(*MM lb*)

Company	*1967*[1]	*1972*	*Process*
Allied (Hopewell, Va.)	20	20	Phenol
Celanese (Bay City, Tx.)	35	90	Cyclohexane
DuPont (Belle, W.Va.)	210	300	Cyclohexane
(Orange, Tx.)	210	210	Cyclohexane
(Victoria, Tx.)	60	60	Cyclohexane
El Paso (Odessa, Tx.)	40	80	Cyclohexane
Monsanto (Pensacola, Florida)	350	440	Cyclohexane
(Luling, La.)	35	50	Phenol
Rohm & Haas (Louisville, Ky.)	20	20	Cyclohexane
Total	980	1,270	

[1] Oil, Paint & Drug Reporter Profile, October 1, 1969.

Cyclohexane can be oxidized under moderate pressure in the liquid phase using cobalt acetate as a catalyst. Partially oxidized products, such as cyclohexanone and cyclohexanol, can be recycled to the reactor and oxidized with fresh cyclohexane to more adipic acid. Yields on conversion are 70 to 90% of theory.

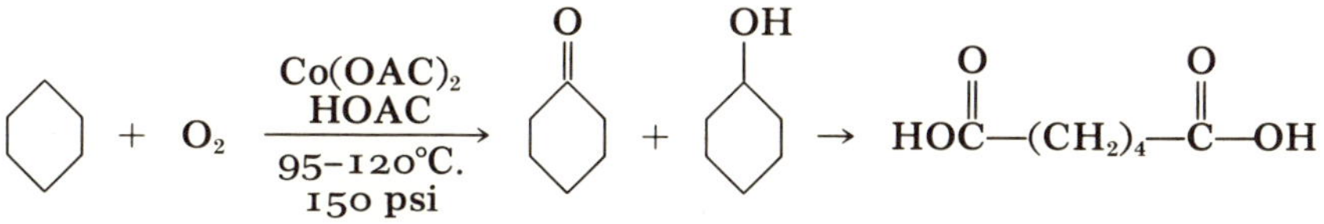

Aside from small amounts of adipic acid used as esters for lubricants and plasticizers, this chemical is nearly all used for nylon-66 manufacture. Manufacture of this well-known fiber forming resin is achieved by condensation of adipic acid with hexamethylene diamine, viz.,

$$HO\overset{O}{\overset{\|}{C}}-(CH_2)_4-\overset{O}{\overset{\|}{C}}OH + H_2N-(CH_2)_4-NH_2 \rightarrow$$

$$(-\overset{O}{\overset{\|}{C}}-(CH_2)_4-\overset{O}{\overset{\|}{C}}-NH-(CH_2)_4-NH-)_n$$

Production of adipic acid has grown from 670 to 1,220 MM lb annually in the 1963–69 period as shown in Table 6.27. Price has also declined

TABLE 6.27
Adipic acid production[1]

Year	*MM lb*	*Unit value, ¢/lb*
1961	na	28
1962	na	25
1963	670	25
1964	na	24
1965	866	23
1966	964	21
1967	971	19
1968	1,163	17
1969	1,220	16
1970	na	19[2]
1971	na	19[2]

[1] US Tariff Commission.

[2] Oil, Paint & Drug Reporter posted price; resin grade; carload lots (delivered).

with market growth to the low level of 16¢/lb in 1969. Only posted prices are available for 1970 and 1971. Approximately 88% of adipic acid is used for nylon-6,6 fiber.

Hexamethylenediamine, the counterpart of adipic acid in nylon-6,6, has been made historically from adipic acid per se. At 300°C. in the presence of phosphoric acid, adipic acid reacts with ammonia to yield adiponitrile.

$$2NH_3 + HO\overset{\overset{O}{\|}}{C}\text{—}(CH_2)_4\text{—}\overset{\overset{O}{\|}}{C}OH \rightarrow NC\text{—}(CH_2)_4\text{—}CN + 4H_2O$$

Liquid phase hydrogenation of adiponitrile leads to the desired hexamethylenediamine.

$$NC\text{—}(CH_2)_4\text{—}CN + 4H_2 \xrightarrow[\substack{300\ \text{atm} \\ \text{Co}}]{200^\circ\text{C.}} H_2N\text{—}(CH_2)_4\text{—}NH_2$$

Recently, new technology has been developed which challenges cyclohexane based processes for adipic acid and hexamethylenediamine. Monsanto has a commercial process for adiponitrile based on electrolytic coupling of acrylonitrile, derived from propylene. Hydrolysis converts adiponitrile to adipic acid, while hydrogenation yields hexamethylenediamine.

$$CH_2{=}CH\text{—}CH_3 + NH_3 \rightarrow CH_2{=}CH\text{—}CN \rightarrow NC\text{—}(CH_2)_4\text{—}CN$$

$$NC\text{—}(CH_2)_4\text{—}CN \rightarrow HO\overset{\overset{O}{\|}}{C}\text{—}(CH_2)_4\text{—}\overset{\overset{O}{\|}}{C}OH$$

$$NC\text{—}(CH_2)_4\text{—}CN \rightarrow H_2N(CH_2)_4\text{—}NH_2$$

In addition, both Esso and DuPont have reported butadiene-based routes for adiponitrile. In the Esso process, butadiene reacts with iodine and cuprous cyanide to yield 1,4-dicyanobutene, which can be hydrogenated to adiponitrile or hexamethylenediamine. Hydrolysis of the adiponitrile again leads to adipic acid.

Adiponitrile technology, which is discussed more fully in Chapter 11, could be a formidable challenger to cyclohexane oxidation.

Caprolactam

Caprolactam is used to prepare nylon-6, which competes directly with nylon-6,6. Although many processes exist for caprolactam production, the only one used commercially in the US is the Beckmann process (Table 6.28).

TABLE 6.28
Caprolactam capacity[1]
(MM lb)

Company	*1967*	*1972*	*Process*
Allied Chemical (Hopewell, Va.)	300	300	Beckmann
Columbia NIPRO (Augusta, Ga.)	44	150	Beckmann
Dow Badische (Freeport, Tx.)	90	176	Beckmann
DuPont (Beaumont, Tx.)	50	shut down	Nitrocyclohexane
Union Carbide (Taft, La.)	50	shut down	Caprolactone
Total	534	626	

[1]Based on Oil, Paint & Drug Reporter, July 1, 1969.

The Beckmann process is based on cyclohexanone as an intermediate, which is derived from cyclohexane. Cyclohexanone manufacture is based on boric acid technology which minimizes adipic acid production during cyclohexane oxidation. The overall process can be stated as:

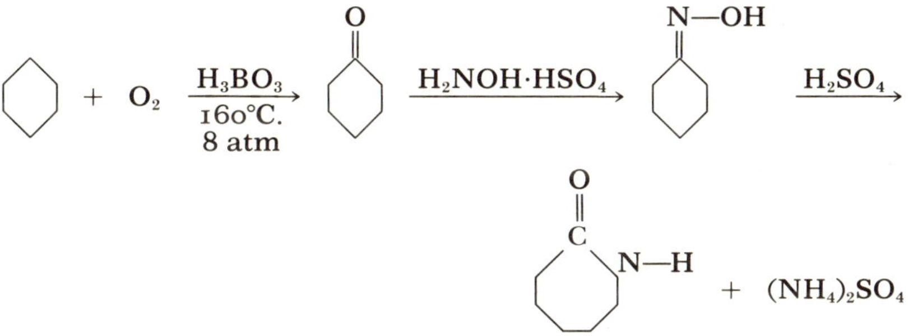

The hydroxylamine sulfate is obtained from ammonium nitrite and sulfur dioxide. The oxime, on contact with concentrated sulfuric acid, is rearranged in nearly quantitative yield to caprolactam. By-product ammonium sulfate is produced during oxime formation and during caprolactam formation. It is a low-value fertilizer whose disposal can be a problem.

The process flow scheme is shown in Fig. 6.5.

Union Carbide introduced an innovation employing caprolactone produced by oxidation of cyclohexanone with peracetic acid. This requires a unit for conversion of acetaldehyde to peracetic acid. By-product acetic acid must also be sold for the process to be competitive. The attractive feature appears to be elimination of $(NH_4)_2SO_4$ production by reacting the caprolactone directly with ammonia to produce caprolactam. Overall yield based on cyclohexane is about 75%.

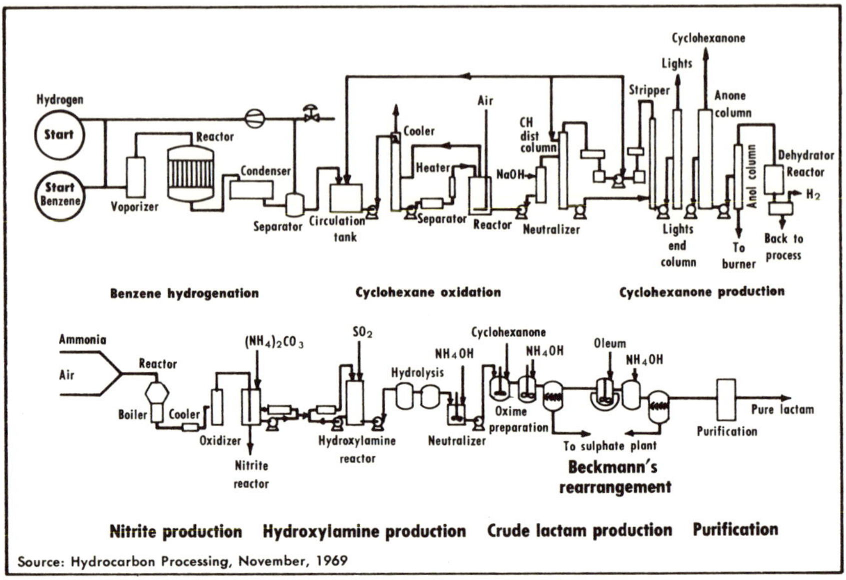

FIG. 6.5 Beckmann process for caprolactam

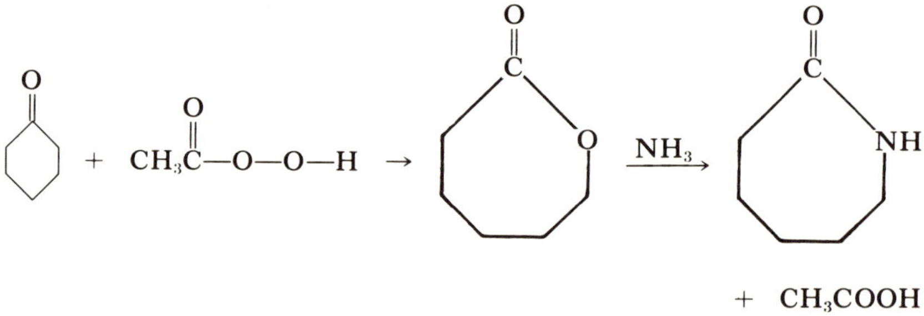

Neither the Toyo Rayon process (used in Japan) nor the Snia Viscosa process (used in Italy) have thus far found acceptance in the US. The Snia Viscosa route is based on toluene and proceeds through a benzoic acid intermediate.

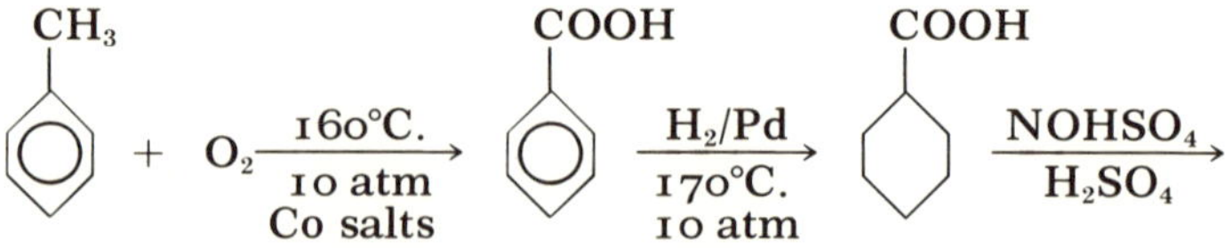

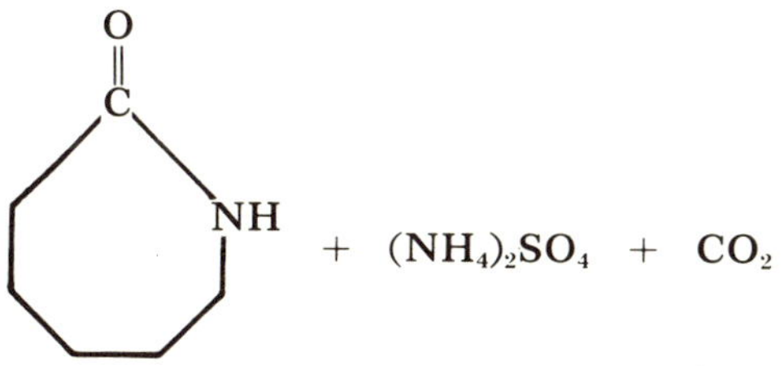

Overall yield based on toluene is about 82%. The nitrosylsulfuric acid is prepared *in situ* by bubbling N_2O_3 into the cyclohexane carboxylic acid dissolved in oleum. Sizable amounts of ammonium sulfate are produced as in the Beckmann process.

The Toyo Rayon route has elicited the greatest amount of interest. It differs from the Beckmann route by producing cyclohexanone oxime directly from cyclohexane. Thus, intermediate cyclohexanone and oximation are eliminated. Nitrosyl chloride reacts directly with cyclohexane under the influence of 60-kw mercury lamps.

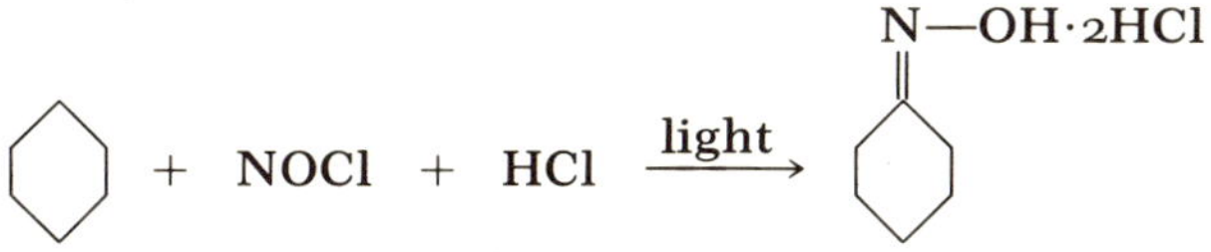

The nitrosyl chloride is produced from nitrogen trioxide which in turn is obtained from ammonia oxidation.

$$2NH_3 + 3O_2 \rightarrow N_2O_3 + 3H_2O$$
$$2H_2SO_4 + N_2O_3 \rightarrow 2HNOSO_4 + H_2O$$
$$HNOSO_4 + HCl \rightarrow NOCl + H_2SO_4$$

Once the oxime has been obtained, it is treated with H_2SO_4 to form caprolactam just as in the convention Beckmann process. The process possesses the advantage that ammonium sulfate is formed in the final step only and not during oximation.

The relative economics for the Beckmann, caprolactone and Toyo Rayon processes are summarized in Table 6.29. The Beckmann route has the lowest investment cost. Because of the need for mercury lamps, the utilities cost of the Toyo Rayon process is quite high, and its investment cost is somewhat higher than the Beckmann. Should ammonium sulfate disposal become a problem, the production costs of the Beckmann and Toyo Rayon routes will be closer.

With increasing market growth, caprolactam prices have steadily

TABLE 6.29
Estimated economics for caprolactam
(150-MM-lb plant; 1972 construction)

Total fixed capital

Process	*$ MM*
Beckmann	37.4[1]
Caprolactone	39.8[2]
Toyo Rayon	40.0

[1] Investment includes cyclohexanone and oximation.

[2] Investment includes peracetic acid and caprolactone units.

Production cost

	¢/lb caprolactam		
	Beckmann	*Caprolactone*	*Toyo Rayon*
Raw materials	11.43[1]	10.71[2]	9.14[3]
Utilities	1.60	1.91	2.25
Labor	0.58	0.40	0.36
Maintenance (6% ISBL + 3% OSBL)	1.20	1.28	1.28
Overhead (45% of maint. & labor)	0.80	0.76	0.74
Taxes & insurance (1.5% of inv.)	0.38	0.41	0.41
Depreciation (10 yr)	2.50	2.66	2.66
Total	18.49	18.13	16.84
By-product credit	4.44	6.22	1.58
Net	14.05	11.91	15.26

[1] Includes cyclohexane (0.88 lb at 3.3¢/lb), NH_3 (1 lb at 2¢/lb) and oleum (1.7 lb at $36/ton). Ammonium sulfate credit at $23/ton.

[2] Includes cyclohexane (1.0 lb at 3.3¢/lb) and acetaldehyde (0.62 lb at 8.0¢/lb). Acetic acid credit at 6¢/lb.

[3] Includes cyclohexane (0.95 lb at 3.3¢/lb), ammonia (1 lb at $40/ton) and oleum (1.7 lb at $36/ton). Ammonium sulfate credit at $23/ton.

declined (Table 6.30). In 1969, they were less than half the price that they were only seven years earlier. It is apparent from Table 6.30 that more than half the caprolactam market is captive for nylon-6 production.

TABLE 6.30
Caprolactam[1]
(MM lb)

Year	*Production*	*Sales*	*Unit value (¢/lb)*
1962	164	69	41
1963	188	90	37
1964	246	122	34
1965	290	134	31
1966	na	142	25
1967	329	165	20
1968	na	na	na
1969	na	450	18
1970	na	na	24.5[2]
1971	na	na	24.5[2]

[1] US Tariff Commission.
[2] Oil, Paint & Drug Reporter posted price. Molten in tanks, freight equalized.

Maleic anhydride

Although maleic anhydride is one of the smaller uses for benzene, it is nonetheless, a rapidly growing market. Consumption of benzene for this purpose has grown 9% annually in the last decade to 286 MM lb in 1971 (Table 6.12). Installation of new capacity has also increased rapidly in order to keep pace with demand. Although there is worldwide interest in use of butene as a raw material, all US capacity for maleic anhydride is based on benzene (Table 6.31).

TABLE 6.31
Maleic anhydride capacity[1]
(MM lb)

Company	*1967*	*1970*	*1972*
Allied Chemical (Moundsville, W.Va.)	20	20	20
Koppers (Bridgeville, Pa.)	20	20	30
Monsanto (East St. Louis, Ill.)	50	75	105
Petro-Tex (Houston, Tx.)	15	30	110
Reichhold (Elizabeth, N.J.)	20	30	30
(Morris, Ill.)	—	—	60
Tenneco (Fords, N.J.)	22	22	22
USS Chemicals (Neville Island, Pa.)	20	40	40
Total	167	237	417

[1] Based on Oil, Paint & Drug Reporter, August 25, 1969 and earlier profiles.

The benzene route comprises a fixed bed catalyst over which a mixture of benzene and air are passed:

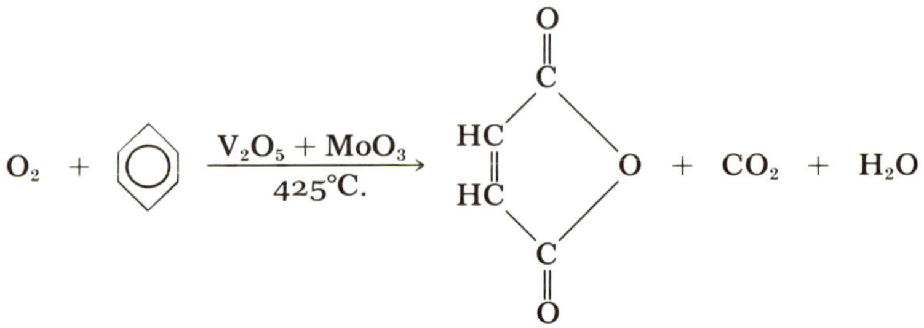

The process possesses the inherent difficulty that two carbon atoms must be discarded for each mole of maleic anhydride produced. As a consequence, theoretical yields are 1.26 lb of maleic anhydride/lb of benzene. In practice, yields are generally 60 to 65 mole %. This has prompted interest in Mitsubishi, Bayer and BASF to consider butene-based processes.

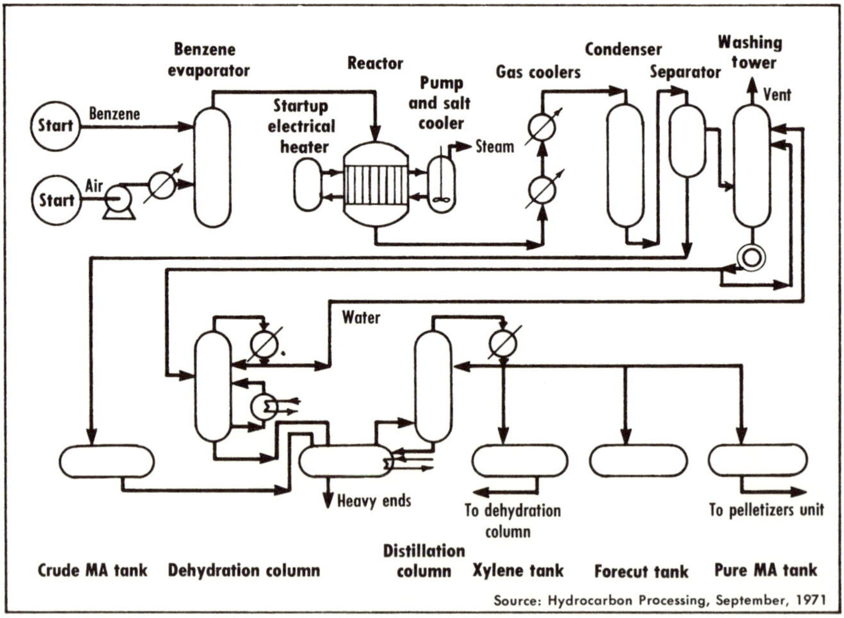

FIG. 6.6 Benzene-based maleic anhydride

A butene-based route would have the theoretical advantage of producing 1.75 lb of maleic anhydride/lb of reactant. Process technology would be similar to the benzene route outlined in Fig. 6.6.

As shown in Table 6.32, should butenes be available in the US at 2.65¢/lb, a butene-based route would have a lower production cost than one based on benzene. This appears to be true even though the yield based on butene is lower (65% vs. 45%). Should butene yields improve, not only would capital investment decline, but so would production cost.

TABLE 6.32
Estimated economics for maleic anhydride
(30-MM-lb plant; 1972 construction period)

Fixed investment cost

Process	*$ MM*
Benzene	5.85
Butene	6.75

Production cost

	¢/lb maleic anhydride	
	Benzene	*Butene*
Raw materials	5.16[1]	4.30[2]
Labor	0.63	0.63
Utilities	0.48	0.61
Maintenance (6% ISBL + 3% OSBL)	0.94	1.09
Overhead (45% labor & maint.)	0.71	0.77
Taxes & insurance (1.5% of investment)	0.39	0.45
Depreciation (10%/yr)	1.95	2.24
Total	10.26	10.09

[1] 1.21 lb benzene/lb at 3.4¢/lb + catalyst and chemicals.
[2] 1.27 lb mixed n-butenes/lb at 2.65¢/lb + catalyst and chemicals.

Maleic anhydride growth continued apace through the 1969–1971 economic downturn to an all time high of 237 MM lb. Prices have remained at the 12 to 15¢/lb level since 1963 (Table 6.33).

TABLE 6.33
Maleic anhydride production[1]

Year	*MM lb*	*Unit value, ¢/lb*
1960	90	24
1961	78	22
1962	84	17
1963	87	13
1964	118	12
1965	128	12
1966	169	13
1967	168	13
1968	182	12
1969	201	15
1970	216	16[3]
1971	237[2]	14[3]

[1] US Tariff Commission.
[2] Based on 7 months' data.
[3] Oil, Paint & Drug Reporter posted price in tanks (delivered).

The estimated 1971 distribution of maleic anhydride uses can be summarized as:

End use	*MM lb*[1]
Unsaturated polyesters	119
Fumaric acid	43
Pesticides	24
Alkyd resins	12
Miscellaneous	39
Total	237

[1] Based on Oil, Paint & Drug Reporter, Jan. 1, 1970.

Aniline

Aniline is produced from benzene via a two-step process comprising nitration and catalytic hydrogenation. Yield on each of the steps is 98%. Originally used primarily for the production of dyestuffs, the aniline market has diversified and grown. Production in 1971 was 332 MM lb, or nearly three-fold greater than a decade earlier (Table 6.34). This 10% annual growth is currently accommodated by six producers (Table 6.35).

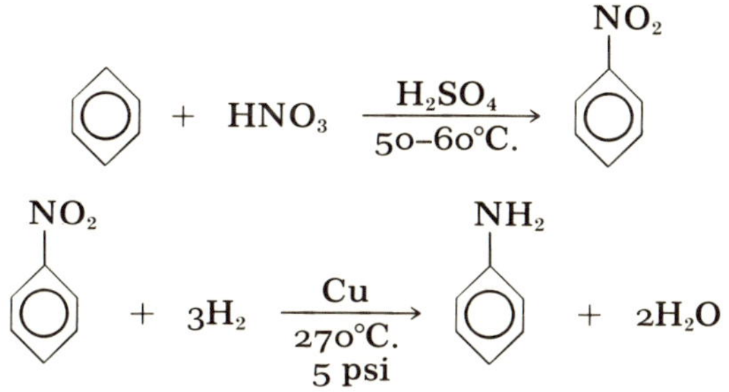

TABLE 6.34
Aniline production[1]

Year	MM lb	Unit value, ¢/lb
1961	123	16
1962	138	13
1963	155	14
1964	170	14
1965	196	13
1966	239	12
1967	226	11
1968	263	11
1969	334	11
1970	372	14[3]
1971[2]	332	15[3]

[1] US Tariff Commission.
[2] Based on 7 months' data.
[3] Oil, Paint & Drug Reporter posted price in tanks (freight allowed).

TABLE 6.35
1972 aniline capacity

Company	MM lb
Allied (Moundsville, W.Va.)	60
American Cyanamid (Bound Brook, N.J.)	60
(Willow Island, W.Va.)	40
DuPont (Gibbstown, N.J.)	130
(Beaumont, Tx.)	200
First Chemical (Pascagoula, Miss.)	35
Mobay (New Martinsville, W.Va.)	70
Rubicon (Geismar, La.)	40
Total	585

A substantial portion (54%) of aniline demand is currently due to derivatives used as anti-oxidants and vulcanization accelerators in the rubber industry. Dyestuffs are currently estimated to account for only 10% of total demand. A more prominent area of growth is use of aniline as an intermediate for isocyanates, specifically methylene-phenylene diisocyanate (MDI) and the related polymeric isocyanates. This market has grown rapidly in recent years and represents a serious challenge to long dominant TDI. Current producers are listed in Table 6.36.

TABLE 6.36
1972 MDI and polymeric isocyanate capacity[1]

Company	*MM lb*
Kaiser (Gramercy, La.)	20
Mobay (New Martinsville, W.Va.)	80
Rubicon (Geismar, La.)	80
Upjohn (La Porte, Tx.)	200
Total	380

[1] Jefferson Chemical may add 100 MM lb in 1973. Kaiser Chemical and Mobay will add 30 and 50 MM lb respectively in that year.

MDI and the related polymerics are principally used for manufacture of rigid polyurethane foams.

Synthesis of MDI and the polymeric isocyanates is achieved by acid catalyzed condensation of aniline with formaldehyde.

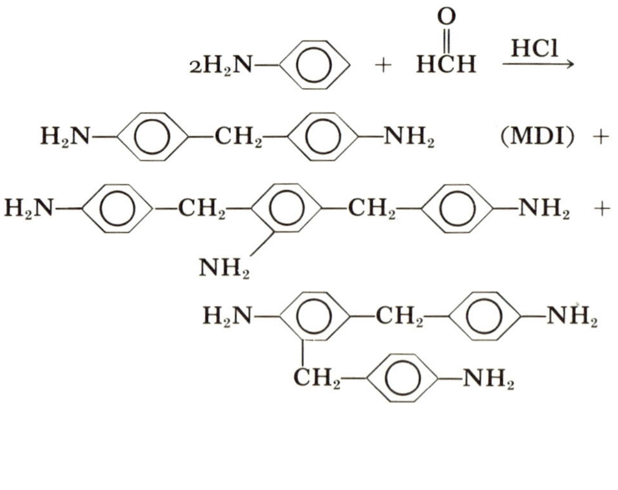

New technology for aniline production has been developed by Halcon International. This is based on ammonolysis of phenol. Lower capital costs are claimed for this route and safety problems are said to be less than the corresponding nitrobenzene process. The first plant is being built by Mitsui in Chiba, Japan. The apparent reaction is:

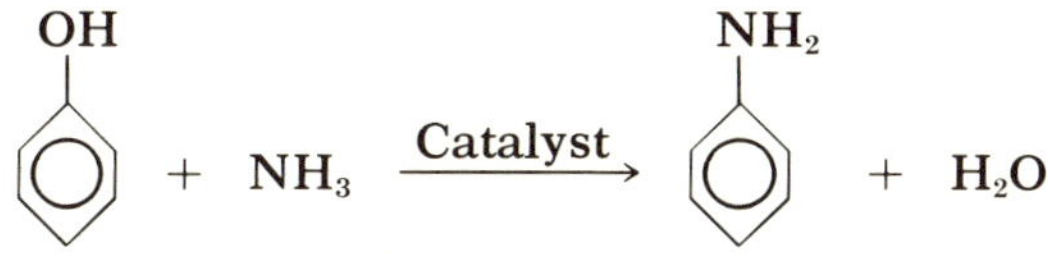

Cresols

Cresol production ranks high as a future use for toluene or benzene. Cresols are used primarily for synthesis of phenolic resins and phosphate esters. The esters are used as plasticizers and fuel additives.

According to some producers, synthetic cresols will take over from the natural variety as a major source of supply before very long. This speculation is summarized in Table 6.38. Accordingly, there has been a dramatic increase in synthetic cresol capacity in recent years. Current synthetic producers are listed in Table 6.37. Cresols are monomethyl phenols while cresylic acids are mixtures of cresols and xylenols. The latter are isomeric dimethylphenols.

TABLE 6.37
Synthetic cresol producers (1972)

Company	*MM lb*	*Process*
Hercules, Inc. (Gibbstown, N.J.)	6	p-cymene oxidation
Koppers (Follansbee, W.Va.)	10	phenol and methanol
Pitt-Consol (Newark, N.J.)	80	phenol and methanol
Sherwin Williams (Chicago, Ill.)	10	toluene sulfonation
Total	106	

Petroleum-based cresols are obtained from the spent caustic liquor used to wash petroleum distillates. The other main source of natural material is distillate from coke ovens. Changing technology is expected to lessen availability from both these sources.

Synthetic processes which have received the greatest attention are vapor phase methylation of phenol over alumina catalysts, and liquid phase oxidation of meta- and para-cymene.

TABLE 6.38
Cresylic acid demand[1]
(MM lb)

Supply	*1968*	*1969*	*1970*	*1971*	*1972*
Coal tar	63	57	51	46	41
Petroleum	69	65	59	52	45
Imports	65	14	8	7	7
Total	197	136	118	105	93
Demand	155	170	183	195	212

Cresylic acid end use pattern[1]

End use	*1968*	*1969*	*1970*	*1971*	*1972*
Phosphate esters	63	68	74	80	87
Phenolic resins	50	54	57	61	65
Wire enamel solvent	17	22	24	25	29
Other	25	26	28	29	31
Total	155	170	183	195	212

[1] Based on Chemical and Engineering News, May 20, 1968, p. 23.

Vapor phase methylation leads principally to o-cresol and 2,4- and 2,6-xylenol, among other things.

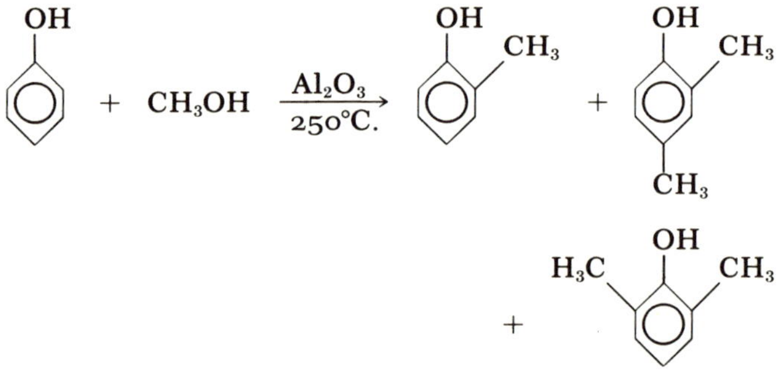

The cymene process is analogous to the cumene-based phenol process.

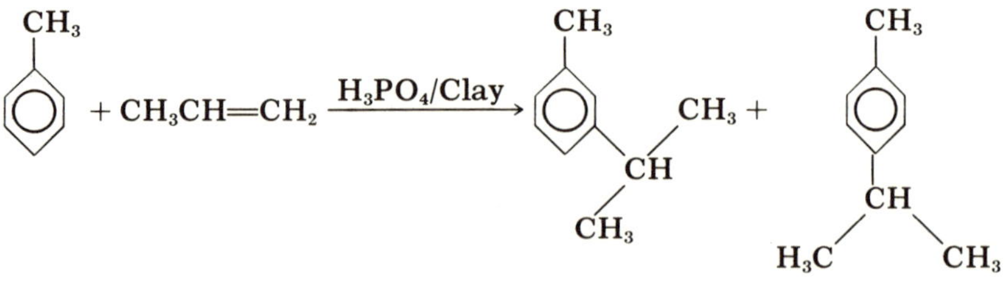

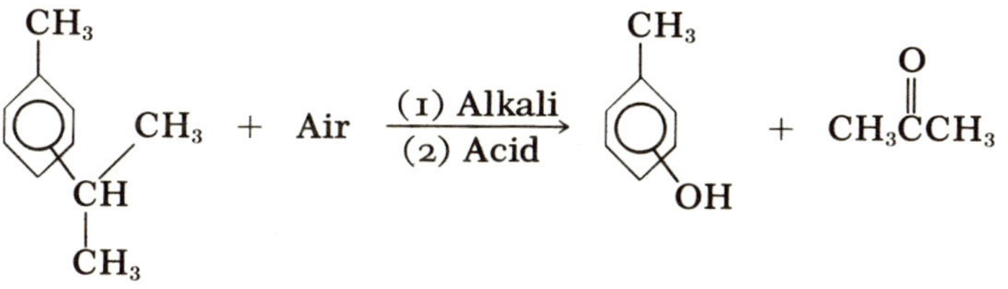

Sumitomo Chemical has commercialized the cymene route in Japan on a large scale. The relative economics of the two processes is summarized in Table 6.39. There would appear to be an advantage to the cymene route depending upon the by-product credit allowed for xylenols.

TABLE 6.39
Estimated cresol economics
(50-MM-lb plant; 1972 construction)

Total fixed capital

Process	*$ MM*
Phenol methylation	7.75
p-cymene	9.55

Production cost

	¢/lb	
	Phenol methylation	*p-cymene*
Raw materials	8.67[1]	4.01[2]
Utilities	0.47	1.22
Labor	0.43	0.43
Maintenance (6% ISBL + 3% OSBL)	0.76	0.92
Overhead (45% of maint. + labor)	0.54	0.61
Taxes & insurance (1.5% of invest.)	0.23	0.29
Depreciation (10%/yr)	1.55	1.91
Total	12.65	9.39
By-product credit	1.92[3]	2.48[4]
Net	10.73	6.91

[1] Includes 0.97 lb phenol/lb at 7.50¢/lb and 0.40 lb methanol/lb at 2.5¢/lb.

[2] Includes 0.95 lb toluene at 2.5¢/lb and 0.48 lb propylene/lb at 2.0¢/lb (propane-propylene stream).

[3] 0.12 lb xylenols at 16¢/lb.

[4] 0.54 lb acetone/lb at 4.6¢/lb.

7

ARTHUR M. BROWNSTEIN

The xylenes

MORE than 95% of the xylenes are obtained from catalytic reformate, in which they account for about 18% of the total. Although they are available in such great abundance, mixed xylenes have little chemical application. Chemical markets are based on the individual isomers, whose means of separation were described in Chapter 4. Of these xylenes, the ortho- and para isomers have the greatest chemical use. Interestingly, the meta isomer, which constitutes about 50% of total xylenes, has only an embryonic chemical market.

PARA-XYLENE

Terephthalic acid (TPA) and dimethylterephthalate (DMT) constitute virtually the sole use for p-xylene. Both TPA and DMT in turn have but one use, the synthesis of polyethylene terephthalate fiber and film.

TPA and DMT have been known since the latter part of the nineteenth century. They were first used for the preparation of polyethylene terephthalate in the late 1930's, but it was not until after World War II that full scale commercial production began. This polyester was first manufactured in England by ICI in 1949, and in the United

States by DuPont in 1953. Both companies used p-xylene which was converted to TPA by a dilute, aqueous nitric acid process originally developed by ICI. Modifications of this technology have been in continual use by both these firms ever since. The process can be summarized by this equation:

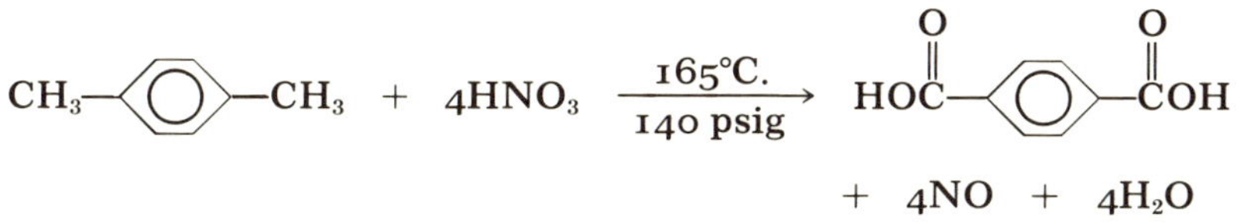

The reaction occurs in the liquid phase (30 to 40% aq HNO_3). Oxygen or air is passed into the reactor where oxidation of p-xylene and of by-product NO occurs. The NO is readily re-oxidized to NO_2 and can therefore be used for additional HNO_3 production. This reduces HNO_3 consumption to a few tenths to 2.0 lb/lb of p-xylene.

Despite any refinements to the original HNO_3 process, there are still three inherent disadvantages: (1) an auxiliary HNO_3 plant is required to use the lower nitric oxides produced in the oxidation; (2) temperature and HNO_3 concentrations necessary for high yields (88 mole %) of TPA make the process susceptible to explosions; and (3) the TPA so produced contains nitrogen impurities which are difficult to remove.

This last point is the reason that DMT, not TPA, became established as the original commercial monomer for polyester manufacture. Conversion of TPA to DMT offered an ease of purification by crystallization or distillation not readily attainable with TPA per se:

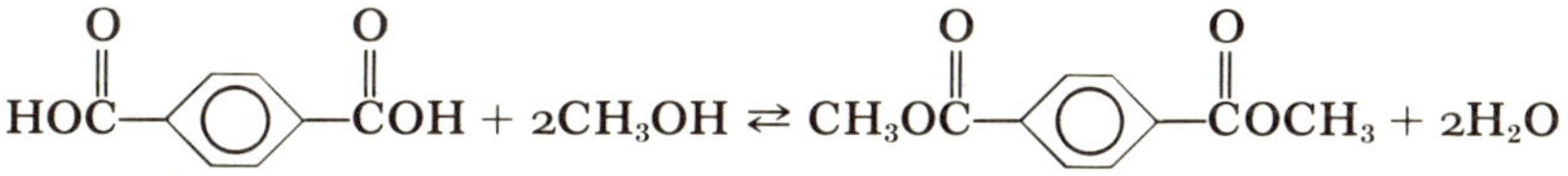

Thus, the industry had a suitable monomer for polyester fiber production, which has grown rapidly as shown in Fig. 7.1.

The data in Fig. 7.1 include only polyester staple and tow, and account for 75% of the total. Polyester filament for which data is unavailable comprise the balance of the fiber market. It is also estimated that polyester film accounts for about 10% of total fiber. Hence, total polyester resin production in 1971 was about 1,650 MM lb. This has been forecast to double by 1980.[1]

[1] H. S. Bryant et al.; Chem. Eng. Prog. 67, 69 (1971).

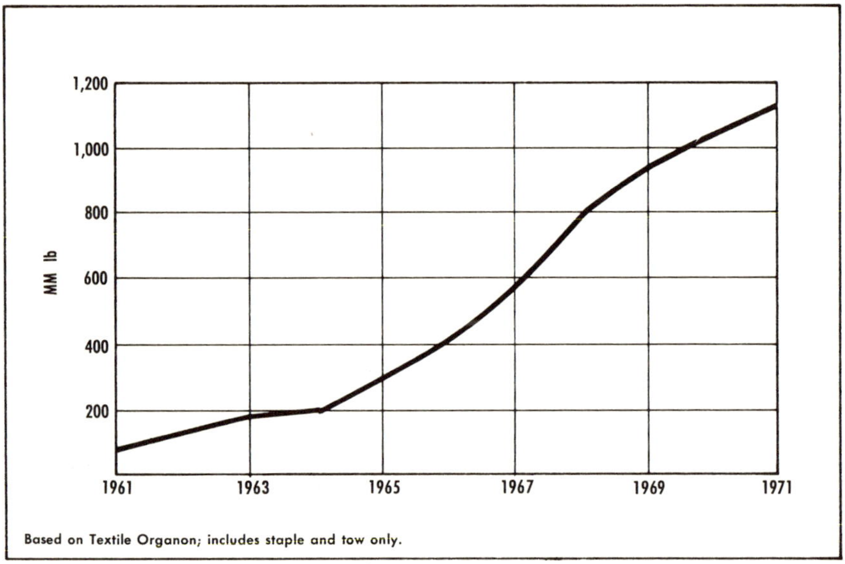

FIG. 7.1 Polyester fiber production

Since the original commercialization of DMT by ICI and DuPont, a number of firms have entered the market (Table 7.1), each of which has its own technology.

TABLE 7.1
US DMT-TPA capacity (1972)[1]

Company	*MM lb*	*Process*
Amoco (Decatur, Ala.)	900	Mid-Century
(Joliet, Ill.)	150	" "
DuPont (Gibbstown, N.J.)	250	Nitric acid
(Old Hickory, Tenn.)	250	" "
Eastman (Kingsport, Tenn.)	300	Acetaldehyde
Hercules (Burlington, N.J.)	100	Imhausen
(Wilmington, N.C.)	500	"
Hoechst (Spartanburg, S.C.)	100	Imhausen
Mobil (Beaumont, Tx.)	170	Methylethyl ketone
Total	2,720	

[1] Based on Oil, Paint & Drug Reporter, April 1, 1971.

All these other processes appear to be economically superior to the nitric acid route. They have one basic similarity in that thay are all liquid phase oxidations using cobalt acetate as a catalyst. They differ essentially in the nature of the promoter used in conjunction with the catalyst.

The Mid-Century process employed by Amoco uses a combination cobalt acetate and manganese acetate catalyst in conjunction with a source of bromine, eg NaBr or HBr, as a promoter:

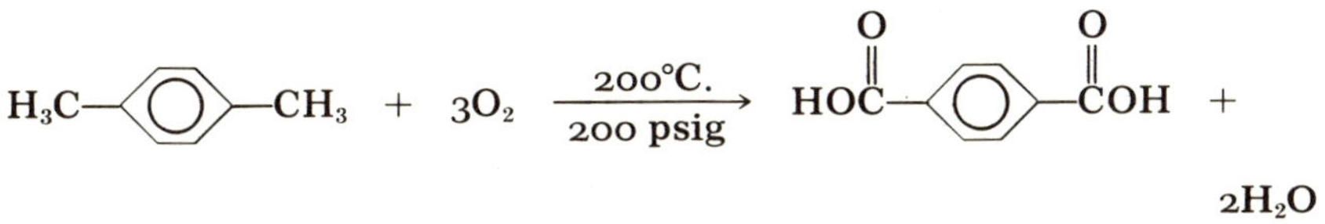

The reaction proceeds in acetic acid as solvent. In one hour's time, the yield is about 95% on conversion. The catalysts are largely recoverable and reusable. Under the reaction conditions, 4-carboxybenzaldehyde (CBA) is formed as a major impurity. In order to form a satisfactory grade TPA for polymerization, this material is removed by hydrogenating the crude terephthalic acid. This converts the CBA to p-toluic acid which can be converted to additional TPA.

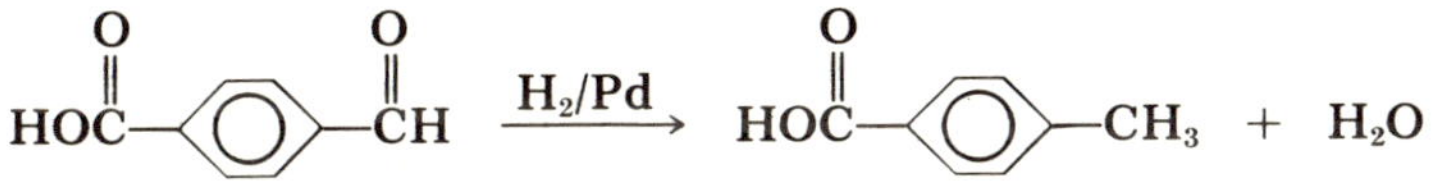

The TPA which is now free from CBA is converted to fiber grade product by crystallization (Fig. 7.2).

By use of this technology, Amoco is capable of producing DMT or TPA which is suitable for fibers. Both Amoco and Mobil have campaigned for use of TPA in place of DMT in the synthesis of polyethylene terephthalate, and have been largely successful.

The Mobil process is reminiscent to Amoco's with the distinction being that methylethyl ketone is used as a promoter in place of bromine compounds. Yields are the same for the two processes. The Mobil route possesses the slight disadvantage that about 20% of the methylethyl ketone is converted to acetic acid. Consequently, sale of the acetic acid is necessary in order to be competitive with the Amoco route. The process flow scheme is illustrated in Fig. 7.3. The crude TPA can be purified by recrystallization or via sublimation.

Eastman technology closely resembles that of Mobil with the distinction being that acetaldehyde is the promoter in place of methylethyl ketone. Acetaldehyde content is about 22% based on p-xylene. Once again, acetic acid is produced as a byproduct. This arrangement

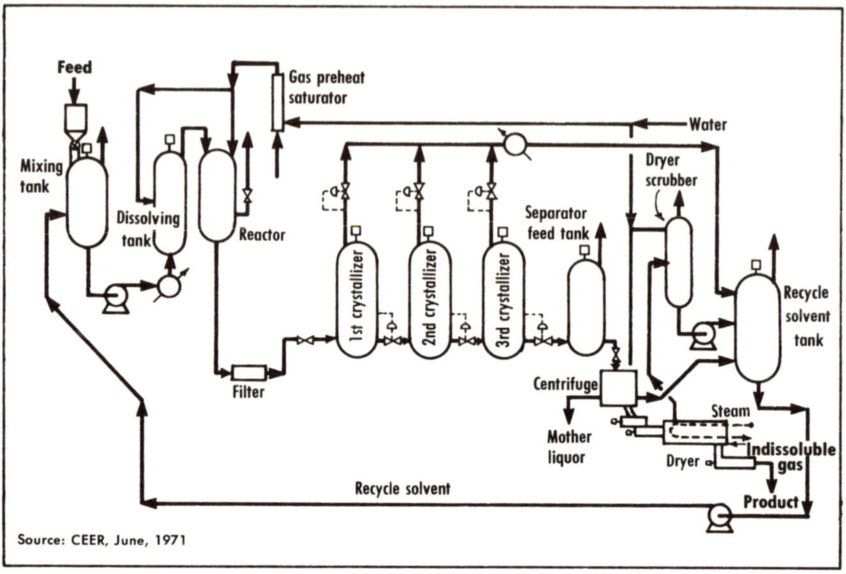

FIG. 7.2 TPA purification

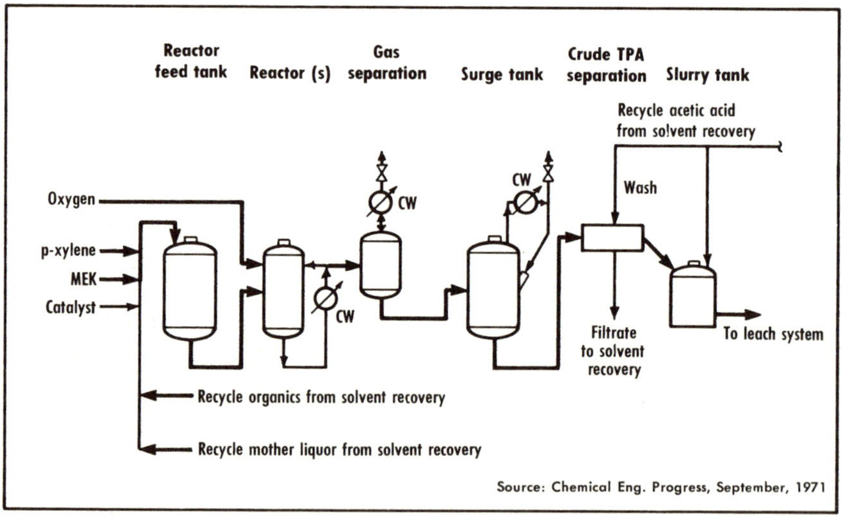

FIG. 7.3 Mobil TPA process

may be very convenient to Eastman, however, since they possess a Wacker license for inexpensive acetaldehyde (Chapter 10), and they have a captive use for acetic acid in cellulose acetate production. The Eastman technology is available for license through Foster-Wheeler. Eastman is understood to produce DMT, not TPA.

Hercules and Hoechst's technology is often referred to as the Witten-Imhausen process. The synthesis comprises the liquid phase oxidation of p-xylene in acetic acid at 165°C. and 65 psig pressure with a cobalt acetate or naphthenate as a catalyst. Reaction time is about 4 hr and the product is p-toluic acid. This is subsequently esterified with methanol, and then again oxidized for 5 hours at 200°C. and 400 psig pressure to produce methyl hydrogen terephthalate. The latter is finally esterified with methanol to form DMT in an overall yield of about 80%. This route can be summarized as:

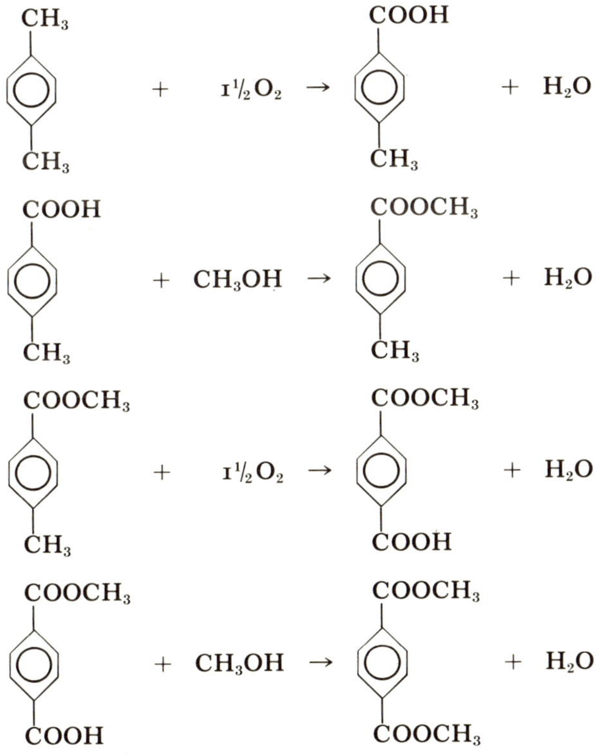

At present, only Amoco and Mobil are suppliers of fiber grade TPA, although Hercules will soon be an additional supplier. Presumably, Hercules will hydrolyze DMT to obtain TPA.

Until about 1963, all polyethylene terephthalate was manufactured from DMT. There is, however, a basic cost disadvantage in using DMT as a monomer. Two moles of methanol/mole of TPA are required for DMT synthesis. Methanol does not add to the final polymer, and must be recovered or sold for satisfactory economics. Hence, each pound of DMT only contains 0.86 lb of usable monomer. The relative advantages of TPA vis a vis DMT are illustrated in Table 7.2.

The technical innovations provided by catalytic, liquid-phase

TABLE 7.2
Estimated terephthalic acid economics
(400-MM lb plant—1972 construction)

Investment cost

Process	*$ MM*
Amoco	52.9
Mobil	58.6
Hercules	57.9

Production costs

	¢/lb		
	Amoco[1]	*Mobil*[2]	*Hercules*[3]
Raw materials	6.62	7.57	5.74
Utilities	0.65	0.88	0.47
Labor	0.09	0.09	0.09
Maint. (6% ISBL + 3% OSBL)	0.64	0.71	0.70
Overhead (45% maint. + labor)	0.33	0.36	0.36
Taxes & insurance (1.5% of invest.)	0.20	0.22	0.22
Depreciation (10 yr)	1.32	1.47	1.45
Total	9.85	11.30	9.03 (10.50)
By-product credit	—	1.20	(0.80)
Net	9.85	10.10	(9.70)

[1] Includes 0.67 lb p-xylene at 6.5¢/lb.

[2] P-xylene at 6.5¢/lb and methylethyl ketone at 10¢/lb; 0.67 lb p-xylene/lb and 0.25 lb MEK/lb; 0.20 lb acetic acid at 6.0¢/lb as by-product credit.

[3] All data expressed as ¢/lb DMT. Parenthetic data are equivalent costs as ¢/lb TPA. By-product credit assigned to 90% of the methanol at 2.0¢/lb; raw materials include p-xylene 0.68 lb/lb at 6.5¢/lb and methanol 0.44 lb/lb at 2.6¢/lb.

oxidation have facilitated a price decline of more than 50% for DMT and TPA in the last decade (Table 7.3). This has contributed greatly to the growth in polyester resins. No breakdown is available of TPA and DMT production and sales. Both these products sell at the same price.

Concomitant with polyester growth and DMT-TPA price decline has been a rapid decline in p-xylene prices (Table 7.4). In addition to domestic demand, p-xylene enjoys a sizeable export market which has

TABLE 7.3
Dimethylterephthalate and terephthalic acid production[1]

Year	*Production*	*Sales*	*¢/lb*
1961	na	70	36
1962	na	65	32
1963	331	na	na
1964	356	na	na
1965	545	219	24
1966	797	279	21
1967	936	353	20
1968	1,309	543	18
1969	1,537	752	15
1970 & 1971	na	na	na

[1] US Tariff Commission.

TABLE 7.4
P-xylene statistics[1]
(MM lb)

Year	*Production*	*Sales*	*¢/lb*
1961	248	219	13
1962	256	272	12
1963	263	247	9
1964	296	313	9
1965	396	375	9
1966	519	460	8
1967	757	635	8
1968	1,316	991	8
1969	1,629	1,207	7
1970	1,588	na	6.75[2]
1971	1,570	na	6.25[2]

[1] US Tariff Commission; 1971 based on 9 months' data.
[2] Oil, Paint & Drug Reporter posted price in tanks (works, freight equalized).

ranged from 270 to 350 MM lb annually in the last several years. Most of the product is exported to Western Europe where it is also used for DMT and TPA production (Table 7.5).

TABLE 7.5
Estimated 1971 p-xylene exports

Importing country	*MM lb*
United Kingdom	9.7
Netherlands	109.9
W. Germany	93.8
Japan	52.9
Total[1]	266.3

[1] Based on 7 months' US Census Bureau Reports.

US p-xylene producers are summarized in Table 7.6. As can be noted from these data and those provided in Table 7.4, the market is geared for about 75% merchant sales. At present, all p-xylene capacity in the United States is based upon crystallization technology. Neither the Parex nor the Japan Gas process have yet been commercialized here.

TABLE 7.6
1972 US p-xylene capacity[1]

Company	*MM lb*
Amoco (Decatur, Ala.)	275
(Texas City, Tx.)	275
Arco (Houston, Tx.)	300
Chevron (El Segundo, Calif.)	90
(Pascagonla, Miss.)	250
(Richmond, Calif.)	110
Cities Service (Lake Charles, La.)	35
Cosden (Big Spring, Tx.)	5
Enjay (Baytonne, Tx.)	200
Hercor (Guayanilla Bay, P.R.)	240
Phillips (Guayama, P.R.)	75
Shell (Houston, Tx.)	100
Signal (Houston, Tx.)	20
Sun Oil (Corpus Christi, Tx.)	170
Tenneco (Chalmette, La.)	100
Total	2,245

[1] Based on Oil, Paint & Drug Reporter, Oct. 1, 1969.

Polyethylene terephthalate is mainly produced by batch condensation of DMT with ethylene glycol to yield principally bis(hydroxyethyl) terephthalate plus some oligomeric material.

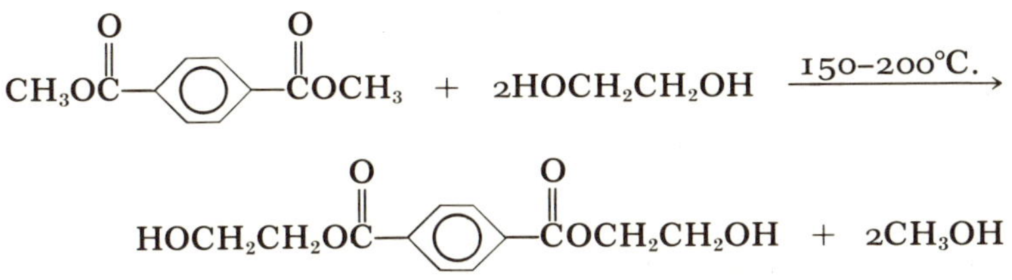

A 5 to 10% molar excess of ethylene glycol is employed, and methanol is continuously evolved and removed by fractionation. It can be returned to the DMT producer and credited against the production cost of the resin or the DMT as shown in Table 7.2. Acetates, carbonates and oxides of Zn, Ca, Mg, Mn, and Sb are commonly used as transesterification catalysts in amounts of 0.05 to 0.1 wt % of the DMT.

Once this bis(hydroxyethyl) terephthalate has been obtained, the pressure of the system is dropped to about 0.5 mm Hg and the temperature is raised to 270 to 280°C. Polymerization now proceeds rapidly to high polymer:

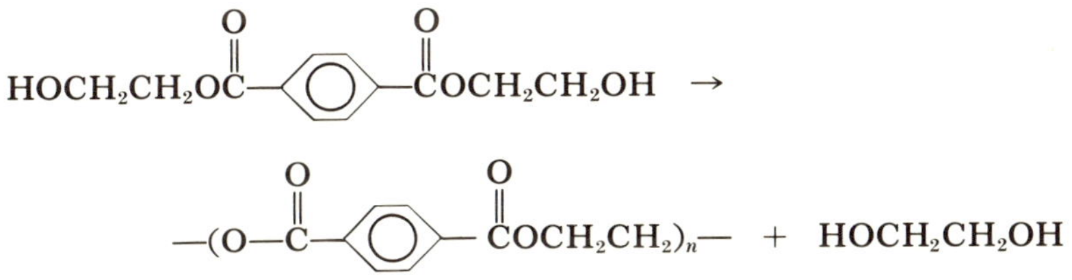

Since the mid-1960's, a continuous polymerization and spinning process for polyester has gained broad acceptance. This was introduced by Vickers-Zimmer and consists of polycondensation of DMT and ethylene glycol (1 to 1.7 mole ratio) in a horrizontal pre-condenser at 160° to 180°C. Ester interchange is complete in 4 hr at 245°C. This product is then passed through two successive polymerization reactors, each divided into individually stirred cells. The first reactor operates at 270°C. and 20 mm Hg, and the second at 280°C. and 0.5 to 1 mm Hg. Fiber-grade polymer of 0.7 intrinsic viscosity is obtained which can be fed to an integrated multi-cell spinning plant.

In view of the cost advantages inherent in the use of TPA over DMT, most new plants will probably employ the former as the monomer in conjunction with continuous polymerization processes. Most existing plants will continue to employ DMT. According to J. G. Lambertsen[1] of Amoco Chemicals, TPA's share of the polyester market increased rapidly from 10% in 1966 to 24% in 1968 and is probably much higher today.

The most recent innovation in polyester technology has been a development pursued by several Japanese companies: Nippon Soda, Mitsui, Toyo Spinning, and Teijin. In these processes, crude or high purity TPA is reacted with ethylene oxide (not the glycol) to yield bis(hydroxyethyl) terephthalate. The latter is then converted to high polymer as already described. In the Nippon Soda version, the reaction between ethylene oxide and TPA is heterogeneous, with high-purity TPA being the dispersed solid. The reaction occurs at 90° to 130°C. in 4 hr. The scheme is shown in the flow diagram of Fig. 7.4.

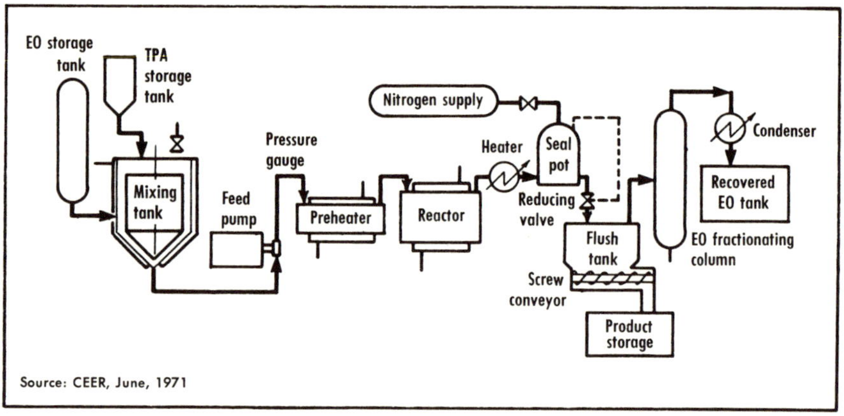

FIG. 7.4 TPA-ethylene oxide polyester process

Typical analysis for fiber grade TPA is shown in Table 7.7.

The attractive properties of polyester fiber are its outstanding crease resistance and work recovery, and its low moisture absorption. These properties result from polymer chain stiffness and the resulting high modulus. Consequently, garments made from polyester fiber are very resistant to wrinkling and can be washed repeatedly without subsequent ironing. Polyester fibers are seldom used alone, but are blended with cotton and wool. There are currently 13 producers in the US (Table 7.8).

[1] Commercial Chemical Development Assoc. Meeting, Dec., 1969.

TABLE 7.7
Typical analysis of polymer–grade TPA[1]

Analysis	Value
p-Carboxybenzaldehyde, ppm	20–50
DMF color	5–10
Ash, ppm	3
Water, wt %	<0.1
Metals, ppm	
Aluminum	<0.1
Silica	0.5
Magnesium	<0.1
Calcium	<0.1
Potassium	<0.1
Sodium	<0.1
Iron	0.2
Cobalt	0.2
Nickel	<0.1
Chromium	<0.1
Molybdenum	<0.1
Titanium	<0.1

[1] Chem. Eng. Prog. *67,* 69 (1971).

TABLE 7.8
1972 US polyester fiber capacity[1]

Company	MM lb
Allied Chemical (Moncure, S.C.)	80
Akzona (Lowland, Tenn.)	130
Dow Badische (Anderson, S.C.)	30
DuPont (Wilmington, N.C.; Kinston, N.C.; and Old Hickory, Tenn.)	700
Eastman (Columbia, S.C.)	175
(Kingsport, Tenn.)	200
FMC Corporation (Lewiston, Pa.)	120
Fiber Industries (Salisbury & Shelby, N.C.)	630
Goodyear (Point Pleasant, W.Va. and Scottsboro, Ala.)	30
Hoechst (Spartanburg, S.C.)	100
Monsanto (Decatur, Ala.)	120
Phillips (Rocky Mount, N.C.)	40
Total	2,355

[1] Based on Chem. & Eng. News, May 31, 1971.

Polyethylene terephthalate surpassed nylon in 1970 as the largest volume synthetic fiber produced in the United States (Fig. 7.5).

A new use for TPA, that of a 1,4-butanediol polyester, was announced by Eastman during 1971. The resin is an engineering plastic that is expected to compete with polyacetals et al.

ORTHO-XYLENE

Ortho-xylene, like p-xylene, has a very narrow end-use pattern. For all practical purposes, it is used solely for the manufacture of phthalic anhydride. Because of this, o-xylene competes directly with naphthalene, whose principal end use is also production of phthalic anhydride.

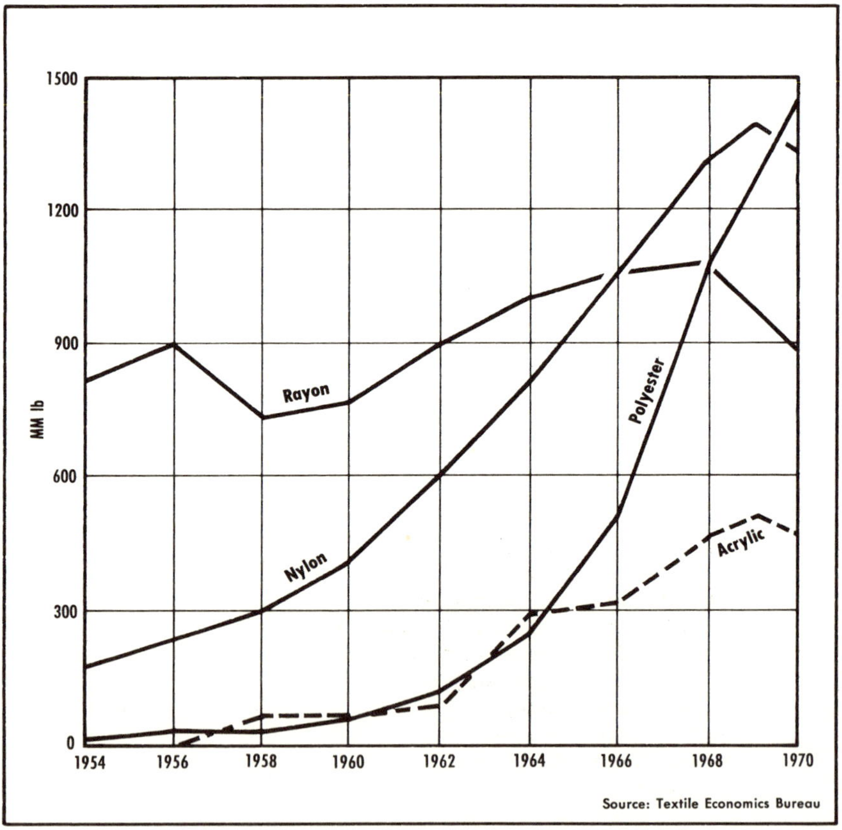

FIG. 7.5 US man-made fiber production

Until 1959, nearly 100% of the phthalic anhydride produced in the US was based on naphthalene. Since that time, o-xylene has made considerable inroads on this market until, in 1972, it shares it nearly evenly with naphthalene. A major factor in the use of o-xylene as feedstock in place of naphthalene is the ultimately greater yield of phthalic anhydride that might be obtained. The theoretical yield of phthalic anhydride from o-xylene is 1.40 lb/lb, whereas it is only 1.16 lb/lb of naphthalene. Further discussion of o-xylene will be deferred until Chapter 8, where it will be considered in context with naphthalene.

A substantial factor affecting US production of o-xylene has been the export market. This has steadily accounted for more than 50% of total production for the last decade (Table 7.9).

TABLE 7.9
Ortho-xylene production, sales and exports
(MM lb)

Year	*Production*[1]	*Sales*[1]	*Exports*[3]	*Unit value,*[1] *¢/lb*
1965	351	344	198	2.7
1966	403	397	218	2.6
1967	478	465	265	3.0
1968	944	768	535	4.4
1969	850	733	596[4]	3.6
1970	825	na	536[4]	2.3[2]
1971	750	na	440[4]	na

[1] US Tariff Commission Reports; 1971 based on 9 months' data.
[2] Oil, Paint & Drug Reporter posted price in barges (Gulf Coast).
[3] European Chemical News Suppl., Sept. 26, 1969.
[4] US Census Bureau.

The bulk of these exports has been to Western Europe. This trade has been declining steadily since its high point in 1969 as new foreign capacity came on stream. It has been forecast by C. Wickam-Jones[1] of ICI that the present Western European deficit will change to a surplus by the early 1970's. Should the export market for o-xylene dwindle away, there would be a substantial excess of domestic capacity (Table 7.10).

[1] *European Chemical News,* Large Plant Supplement, Sept. 26, 1969, pp. 10–12.

TABLE 7.10
1972 ortho-xylene capacity[1]

Company		*MM lb*
Arco	(Houston, Tx.)	210
Chevron	(El Segundo, Calif.)	15
	(Richmond, Calif.)	130
Cities Service	(Lake Charles, La.)	120
Commonwealth	(Penuelas, P.R.)	140
Cosden	(Big Spring, Tx.)	18
Crown Central	(Houston, Tx.)	40
Enjay	(Baytown, Tx.)	100
Monsanto	(Alvin, Tx.)	30
Phillips	(Guayamas, P.R.)	130
Sun	(Corpus Christi, Tx.)	160
Tenneco	(Chalmette, La.)	155
	Total	1,248

[1] Based on Oil, Paint & Drug Reporter Profile, Jan. 1, 1970.

A breakdown of 1971 exports of o-xylene is given in Table 7.11.

TABLE 7.11
US ortho-xylene exports[1]

Importing country	*MM lb*
Argentina	23.4
Brazil	34.3
Canada	33.3
Colombia	3.8
India	6.4
Italy	82.5
Japan	11.8
Netherlands	156.0
Republic of South Africa	2.3
Sweden	6.5
Venezuela	7.8
West Germany	71.9
Total	440.0

[1] US Census Bureau; based on 7 months' data.

META-XYLENE

Use of this material for large-volume chemical synthesis appears to be just beginning. Only two products of note exist at this time, isophthalic acid (IPA) and m-xylylene diamine. Only IPA is produced domestically.

IPA was first introduced commercially to the chemical industry by Chevron Chemical from a plant in Richmond, Calif. in 1956. For a number of years, Chevron produced IPA by a proprietary process. Meta-xylene was oxidized with sulfur in an aqueous ammonia system at 250 to 300°C. and 1,000 to 2,000 psig to yield isophthamide.

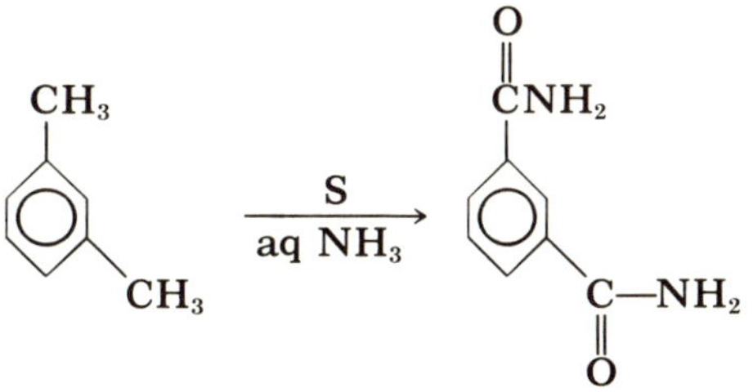

This intermediate was subsequently hydrolyzed with aq H_2SO_4 to yield IPA.

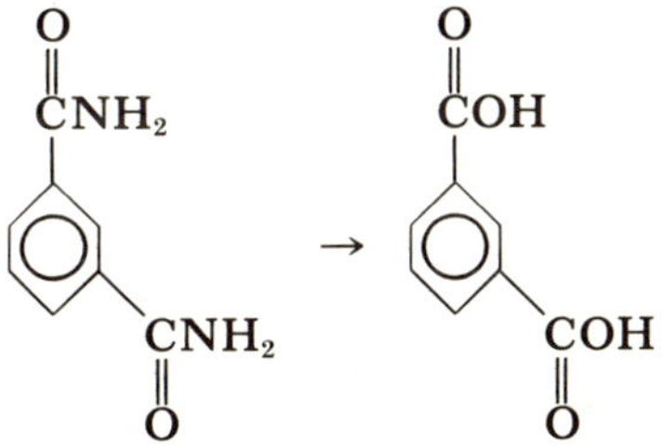

Amoco entered this market in 1960, shortly after Chevron. Its product was obtained by the superior Mid-Century process which has already been described. Chevron evidently was unable to compete and shuttered its 35-MM-lb plant in 1967, thus leaving the entire market to Amoco.

Since that time, the IPA market has grown, such that there are now three sizable producers in the world as shown in Table 7.12.

TABLE 7.12
International IPA capacity (1972)

Company	MM lb
Amoco (Joliet, Ill.)	100
AG International (Mizushima, Japan)	80
Arco (Channelview, Tx.)	35
Total	215

AG International is a joint venture between Amoco and Japan Gas Co., and as would be expected, also uses Mid-Century technology. The process used by Arco Chemical was originally developed by Sinclair. Its details have never been made public.

Isophthalic acid is produced in three grades by the industry. Amoco sells an IPA-85 and an IPA-95 which is composed of 85% and 95% IPA respectively. The balance of the material is essentially TPA. Arco Chemical and AG International each produce a 99+% IPA.

IPA purity is largely a function of the purity of the m-xylene feedstock. In the US, this is generally 85% to 95%. Although m-xylene can be produced by nearly any oil company, there are now only 5 producers (Table 7.13). Arco is believed to produce its high purity m-xylene via its disproportionation process. The high purity material used by AG International is via the Japan Gas BF_3 process.

TABLE 7.13
US meta-xylene producers[1]

Arco Chemical
Chevron Chemical
Cosden Oil
Sun Oil
Tenneco Oil

[1] Hydrocarbon Processing, Aug., 1971.

IPA of 85% purity sells for about 13½¢/lb vs. 16½¢/lb for 95% purity and higher.

According to N. E. Ockerbloom, unsaturated polyesters currently constitute the major market for IPA, and will continue to be the premier market for the foreseeable future. His estimates of the market are given in Table 7.14.

TABLE 7.14
Isophthalic acid market use[1]
(MM lb)

	1970	*1975*
Unsaturated polyesters	40–50	80–100
Alkyds and plasticizers	20–25	23–30
Polymers and miscellaneous	10–15	20–30
Export	28.9	10
	100–120	133–170

[1] Hydrocarbon Processing, Aug., 1971.

Unsaturated polyesters differ distinctly from the saturated type used for polyester fiber and film. The former are thermoset resins comprised of maleic anhydride and mono- or dipropylene glycol. Phthalic anhydride or isophthalic acid, among others, are often added as modifiers, and the entire system is cross-linked by the addition of an olefin, such as styrene. IPA polyesters find their largest outlets in fiber glass-reinforced plastics. These are used in boats, gasoline storage tanks, luggage, etc.

A recent use for m-xylene is production of m-xylylene diamine. This is an innovation of Showa Denko in Japan. The first commercial plant was a 2.5 MM-lb/yr unit at Kawasaki which came on stream in late 1966. It has since been expanded to 3.5 MM lb. There are no such units as yet in the United States.

The xylylene diamine is a 70/30 mixture of meta/para isomers. It is produced by passage of a 70/30 mix of m- and p-xylene with NH_3 and air over a fixed bed V_2O_5 catalyst, supported on α-alumina at 350 to 450°C. The isophthalonitrile-terephthalonitrile mixture so obtained is then catalytically hydrogenated to the diamines:

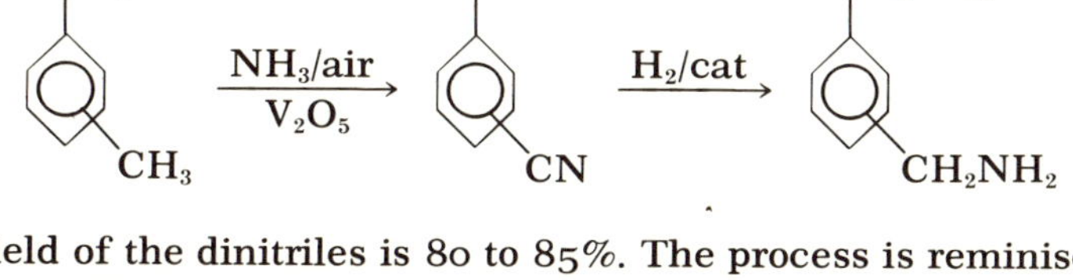

Yield of the dinitriles is 80 to 85%. The process is reminiscent of the ammoxidation route used for acrylonitrile. Because of the small plant size now in use, production costs are high, about 70¢/lb for isophthalonitrile. Showa Denko estimates that a larger plant of about 18 MM lb/yr would drop isophthalonitrile costs to about 30¢/lb. The flow scheme for the ammoxidation portion of this process is illustrated in Fig. 7.6.

Xylylene diamines have application as diisocyanates for non-yellowing polyurethane resins. Takeda of Japan has pioneered in this area. They can also be used with adipic acid for polyamide synthesis.

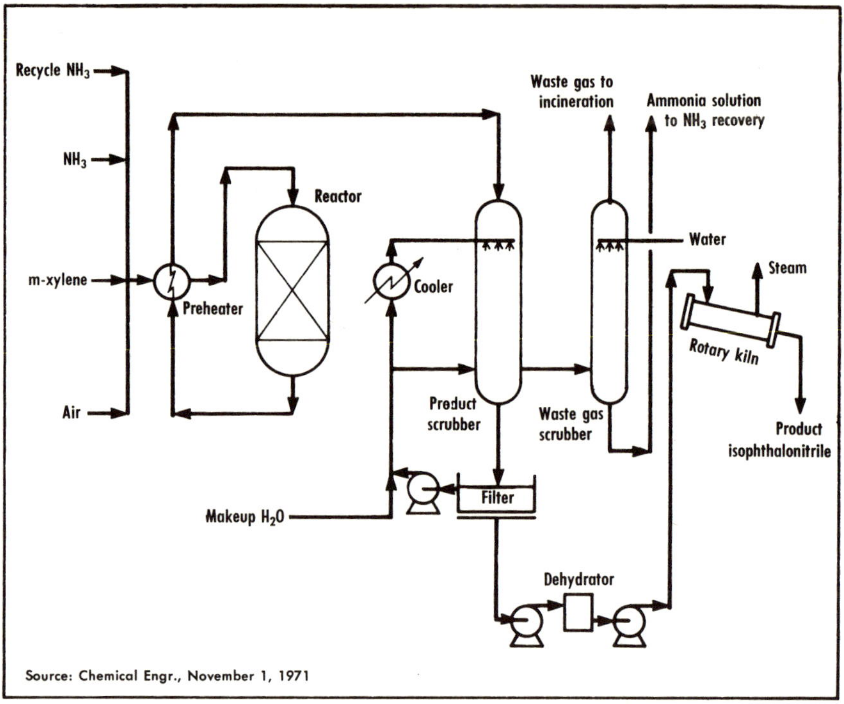

FIG. 7.6 Isophthalonitrile process

8

NELSON E. OCKERBLOOM

Naphthalene and higher aromatics

NAPHTHALENE is a white flake melting at 80°C. and is characterized by a typical moth ball odor. It is the last (in point of time) major aromatic compound to be produced from petroleum. The inability of coal-tar producers to supply the market in the early 1960's gave rise to the development of processes which produce naphthalene from petroleum. Today, 41% of United States naphthalene production is petroleum derived.

The historic source of naphthalene is tar distillation and coke ovens. The high-temperature carbonization of coal in coke ovens yields coke-oven gas, Fig. 8.1. This gas, upon cooling, yields a coal tar of this approximate composition (weight percent): naphthalene 9.0, creosote 26.3, tar acids 1.8, pitch 59.0, and others, 3.9. The yield of tar varies from plant to plant and depends upon the type of coal processed and coking temperatures.

The resulting tar is fractionated into a middle oil (creosote oil or chemical oil), creosote, and pitch (the residue). The naphthalene is contained in the middle oil (bp, 200 to 250°C.) in about 40% concentration. The tar bases are removed with sulfuric acid, and the tar acids with caustic soda. The product is then crystallized yielding a product melting at 68°C., or pressed to yield naphthalene melting at 74 to 76°C.

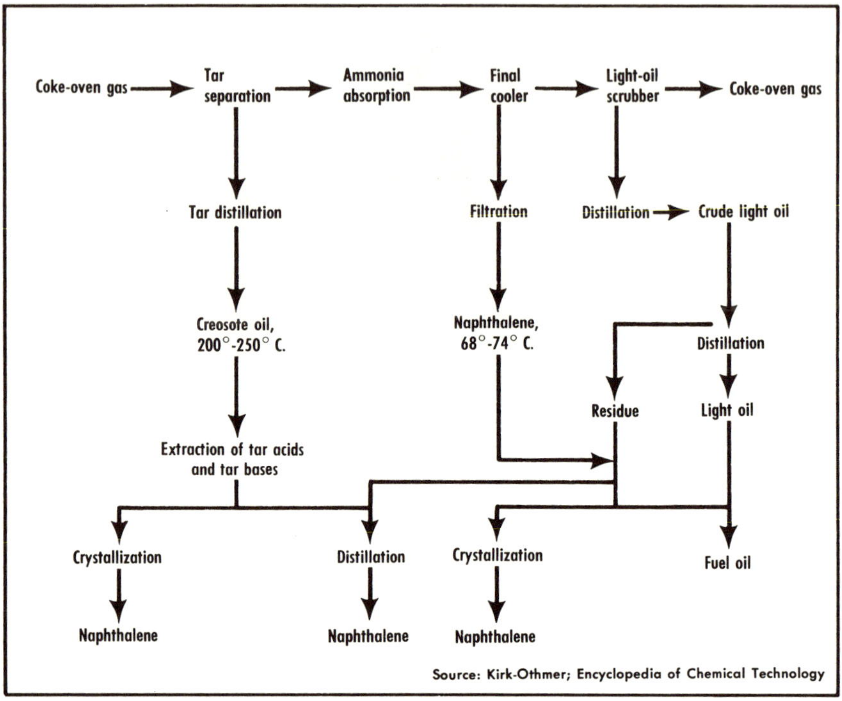

FIG. 8.1 Naphthalene recovery from coke-oven gas

Instead of purification by crystallization or pressing, distillation can be utilized and a 78°C. product obtained. Further treatment with sulfuric acid, washing, and redistillation yields a 79.6°C. product. Pure naphthalene melts at 80.2°C.

High purity 79.6°C. + product may also be obtained by removal of sulfur and impurities by hydrogenation. Costs to hydrogenate a 78°C. product at economic levels are estimated to be greater than 0.5¢/lb.

Availability of coal tar naphthalene depends upon the quantity of coke required for steel making and disposition of the coal tar. During the 1960 to 1969 period, the amount of coke required to produce a ton of pig iron decreased 17%, while United States steel production increased 3.6%. The result was an increase in coal tar production of only 1.1%. A further decline in coke requirement and a negligible increase in coal tar production is expected in the 1970's. Whether or not the coal tar is distilled depends upon two factors: (1) the value of the coal tar as an energy source, and (2) the chemical values.

Recent data on pig-iron–coal-tar production and disposition are:

	Pig-iron production MM ton	*Coal-tar production MM gal*	*Coal-tar distilled MM gal*	*Coal-tar burned MM gal*
1965	88.8	802.7	615.8	122.9
1967	87.6	780.3	594.6	129.0
1969	95.5	768.8	667.2	98.1
1970	91.0	760.9	657.7	108.0

Current coal tar producers are:

	Product quality	
Producer	*74°C.*	*74 to 79°C.*
Allied Chemical		x
Armco Steel	x	
Bethlehem Steel	x	
Colorado Fuel & Iron	x	
Granite City Steel	x	
Interlake Steel	x	
Koppers Company	x	x
Republic Steel	x	
US Steel	x	x
Woodward Iron		x

Sources for both of these tables are US Bureau of Mines and US Tariff Commission.

In 1961, Ashland at Cattlettsburg, Kentucky, brought onstream the world's first petroleum naphthalene unit. This unit was soon followed by those of Sun Oil Company, Getty-Collier, Monsanto, and Cities Service. All except Cities Service are operating.

Each of the current producers uses a different specific process, but in general the processes are similar.

1. All are hydrodealkylation processes—catalytic or non-catalytic.

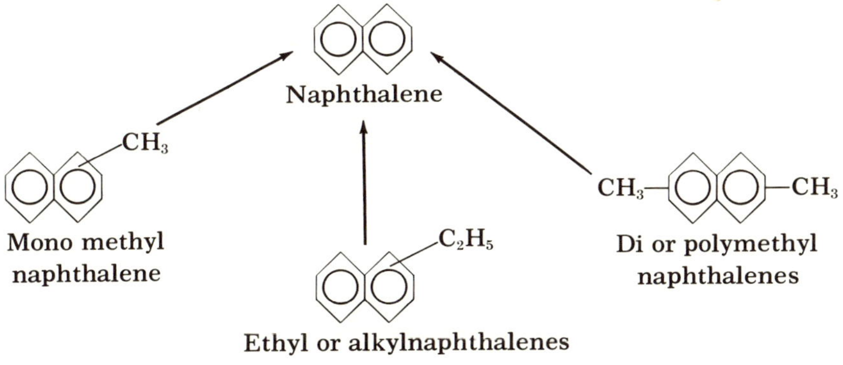

2. All can charge reformer bottoms, or low-sulfur methyl naphthalenes.
3. All produce high-purity 80°C.+ refined naphthalene.

Ashland uses a process developed jointly with Universal Oil Products. These units can utilize reformer bottoms, high-sulfur methyl naphthalenes, or crude coal-tar naphthalene. The process consists of catalytically dealkylating alkyl naphthalenes, using sulfur as the moderator in the presence of hydrogen, and purifying by distillation and clay treating.

The Getty-Collier process (developed by Union Oil) utilizes reformer bottoms, is catalytic in nature, and utilizes centrifuges to purify the product.

Sun Oil operates its own process—a thermal (non-catalytic) process utilizing heart cut refinery catalytic cycle oil as charge stock.

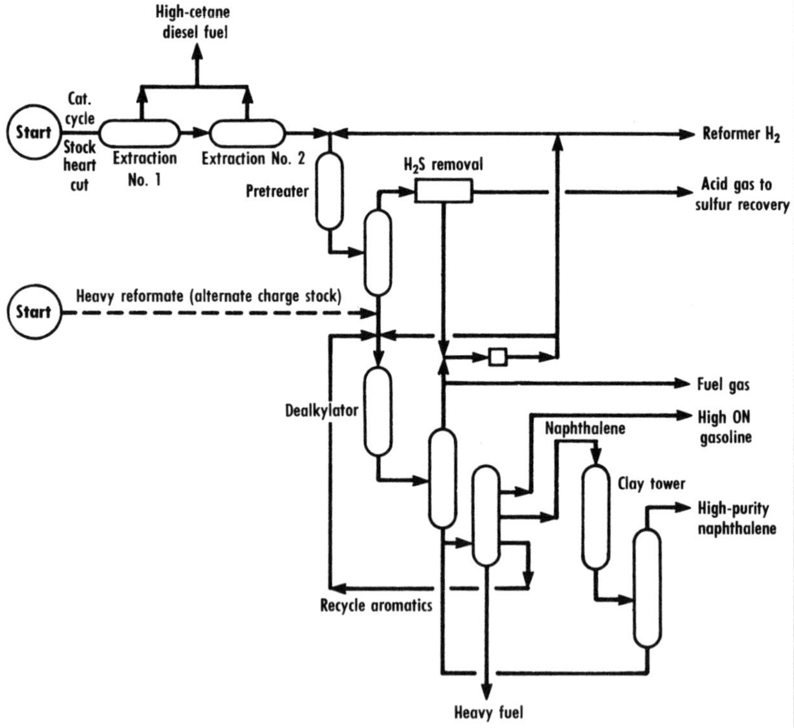

Source: Hydrocarbon Processing, Vol. 45, March 1966, p. 150

FIG. 8.2 Sun Oil's naphthalene process

The process can also utilize reformer bottoms, Fig. 8.2. This process also utilizes a different purification system.

The process consists of three sections: the feed-preparation section, the conversion, and the purification section.

The feed-prep section consists of distillation and then extraction from the catalytic gas oil of all the methyl naphthalenes in the 440° to 515°F. cut (Hydrocarbon Processing, May 21, 1962, p. 123).

The conversion section consists of desulfurization and thermal dealkylation in a hydrogen atmosphere. The product then passes to the purification section where the partially converted naphthalene formers (methyl naphthalenes obtained from dimethylnaphthalene) and impure naphthalene from the distillation tower are returned to dealkylation. Pure product is then clay treated and distilled.

Typical yields for a charge oil (75 vol % fresh feed and 25% recycle) are shown in Table 8.1.

TABLE 8.1
Hydrodealkylation of aromatic concentrate

Charge properties:	
75 vol % fresh aromatic extract	
25 vol % liquid recycle	
Gravity, °API at 60°F.	12.9
Naphthalene, wt %	9.0
β-Methylnaphthalene, wt %	24.1
α-Methylnaphthalene, wt %	10.1
C_{12} naphthalenes, wt %	23.2
Other, wt %	33.6
Yields, wt % of charge oil:	
Gas, C_1 to C_4	19.2
400°F. e.p. gasoline	9.4
400° to 450°F. fraction	50.3
450° to 480°F. fraction	14.7
480°F.+ fraction	6.4
Total	100.0
Naphthalene	39.9
Hydrogen consumption, scf/bbl of charge oil	1,600

Reference: R. Barbor, Sun Oil Co.—The Oil and Gas Journal, May 21, 1962, page 125.

Onsite investments and utilities are shown in Table 8.2. These are 1962 cost data, and today's costs would be higher.

Monsanto produces petronaphthalene by a self-developed process from methyl naphthalene fractions obtained from its Chocolate Bayou Complex. It is suspected that Monsanto's process is similar to Ashland's.

TABLE 8.2
Onsite investment and utility requirements for processing 50 MM lb/yr of naphthalene from catalytic gas oil

Investment, $ MM	3.8
Utilities:	
Electric power, kw	450
Fuel, BTU/hr (fired)	75,000,000
Steam, lb/hr	7,500
Circulating water, gpm (35°F.ΔT)	2,500
Boiler feed water, gpm	70
Chemical, dollars/month	2,150

The major factor in petronaphthalene production economics is charge stock value as shown in Table 8.3.

TABLE 8.3
Petronaphthalene economics

Basis: *75-MM-lb/yr unit—Gulf Coast* (*1962*)

	¢/lb	
Value of raw material	2.0	
Cost of raw material	5.8	
Hydrogen at fuel value	0.1	
Benzene credit at 21¢/gal	(1.3)	
Heavy product credit 0.8¢/lb	(0.1)	
Off gas credit at 20¢/MM BTU	(0.6)	
Net cost of raw material		3.9
Direct operating costs		0.7
Out-of-pocket costs		4.6
Depreciation @ 10% bl		0.4
		5.0
Sales and admin. cost		0.3
Grand total, no return on investment		5.3

Reference: Hydrocarbon Processing, Vol. 45, No. 3, p. 154, 1966.

Since these economics were published raw material values, fuel value, labor, and other costs have increased. No new units have come onstream since 1963.

Properties of several petroleum and coal tar naphthalenes are shown in Table 8.4.

TABLE 8.4
Properties of naphthalenes

		Coal tar	
	Petroleum naphthalene	*78°C.*	*79.6°C.*
Solidification point, °C.	80.0 min	78.0	79.6°
Benzothiophene, wt %	0.04 max	2.0	1.2
Sulfur, wt %	0.005 max	0.52	0.01
Indene, wt %	0.02 max	0.20	nil
Non volatile matter, wt %	nil (0.01 max)	0.40	nil
Beta and alpha methyl naphthalenes, wt	0.30 max	–	–
Color, apha (30% in benzene)	25 max white	off white	off white

Production capabilities of current naphthalene producers are shown in Table 8.5.

TABLE 8.5
Naphthalene production capabilities

Producer	*Capacity, MM lb*	*Type*
Allied Chemical	200	CTN
Ashland Oil	90	PN
Getty-Collier	100	PN
Koppers	185	CTN
Monsanto	100	PN
Productol	12	CTN
Reilly Tar	12	CTN
Sun Oil	150	PN
US Steel	130	CTN
Total	979	

CTN = Coal-tar naphthalene.
PN = Petroleum naphthalene.
Reference: Oil, Paint and Drug Reporter, April 1, 1970.

Historical output and sales of coal tar naphthalene (MM lb/yr) are shown in Table 8.6.

Coal-tar naphthalene production data contain statistical duplications, and the total is incorrect. This is due to the fact that the less than 74°C. product is sold to a producer of 74 to 79°C. product for upgrading, and both report production to the US Tariff Commission.

TABLE 8.6
Naphthalene production and sales

	74°C.		*74 to 79°C.*		*Total*	
Year	*Production*	*Sales*	*Production*	*Sales*	*Production*	*Sales*
1965	81.9	na	382.0	na	463.9	na
1966	85.7	na	407.9	na	493.6	270.9
1967	84.2	na	436.8	na	520.9	302.6
1968	75.9	59.5	449.9	274.3	525.9	333.8
1969	42.7	32.0	453.2	287.1	495.8	319.2
1970P	19.4	10.2	408.7	na	428.1	na
1971P	na	na	na	na	369.7	na

Basis: US Tariff Commission.
P = Preliminary.

Analysis reveals that 74 to 79°C. production approximates actual (non-duplicated) production.

Historical production of petroleum naphthalene is shown

TABLE 8.7
Petroleum naphthalene
(MM lb)

	Production	*Sales*
1962	167.1	160.7
1965	346.6	269.8
1966	354.1	279.8
1967	376.7	280.9
1968	375.9	243.9
1969	357.6	256.5
1970P	290.5	201.5
1971P	262.8	na

Basis: US Tariff Commission.

Not all petroleum naphthalene production is product derived solely from petroleum. Some product is based upon coal tar methyl naphthalenes and crude naphthalene both domestic product and overseas. Major imports of methyl naphthalenes were obtained from W. Germany and the Netherlands in the years 1967 to 1969. This product was dealkylated to naphthalene in petroleum naphthalene units.

In addition to domestic production, naphthalene has historically been imported into the United States, Table 8.8. The majority of the imports come from W. Germany.

TABLE 8.8
Naphthalene imports
(MM lb)

Year	*Crude naphthalene*[1]	*Refined naphthalene*
1960	41.4	na
1961	101.8	na
1965	15.1	Neg.
1966	20.2	0
1967	5.7	0
1968	12.1	Neg.
1969	18.9	1.385
1970	23.4	1.515
1971	38.9	1.397

[1] Below 79°C.
Reference: US Dept. of Commerce data.

Overall naphthalene availability for domestic use is shown in Table 8.9

TABLE 8.9
US naphthalene availability

Year	*Petroleum 80+°C.*	*Coal tar 78°C.*	*Imports*	*Total*
1965	347	382	15	744
1966	354	408	20	782
1967	377	437	6	820
1968	376	450	12	838
1969	358	453	19	832
1970P	291	409	25	725
1971P	263	350	40	653

P = Preliminary.
Note: Excludes inventory carry over.

The major market for naphthalene, both petroleum and coal tar, is phthalic anhydride. Over 75% of the naphthalene produced in 1970 entered this end use.

PHTHALIC ANHYDRIDE

Phthalic anhydride, the anhydride of 1,2 benzene dicarboxylic acid $C_6H_4(COOH)_2$, is a white flake with a freezing point of 131°C. It is a raw material for plasticizers, alkyds, unsaturated polyesters, and dyestuffs production.

Phthalic anhydride is derived by the oxidation of orthoxylene or naphthalene:

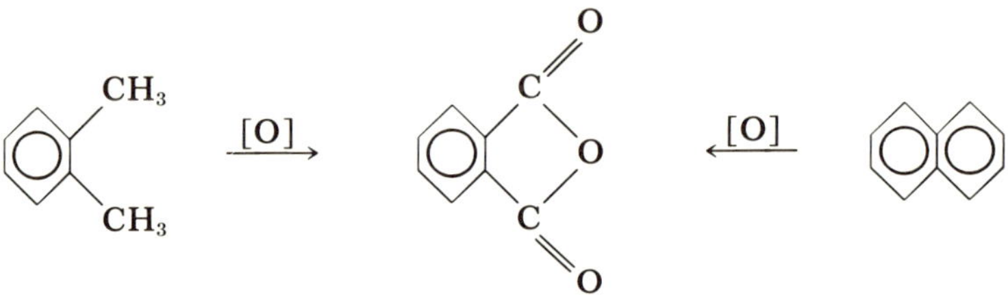

Ortho-xylene Phthalic anhydride Naphthalene

In the United States, capacity based upon each raw material has changed dramatically since 1965—increasing from less than 20% based upon ortho-xylene to over 55% in the early 1970's, Fig. 8.3. The changeover is due to process developments in the 1960's and availability and cost differentials between the two raw materials.

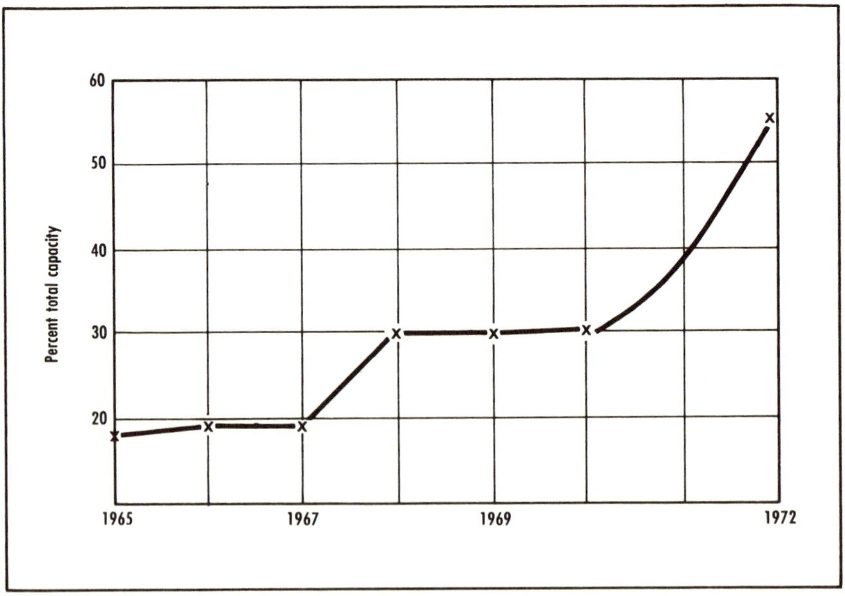

FIG. 8.3 Approximate US phthalic anhydride capacity based on o-xylene

The first patented phthalic process (BASF-1896) utilized naphthalene. Phthalic anhydride was produced by oxidation of naphthalene with sulfuric acid in the presence of mercury salts. In 1917, Gulf and Conover in the United States and Wohl in Germany developed a catalytic vapor phase process. In 1934, Wohl in patent litigation was awarded priority. Processes today are very similar to those of the original inventors; the major improvements having been made in catalysts.

A major change which has occurred is that the raw material choice has gone from coal-tar naphthalene to petroleum naphthalene (in the United States) and now to ortho-xylene worldwide.

Research in Western Europe (Germany) has centered on oxidation at low temperatures, where long-life catalysts giving higher yields but at a slow rate were developed. US research developed catalysts which yield phthalic at high rates and temperatures, but at low yields. Catalysts basically are potassium-modified vanadium pentoxides.

Early processes were fixed-bed systems. Fluid-bed-systems development was pioneered by Sherwin-Williams and Badger. Fluid-bed systems became the process of choice and remained so until BASF made improvements in the fixed-bed process, again making it the preferred process.

A schematic of the BASF process is shown in Fig. 8.4.

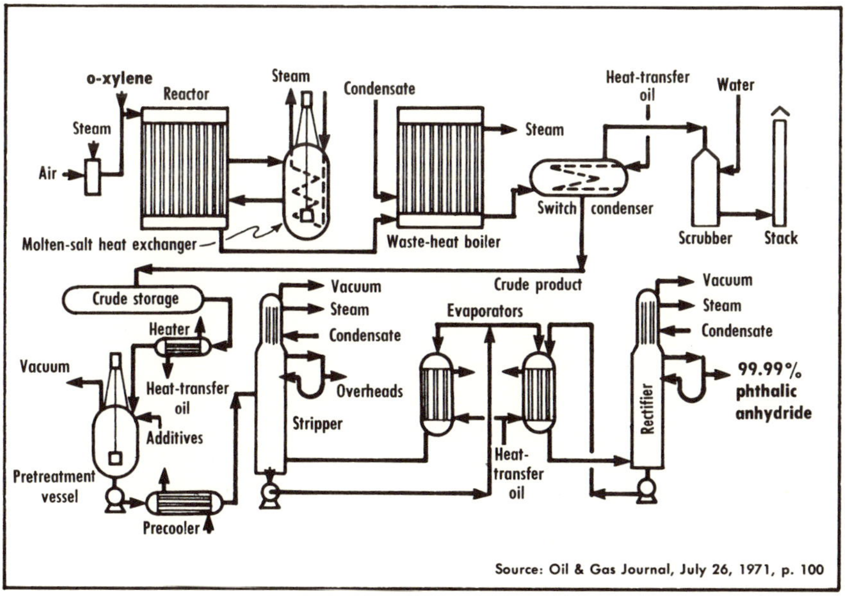

FIG. 8.4 BASF route to phthalic anhydride

Economics for the BASF process and for the Von Heyden Process employing the "F" Type catalyst are shown in Table 8.10:

TABLE 8.10
Phthalic anhydride economics

	BASF	*Von Heyden*
Plant capacity, tons	28,000	22,000
MM lb	61.6	48.4
Battery limits invest	$3.08 MM	$3.25 MM

Reference: Oil and Gas Journal, July 26, 1971, p. 100; European Chemical News, October 8, 1971, p. 40.

Production cost, $/ton

	BASF	*Von Heyden*
Raw materials		
Ortho-xylene	74.7	75.5
Catalysts	2.3	1.9
Utilities	22.0	43.9
Fixed costs		
Labor, maintenance, and overhead	15.4	7.0
Capital charges	22.0	36.9
Total cost		
$/ton	132.0	131.5
¢/lb	6.0	6.0

All phthalic anhydride units worldwide—except for one—are vapor processes. The lone exception is a plant of 15,000-ton/yr capacity owned by Progil and an association of French companies at Villers, St. Paul, France. In the process developed by Progil, orthoxylene is oxidized with air under pressure in liquid phase cascade reactors. Product is then purified by continuous distillation and crystallization. Phthalic acid is dehydrated into the anhydride and redistilled.

Economics are speculated to be as shown in Table 8.11.

The process requires some 1,763 lb of ortho-xylene/2,240 lb of phthalic produced—a high yield—but the process, unfortunately, has a high catalyst and chemical cost which raises the raw material

TABLE 8.11
Liquid-phase phthalic process
(Capacity, 60 MM lb/yr)

Battery-limits capital	$4,600,000
Raw materials*	4.0¢/lb
Utilities	0.8¢/lb
Labor maintenance & overheads	0.8¢/lb
Capital-related @ 20% blcc	1.5¢/lb
Cost of production	7.1¢/lb

*Based on o-xylene at 3.5¢/lb and a catalyst and chemical cost of 1.25¢/lb phthalic produced.

Reference: P. Spitz, Hydrocarbon Processing, November 1968, Vol. 47, No. 11.

cost to the 4.0¢/lb level. It is speculated that when phthalic anhydride units get larger, the liquid-phase process will become a competitive route to phthalic anhydride.

Phthalic anhydride will be produced by eleven firms in the United States by the end of 1972. Capacity and producers are shown in Table 8.12.

As noted in the table, all new capacity expected onstream by the end of 1972 is ortho-xylene based. Outside of stoichiometry (theoretical yields of 0.72 lb ortho-xylene or 0.86 lb naphthalene required 1.0 lb phthalic anhydride) other reasons for the change include new process technology and raw material price, ease of availability and low transportation costs.

Phthalic anhydride demand grew from 573 MM lb in 1965 to 774 MM lb in 1969. In 1970–1971 demand remained at the 750-MM-lb level.

Statistics on phthalic anhydride production, sales, unit value and estimated raw material use are shown in Table 8.13.

Between 42 to 45% of total phthalic anhydride production is captively consumed by the producer. Domestic disposition and foreign trade are shown in Table 8.14.

Three basic markets account for the major outlets for phthalic anhydride: plasticizers, alkyds, and unsaturated polyester resins. Other uses are small in comparison.

The largest outlet for phthalic anhydride is for the production of plasticizers. These esters of phthalic anhydride are mostly used for the

TABLE 8.12
US phthalic anhydride capacity
(MM lb)

	Location	*Estimated '68*[a]	*Expected at end of '72*[b]	*Probable feedstock*	*Process*
Allied Chemical	El Segundo, Calif.	33	35	*o*-X	Chemiebau (V.H.)
	Frankford, Pa.	80	100	Coal tar naph.	Badger
	Ironton, Ohio	36	37	Coal tar naph.	Own
BASF	S. Kearney, N.J.	–	130[d]	*o*-X	BASF
Chevron Chemical	Perth Amboy, N.J.	30	–	*o*-X	Own
	Richmond, Calif.	15	50	*o*-X	Lurgi (V.H.)
Enjay	Baton Rouge, La.	–	90[c]	*o*-X	BASF
W. R. Grace	Fords, N.J. (2 plants)	53	75	naph.	Badger (Sh. W.) Chemiebau
Hooker	Arecibo, P.R.	100	100	*o*-X	Chemiebau (V.H.)
Koppers	Bridgeville, Pa.	70	90	Coal tar naph.	Own (Fluid Bed)
	Stickney, Ill.	–	130	*o*-X	BASF
Monsanto	St. Louis, Mo.	40	–	naph.	Own
	Bridgeport, N.J.	60	80	naph.	Badger (Sh. W.)
	Texas City, Tx.	75	–	naph.	Badger (Sh. W.)
	Chocolate Bayou, Tx.	–	130[d]	*o*-X	BASF
Reichhold	Elizabeth, N.J.	30	–	naph.	Badger (Sh. W.)
Sherwin-Williams	Chicago, Ill.	20	–	naph.	Badger (Sh. W.)
Stepan Chemical	Millsdale, Ill.	45	48	*o*-X	Chemiebau (V.H.)
Union Carbide	Institute, W.Va.	75	75	naph.	Badger (Sh. W.)
USS Chemicals	Neville Island, Pa.	43	125	Coal tar naph.	Badger (Sh. W.)
Witco	Chicago, Ill.	20	–	naph.	Scientific Design
	Perth Amboy, N.J.	30	–	*o*-X	Scientific Design
Total		855	1295		
o-X based, MM lb		253	713		
% total *o*-X based		30	55		
naph., based, MM lb		602	582		

[a] Spitz, Peter, Hydrocarbon Processing, November 1968, Vol. 47, No. 11, p. 162.
[b] Chemical & Engineering News, March 15, 1971, p. 18.
[c] Expandable to 130 MM lb.
[d] Startup 1971.

TABLE 8.13
Phthalic anhydride production, sales

	MM lb			Estimated Raw-material requirement, MM lb	
	Production	Sales	Unit value, ¢/lb	Naphthalene[1]	Orthoxylene[2]
1965	608.3	336.3	8	550	110
1966	675.2	365.3	9	610	120
1967	727.5	403.5	12	640	145
1968	743.8	428.2	13	570	225
1969	760.0	446.4	11	585	228
1970P	734.0	440.9	9	540	219
1971P	766	na	8 OPD	480	306

P = Preliminary.
Reference: US Tariff Commission; Oil Paint & Drug Reporter.
[1] 1.1 lb naphthalene/1.0 lb phthalic through 1969.
[1] 1.05 lb naphthalene/1.0 lb phthalic post 1969.
[2] 1.0 lb ortho-xylene/1.0 lb phthalic.

TABLE 8.14
Phthalic anhydride supply, demand
(MM lb)

	Supply		Demand	
	Production[1]	Imports[2]	Exports[2,3]	Estimated other uses
1965	608.3	0.1	35E	573
1966	675.2	6.0	8.55	671
1967	727.5	3.0	48.3	683
1968	743.8	14.6	20.6	748
1969	760.0	19.5	6.0	774
1970P	734.0	45.6	7.1	773
1971P	766	9.8	7.6	778

P = Preliminary.
[1] US Tariff Commission.
[2] US Department of Commerce.
[3] Includes isophthalic acid prior to 1970.

plasticization of vinyl resins (polyvinyl chloride and copolymers, and others).

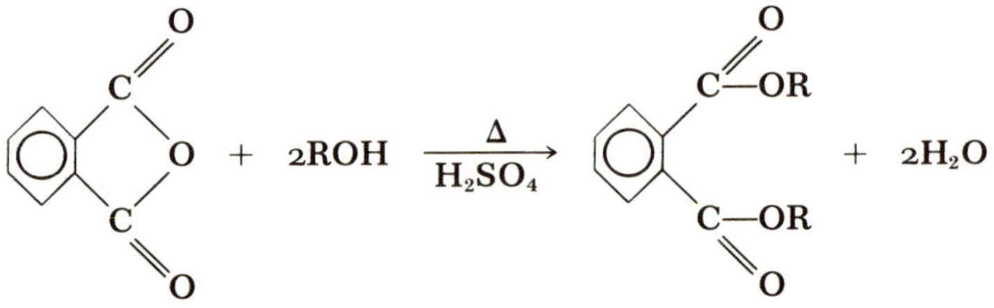

Phthalic Anhydride + Alcohol → Plasticizer

As expected, plasticizer growth has paralled growth in vinyl resins.

Growth in phthalic based plasticizer output since 1965 is shown in Table 8.15.

TABLE 8.15
Phthalic anhydride esters
(MM lb)

	Production	*Sales*	*Unit value ¢/lb*
1965	678.7	646.4	15
1966	754.5	742.4	16
1967	783.9	733.6	16
1968	840.6	785.3	17
1969	883.8	821.4	15
1970P	855.1	814.5	13
1971E	880	—	—

P = Preliminary.
E = Estimate based on output of dioctylphthalate.

Phthalic-based plasticizers are produced by at least 26 firms. The four major phthalates produced in 1970 were di(2 ethylhexyl) phthalate—350 MM lb; diisodecylphthalate 123 MM; diiso-octylphthalate 85 MM; and N-octyl-N decyl phthalate 59 MM lb.

About 0.43 lb of phthalic anhydride is required/lb of phthalate plasticizer produced.[1] This average value does not correct for the isophthalic acid utilized to produce plasticizers (see Chapter 7).

[1] Petrochemical Engineer, February 1963, p. 57.

Domestic phthalic anhydride plasticizer demand and export history are shown in Table 8.16.

TABLE 8.16
Phthalic anhydride esters
(*MM lb*)

	Domestic demand	*Exports*
1965	572	106.5
1966	659	95.2
1967	660	124.1
1968	726	114.5
1969	762	122.0
1970	783	72.5
1971E	818	62

E = Estimated from 6 months' data.

Phthalic plasticizer's future is tied directly to the non-rigid vinyl marketplace where strong growth has been exhibited and is expected to continue.

Phthalic-based alkyd resins are utilized as a prime ingredient in surface coatings. They are used because of low cost, ease of production, and wide versatility. Alkyd resins are widely used in surface coatings for boats, maintenance, exteriors, and appliances, although acrylics and latex paints have taken a large share of trade and industrial markets.

Alkyd resins—a type of polyester resin—are produced by reacting polybasic acids (such as phthalic or maleic acids) with polyalcohols (glycerine or pentaerythritol) and fatty oils and acids.

Phthalic-based alkyds are the major alkyds produced, as shown in Table 8.17.

Approximately 0.37 lb of phthalic anhydride is required/lb of phthalic alkyd produced. This does not take into account the isophthalic acid consumed (see Chapter 7).

The third largest and most rapidly increasing outlet for phthalic anhydride is unsaturated polyester resins. Unsaturated polyester resins are polymers made with unsaturated acids and glycols but cross-linked with such monomers as styrene, vinyl toluene, and diallyl phthalates.

Production data since 1965 are shown in Table 8.18.

TABLE 8.17
Alkyd resins
(MM lb)

	Total	*Phthalic anhydride type*	*Polybasic acid type*	*Sales*	*Unit value ¢/lb*
1965	639.6	562.6	59.4	334.9	26
1966	666.1	584.5	81.5	350.8	26
1967	638.5	549.8	88.7	313.3	27
1968	691.6	581.3	110.2	350.2	26
1969	694.1	585.9	108.2	335.3	28
1970P	635.6	559.6	75.9	332.3	27

Basis: US Tariff Commission data.

TABLE 8.18
Unsaturated polyesters
(MM lb)

	Production	*Sales*	*Unit value, ¢/lb*
1965	398.9	343.6	29
1966	470.0	406.7	30
1967	513.5	449.2	28
1968	615.4	543.3	28
1969	688.3	606.7	27
1970P	569.2	505.9	23
1971[1]	648	740	na

[1]Based on 11 months' data.
Basis: US Tariff Commission data.

Polyester production (70% of which is glass-reinforced) is finding increased use in building, appliances, boats, automobiles, and other uses. The growth rate found between 1965–1969, some 14%/year should continue at least into the mid 1970's.

About 0.22 lb of phthalic anhydride are used/lb of unsaturated polyester resin produced. This approximation does not take into account product made with other modifying agents—adipic, azelaic, isophthalic (see Chapter 7) and other acids.

There are a number of smaller uses for phthalic anhydride which amount to 80 to 100 MM lb. These include dyestuffs, such as anthraquinone, phthalocyanine, and xanthalene dyes (total volume 1970 = 54.8 MM lb); phenolphthalein (phenol-phthalic anhydride condensate); tetrachloro or tetrabromophthalic anhydride (fire resistant retardant—reactants produced by chlorinating or brominating phthalic anhydride)—possibly 10+ MM lb/yr. Lead phthalate (lead salt of phthalic acid, a PVC stabilizer); diallyl phthalate (a phthalic acid ester of allyl alcohol); dibutyl phthalate (an insecticide); and polyester polyols are other uses.

It is difficult to estimate the amount of phthalic entering the other-use area, as no production data are available. Other uses are therefore estimated as being the difference between production and identifiable uses. This approximation may be in error, as it reflects changes in inventory and any errors in other end uses.

Phthalic anhydride's end-use pattern since 1963 is estimated to be that shown in Table 8.19.

TABLE 8.19
Domestic use of phthalic anhydride
(MM lb)

	Plasticizers	*Alkyds*	*Polyesters*	*Misc.*	*Total*	*Export*	*Total demand*
1965	290	190	65	28	513	35	608
1966	320	195	75	81	671	9	680
1967	330	180	80	93	683	48	731
1968	360	195	95	98	748	21	769
1969	380	195	100	99	774	6	780
1970	365	190	80	138	773	7	780
1971E	376	190	100	112	778	8	786

E = Estimate based on 8 months' data.

Miscellaneous uses of 80 to 100 MM lb appear highly probable. Any higher value may be inventory buildup.

Phthalic anhydride exports have been on the decline since 1967, a year when the chemical was short in many parts of the world. Since then, a surplus of capacity has existed in many areas outside of the United States.

Phthalic anhydride is a good example of the reaction of a chemical to the laws of supply-demand not only for the product itself but of its raw material supply.

Phthalic anhydride prices, Fig. 8.5, dropped in the early 1960's due to overcapacity, the availability of petroleum naphthalene at attractive prices, and a surplus of ortho-xylene (also at low prices). With strong world phthalic growth in the 1967 to 1969 period, and the resultant lack of raw material, prices again rose. After 1969,

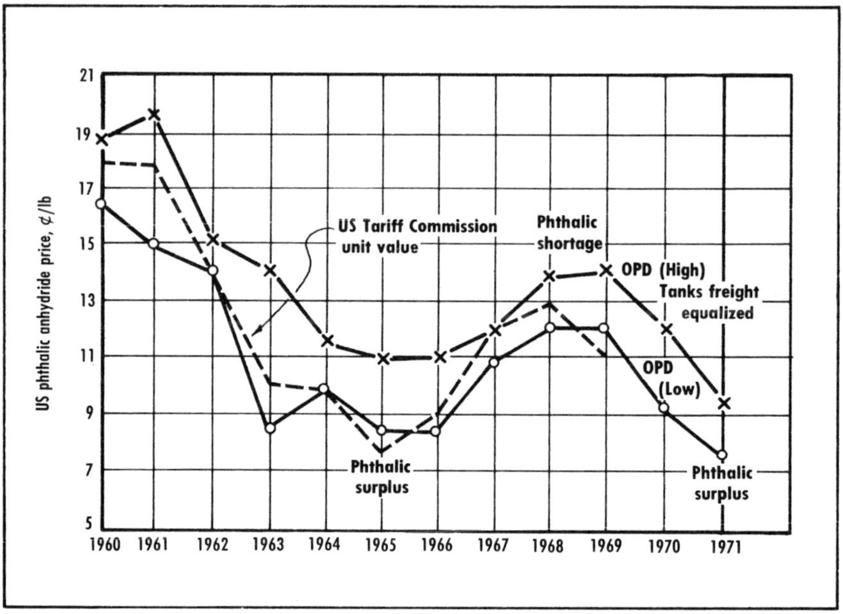

FIG. 8.5 Phthalic anhydride prices

adequate raw material, increased capacity, and the threat of near-term oversupply again dropped prices.

A number of other uses exist for naphthalene:

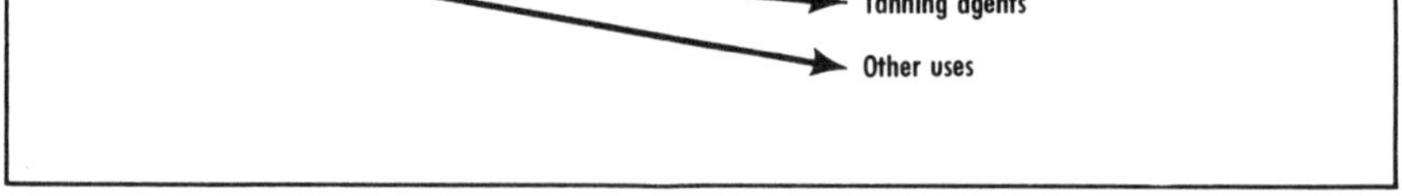

Outside of phthalic anhydride, the largest outlet for naphthalene is for the manufacture of the insecticide *Sevin*® (1-naphthyl N-Methyl carbamate).

Sevin® is thought to be produced by Carbide by this process.

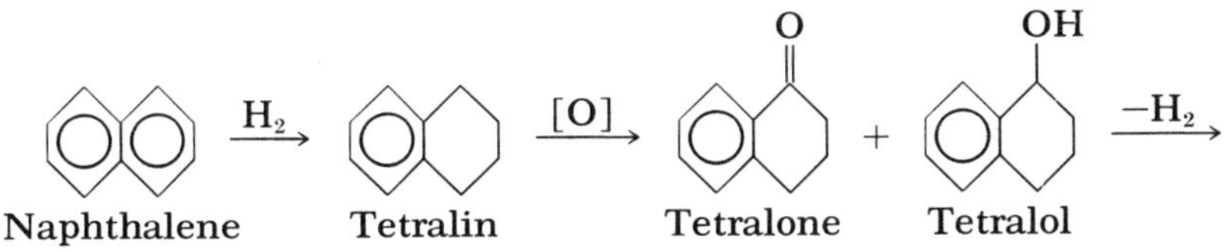

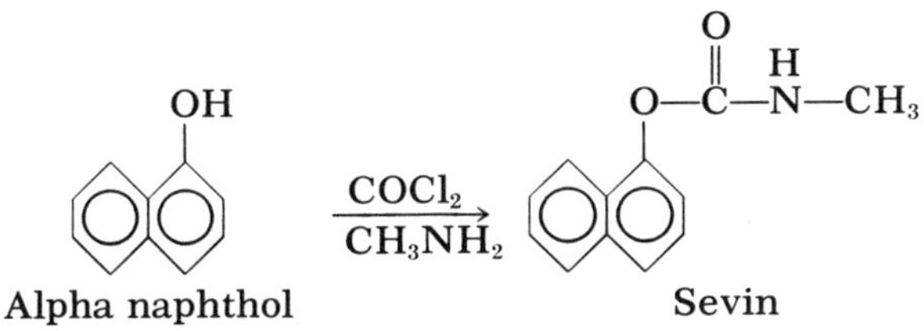

As there is only one producer of Sevin, there are no official production statistics but production is reported under the "all other" category of the U.S. Tariff Commission for pesticides. This category has grown from 30 MM lb in 1960, to 58 MM in 1965, to 76 MM lb in 1970.

Literature sources indicate that all but about 15 MM lb of this category is Sevin. Naphthalene use in 1970 is estimated at the 77 MM-lb level. With the demise of DDT, some market observers expect Sevin to exhibit strong growth in the future.

The third largest outlet for naphthalene is beta naphthol. This dyestuff intermediate is currently produced by the sulfonation of naphthalene, followed by caustic fusion:

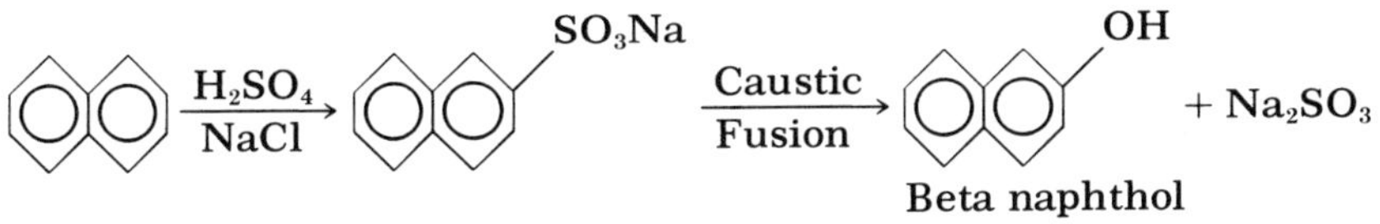

Producers, according to recent (1969) Tariff Commission reports, are American Cyanamid, Sherwin-Williams, and DuPont.

American Cynamid has announced it will construct a new plant to produce beta naphthol from isopropylnaphthalene, in a manner similar to the cumene-phenol process:

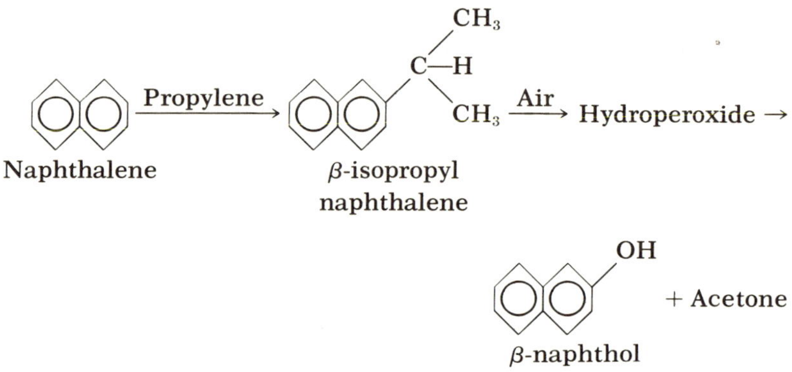

Beta naphthol is used for the manufacture of rubber antioxidants, dyes, and pigments. Major antioxidant outlets include phenyl beta-naphtholamine, 1,1′methylenebis-2 naphthol, and N,N′ di-2 naphthyl-p-phenylenediamine. Beta naphthol dye and pigment use includes such dye intermediates as 3-hydroxy-2 naphthoic acid (BON) and 2 amino-1 naphthalene sulfuric acid (Tobias acid).

As noted in Chapter 2, the Kennedy Round of Tariff changes lowered tariffs on many dyestuffs and intermediates. An example of the negative effects of these tariff lowerings are the increased imports of beta naphthol and derivatives. The increased imports, Table 8.20, are one reason why beta naphthol production may not have increased in recent years.

TABLE 8.20

Imports of beta naphthol and derivatives

(MM lb)

	1966	*1968*	*1970*
Beta naphthol	0.220	2.681	2.719
3-Hydroxy 2 naphthoic acid	0.917	1.043	1.450
Naphthol AS + derivatives	1.557	0.945	1.580

Reference: US Department of Commerce.

According to the literature, 22 MM lb of beta naphthol were produced in 1965. Current production is estimated at the 25-MM-lb level. 1.16 lb of naphthalene are required/lb of beta naphthol produced.

Another naphthalene outlet is synthetic tanning agents or syntans (naphthalene sulfonate–formaldehyde condensates). Syntans are used with vegetable-tanning materials, as dispensing agents for ceramics, rubber, paper pulp, and pigments.

Structually, they are:

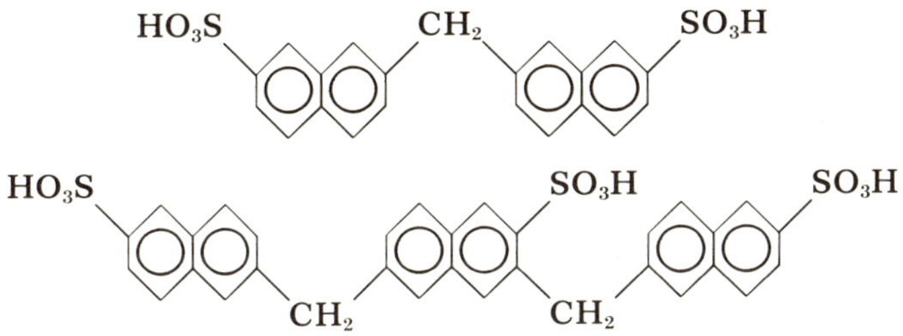

Syntans are produced by Arkansas Co., Diamond Shamrock–Nopco, W. R. Grace–Dewey and Almy Division, Rohm and Haas, and Tenneco Chemicals.

Production and sales history is shown in Table 8.21:

TABLE 8.21
Syntans production and sales
(MM lb)

	Production	*Sales*	*Unit value, ¢/lb*
1965	29.8	28.9	19
1967	30.5	30.2	19
1969	37.5	36.7	21
1971E	38	–	–

E = Estimate based on 8 months' data.
Reference: US Tariff Commission.

About 0.6 lb of naphthalene is consumed/lb of syntan produced.

Alkyl naphthalene sulfonates–naphthalene based surface active agents–are prepared from naphthalene by alkylation with an olefin and subsequent sulfonation. These naphthalene adducts are used as wetting, dispersing, and scouring agents for textiles.

Over 14 firms currently produce these products. Their production and sales history is shown in Table 8.22.

TABLE 8.22
Alkyl naphthalene sulfonates[1]
(MM lb)

	Production	*Sales*	*Unit value, ¢/lb*
1965	9.63	6.78	39
1967	10.62	7.03	39
1969	11.02	8.99	37
1970E	11	—	—

E = Estimate
[1]Butyl, isopropyl, diisopropyl and others.
Reference: US Tariff Commission.

Average naphthalene consumption/lb of sulfonate depends upon the size of the alkyl group, but averages around 0.5 lb/lb of sulfonate.

Moth balls remain a major outlet for naphthalene in spite of inroads by paradichlorobenzene. The volume of naphthalene used as a mothproofing agent and deodorant has been at the 14-MM-lb level for the past five years.

A number of other outlets consume smaller quantities of naphthalene. Other uses include:

- Decalin production by DuPont for solvent and chemical uses.
- Tetralin production by Carbide and DuPont for solvent uses (see earlier discussion on Sevin®).
- Chloronaphthalenes which are used as solvents and plasticizers; produced by GAF and Koppers-Pittsburg Co.
- Alpha naphthol produced by Allied, DuPont, and Carbide for dyestuff use.
- Alpha naphthyl amine used as dye intermediate and rubber-antioxidant raw material produced by Allied and DuPont.

TABLE 8.23
Estimated naphthalene requirement
(MM lb)

	1965	*1970*
Phthalic anhydride	>550	>525
Sevin®	54	77
Beta naphthol	25	25
Syntans	18	21
Surface-active agents	5	6
Moth balls	14	14
Other uses[1]	< 78	< 67
Total	744	735

[1]Obtained by difference.

Naphthalene is also directly consumed in the production of dyes and pigments. Estimated naphthalene consumption by end use 1965 and 1970 is shown in Table 8.23.

Naphthalene's future depends upon growth in non-phthalic anhydride uses.

Naphthalene prices reflect the supply-demand characteristics of the market—shortages in 1962 and 1968 and oversupply in 1965 and 1970. Historical coal tar naphthalene pricing is shown in Table 8.24.

TABLE 8.24
Pricing of coal-tar naphthalene

	Price,[1] ¢/lb			*OPD posted prices*	
	74°C.	*74–79°C.*	*Average*	*Crude domestic*	*78°C., tanks*
1965	1.57	na	na	3.25–3.25	
1966	na	na	4.1	3.25–3.75	
1967	na	na	4.3	4.00–5.50	
1968	4.4	4.7	4.5	6.50	
1969	3.6	4.85	4.7	6.50	
1970	–	–	–	6.50	
1971	–	–	–	6.50	

[1] Reference: US Tariff Commission.

Petroleum naphthalene's price history is shown in Table 8.25.

TABLE 8.25
Pricing of petroleum naphthalene

	Unit value[1] ¢/lb	*OPD posted price*
1962	5.9	–
1965	3.9	–
1966	4.0	4.0
1967	4.9	5.0
1968	7.0	8.25
1969	6.7	7.50
1970	5.2	6.75
1971	–	6.75

[1] US Tariff Commission.

It is suspected that recent posted prices for petroleum and coal-tar naphthalene are not indicative of the market conditions.

POLYMETHYLBENZENES

Polymethylbenzenes, like dimethylbenzenes (xylenes) are obtained from catalytic reformate, naphtha cracking, and coal-tar fractions.

Polymethylbenzenes are characterized by three or more reactive methyl groups and an aromatic ring capable of undergoing substitution reactions at far greater rates than toluene and xylenes. These polyalkylbenzenes have only become of commercial significance during the past five or so years. All eight are available for commercial and development applications. Their structures are:

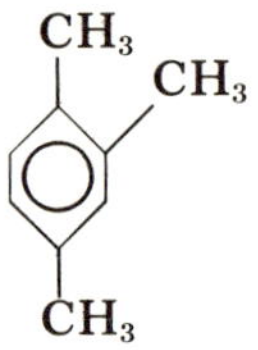

Pseudocumene

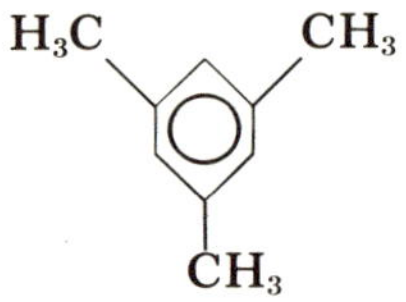

Mesitylene

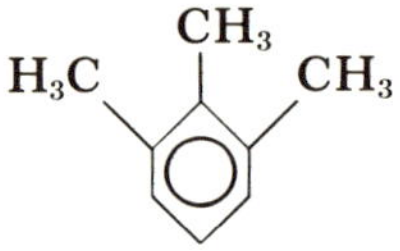

Hemimellitene

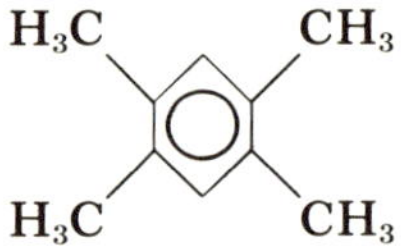

Durene

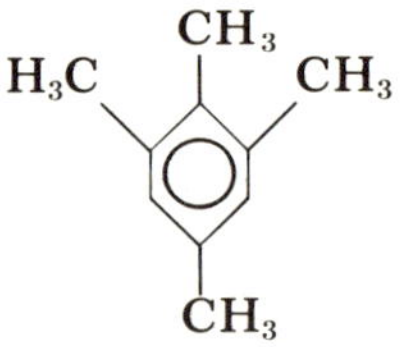

Isodurene

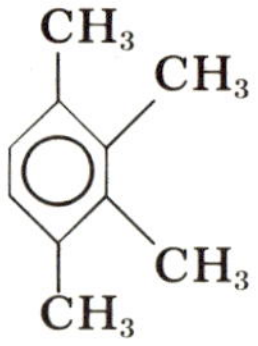

Prehnitene

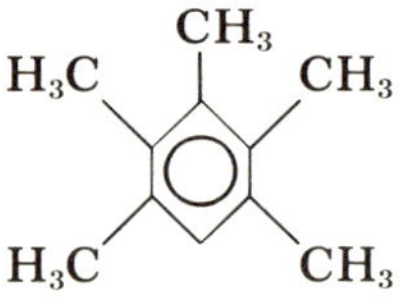

Pentamethylbenzene

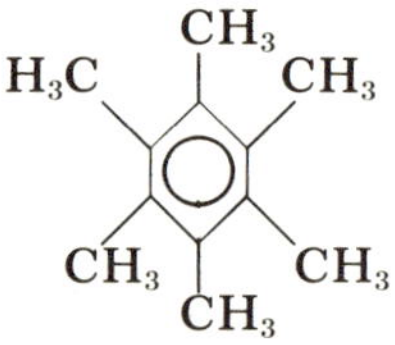

Hexamethylbenzene

Of the eight polymethylbenzenes, only pseudocumene, mesitylene, and durene are of commercial significance.

Pseudocumene (1,2,4 trimethylbenzene) is currently the polymethylbenzene in greatest demand. It is currently produced in the United States only by the Sun Oil Co. and is obtained by fractional distillation of heavy-reformate fractions. Current production capacity is 50 MM lb/yr. Overseas, pseudocumene is extracted from reformate and pyrolysis-gasoline fractions.

The oxidation of pseudocumene yields trimellitic anhydride (TMA):

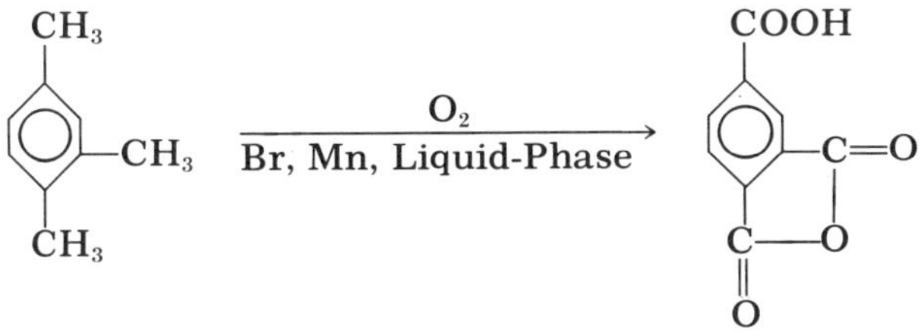

Amoco Chemical is the only trimellitic anhydride producer in the United States, with a plant of 50 MM lb capacity at Joliet, Ill.

The major outlet for trimellitic anhydride is in trimellitate plasticizers. Being a trifunctional compound, it has many advantages over phthalic anhydride as a plasticizer raw material. The advantages are much lower quantities of plasticizer required and low volatility. The latter point is a major reason why it has been used in automotive PVC seat covers where it has helped to eliminate plasticizer "windshield fog."

The plasticizer is produced by the usual reaction with an oxo alcohol:

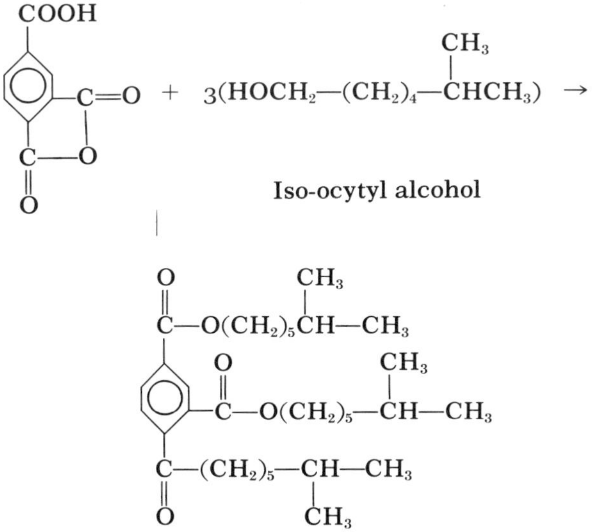

United States production and sales history for trimellitic acid esters is shown in Table 8.26.

TABLE 8.26
Trimellitic plasticizer output
(MM lb)

	Production	Sales	Unit value ¢/lb
1965	–	1.98	40
1966	2.54	2.62	41
1967	6.25	5.69	45
1968	4.73	4.36	58
1969	7.55	5.62	30
1970P	9.70	7.40	29

P = Preliminary.
Reference: US Tariff Commission.

Output of plasticizer by type in 1969 and 1970 is shown in Table 8.27.

TABLE 8.27
Plasticizer output

	Production, MM lb	
	1969	*1970*
Triiso-octyl trimellitate	2.75	3.18
Tri-n-octyl n-decyl trimellitate	2.30	2.51
Trioctyl trimellitate	1.71	2.34
All other trimellitic acid esters	0.78	1.66

US trimellitic plasticizer producers currently include Hooker Chemical, Pfizer, Reichhold and Enjay Chemical.

Another outlet for trimellitic anhydride is in polyamide-imides and polyester-imide resins. These high temperature resistant polymers find outlets as magnet-wire enamels, laminating resins, and stock shapes. According to Stanford Research Institute (Menlo Park, California) Chemical Economic Handbook, 1970, modified polyimide production based upon trimellitic anhydride was slightly over 2.1 MM lb.

Other outlets include use in polyester manufacture, epoxy curing agents, and miscellaneous small uses. Trimellitic anhydride production costs when made by the nitric acid oxidation process, which will be used by Saarbergwerke in a 15,000-ton unit in West Germany, are shown in Table 8.28.

TABLE 8.28
Trimellitic anhydride production costs

Plant size		12,000 tons (26.5 MM lb)
Battery limits investment		\$5.5 MM (1970)
Process		Bergwerksverband Chemiebau
Raw materials	*Unit costs*	*Cost, \$/ton*
Pseudocumene	\$143/ton	132
Nitric acid	\$20/ton	36
Utilities		
Steam	\$2/ton	80
Cooling water	\$0.007/cu m	18
Electricity	\$0.01/kwh	25
Fuel	\$2.10/10^6 kcal	10
Labor maintenance and overhead at 5% Bli		23
Capital charges at 25% Bli		114
Ex-plant production costs		\$738 or 20¢/lb

Reference: European Chemical News, Aug. 21, 1970, p. 28.

It is speculated in the referenced ECN article that costs for the Amoco liquid-phase process could be \$50 to 70/ton lower than those shown for the Bergwerkesverband process. If so, the mill cost by the Amoco process is in the area of 17¢/lb. Trimellitic anhydride is currently sold for 23¢/lb in the United States and 27 to 30¢/lb in Western Europe.

Other uses for pseudocumene include trimethylaniline and Vitamin E manufacture. Pseudocumene can also be methylated to yield durene or isomerized to mesitylene.

World production of pseudocumene is estimated to have been less than 50 MM lb in 1970.

Pseudocumene is produced outside of the United States by Societe d' Hydrocarbons de St. Denis, Deutsche Hydrocarbons, Deutsche Erdol, Maruzen Oil, and Nippon Petrochemicals. Trimellitic anhydride is produced by Saarbergwerke, Japan Catalytic Chemical, Daicel, and Mitsui Toatsu.

Durene (1,2,4,5 tetramethyl benzene) is the second largest volume polymethyl benzene produced. In the United States, Sun Oil Co. produces the product from reformate. Rutgerswerke in West Germany produces from coal tar fractions, and Mitsubishi (Japan), from pyrolysis gasoline. It can also be obtained from pseudocumene by methylation.

Like pseudocumene, durene has only one major outlet today—pyromellitic dianhydride (PMDA), its oxidation product.

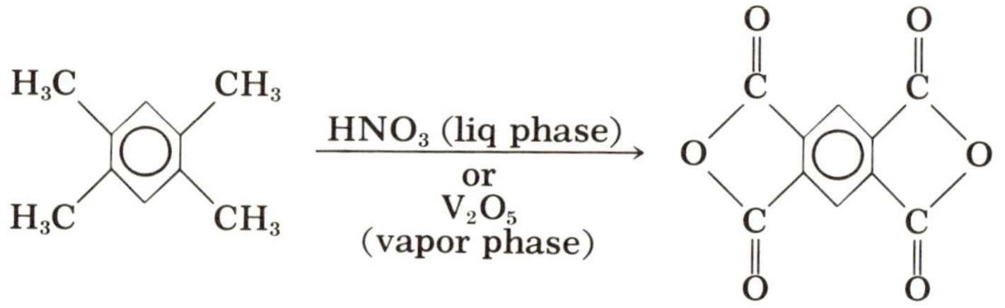

Pyromellitic dianhydride (PMDA) is produced commercially in the United States by either a liquid-phase nitric acid oxidation or by a V_2O_5 vapor phase oxidation. DuPont, the major producer and consumer, practices the former process, and Princeton Chemical Research the latter in a unit smaller than 500,000 lb/yr.

Outside of the United States, pyromellitic dianhydride is produced by Veba Chemie, Furukawa Electic, Tokkasha, and Nippon Shokubai.

PMDA may also be produced from raw materials other than durene, i.e., pseudocumene and paraxylene. The Bergbau Forschung process is based upon chloromethylation of paraxylene, while another process is based upon isopropylpseudocumene.

Uses for PMDA include epoxy curing agents, polyimides, and plasticizer raw materials.

PMDA's largest outlet is polyimide resin, a high-temperature (above 500°F.) resistant polymer whose outlets include film, wire coating, molded parts, laminates, and adhesives.

Polyimides are produced by this reaction:

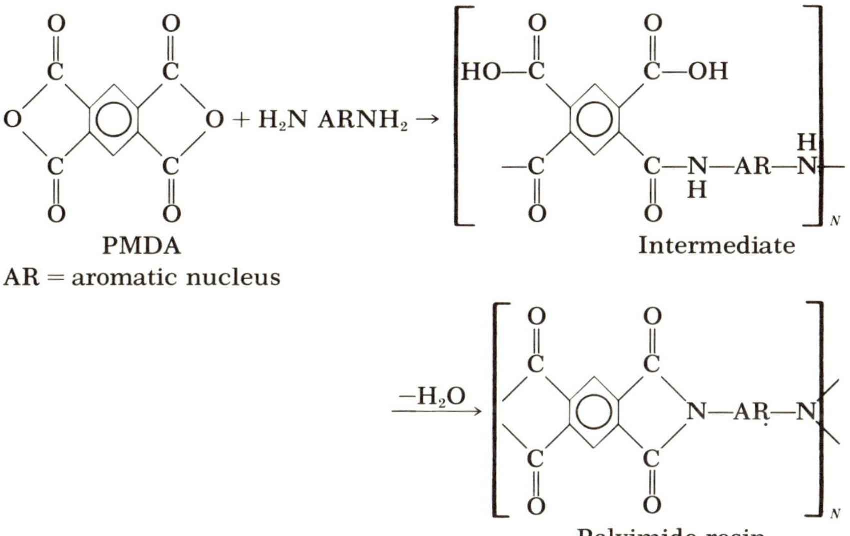

Total output of resins based upon pyromellitic dianhydride was estimated by Stanford Research Institute (Menlo Park, California) Chemical Economics Handbook to be 0.6 MM lb in 1970. Other polyimide resin output (those based on benzophenone tetracarboxylic dianhydride) is estimated to have been 0.3 MM lb.

World polyimide producers include DuPont, 3M, Toray, Raychem, Teijin, Dixon, Rhone Poulenc, General Electric, American Cyanamid, Monsanto, ICI, and Furakawa Electric.

The overall market for PMDA and durene in the United States was in the million-lb range in 1970.

PMDA use breakdown is estimated to be: 95% used in polyimide manufacture, 1% as epoxy resin curing agents, and 4% in miscellaneous outlets.

Mesitylene (1,3,5 trimethylbenzene) like durene and pseudocumene finds a major outlet as an oxidation product – trimesic acid. Amoco chemical produces it in development quantities in this manner:

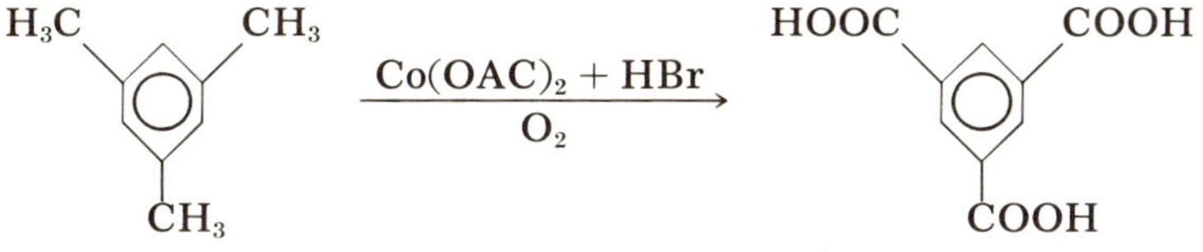

The trifunctional nature of trimesic suggests its possible future use in alkyds, plasticizers, unsaturated polyesters, and as a modifier in synthetic fibers and films.

Commercial uses for mesitylene include its nitration product mesidine (2,4,6 trimethylaniline) a dye intermediate, and Ethyl's Antioxidant 33D, a non-coloring stabilizer for plastics, resins, rubber, and waxes.

Mesitylene is produced in the United States by the Sun Oil Co. and overseas by Rutgerswerke and Toyo Gosei Kogyo. It is obtained from coal-tar fractions or by extraction from C_9 aromatic fraction obtained from naphtha cracking or reformate.

The remaining polymethylbenzenes are available to the world in development quantities. Their structure and reactive character suggest that some day they too will find their way into the high-volume, multi-million-lb specialty-chemical world.

9

ARTHUR M. BROWNSTEIN

Acetylene

ACETYLENE is a low-boiling (−83.6°C.), colorless gas that has been known since 1836. It is an endothermic compound which is unstable at room temperature, particularly in the presence of oxygen. Thermal decomposition can be so rapid that explosions often result.

Acetylene became a major raw material for the chemical industry in the early 20th century and has had its most significant growth since 1940. Its attractiveness centered about the classically easy reactivity of the triple bond. This was conducive to a number of addition reactions that readily led to the commercialization of such compounds as acrylonitrile, vinyl chloride, and vinyl acetate.

Acetylene has both chemical and non-chemical uses. The latter applications are in welding and metal-cutting, and account for about 10% of total acetylene production. As a chemical building block, acetylene is expensive. The majority of acetylene used for this purpose is captive, and price is very difficult to determine. Actual internal transfer prices are probably 50% lower than those of shipments shown in Table 9.1. The probable price range may be 7–8¢/lb depending on product value and the manufacturing process employed. The only years in which the unit value of acetylene was published by the US Tariff Commission were 1963 and 1964. At that time, they were 10.3 and 10.2¢/lb respectively.

TABLE 9.1
Acetylene sales price[1]

Year	*¢/lb*	*Year*	*¢/lb*
1960	17.1	1965	14.6
1961	16.4	1966	14.8
1962	15.7	1967	15.4
1963	15.4	1968	15.9
1964	15.4	1969	16.3

[1]Current Industrial Reports, US Census Bureau.

In the 1950's, new ethylene and propylene technologies were developed which threatened the supremacy of acetylene. For the first time, it became possible to achieve with lower-cost olefins what had hitherto only been possible with acetylene. Although the impact of these new olefin processes would not be felt until the middle 1960's, the seeds had been sowed for the demise of acetylene as a large-volume chemical.

ACETYLENE MANUFACTURE

At the close of World War II, calcium carbide processes represented the sole commercial route for acetylene production. In the early 1950's, more economical processes based on hydrocarbons were introduced. By the beginning of the 1970's, these accounted for 40% of US acetylene capacity as illustrated in Table 9.2.

TABLE 9.2
1972 acetylene capacity[1]

Company	*MM lb*	*Process*
Diamond Shamrock (Houston, Tx.)	40	Partial oxidation
Dow (Freeport, Tx.)	15	" "
Monochem (Geismar, La.)	165	" "
Rohm and Haas (Houston, Tx.)	35	" "
Tennecco (Houston, Tx.)	100	" "
Union Carbide (Seadrift, Tx.)	15	Ethylene byproduct
(Taft, La.)	18	Wulff
(Texas City, Tx.)	80	Partial oxidation
Other	735	Calcium carbide
Total	1,203	

[1]Based on Oil, Paint & Drug Reporter, April 5, 1971, p. 9.

Calcium carbide process

Pyrolysis of a mixture of lime and coke at 2,000°C. in an electric furnace leads to the formation of calcium carbide:

$$CaO + 3C \rightarrow CaC_2 + CO$$

The carbon monoxide byproduct is separated and can be used as fuel or synthesis gas. The CaC_2 is ground under anhydrous conditions and then treated with a limited quantity of water to produce acetylene:

$$CaC_2 + 2H_2O \rightarrow H{-}C{\equiv}C{-}H + Ca(OH)_2$$

The use of a limited amount of water (1 to 1 weight ratio) allows recovery of calcium hydroxide for re-use in CaC_2 production. Addition to water is exothermic (900 BTU/lb of calcium carbide), and this removes water rapidly by vaporization to allow recovery of substantially dry calcium hydroxide.

Methane partial oxidation

In this technique, natural gas is partially oxidized with oxygen, and the evolved heat cracks the hydrocarbon to acetylene:

$$2CH_4 \xrightarrow{1{,}500^\circ C.} HC{\equiv}CH + 3H_2$$

As shown in Table 9.2, this process or its variant is employed in six US plants. Residence time during pyrolysis is 0.001 to 0.01 seconds. The product gases (Table 9.3) are rapidly quenched with water at 40°C. to prevent acetylene decomposition.

TABLE 9.3
Product composition of methane partial oxidation process

Product	*wt %*
Acetylene	8.5
Hydrogen	57.0
Carbon monoxide	25.3
Methane	4.0
Others	5.2
Total	100.0

Acetylene with greater purity than 99% is obtained in 31 mole % yield. The low yield is a principal factor in the relatively high cost of acetylene. The process is summarized in Fig. 9.1.

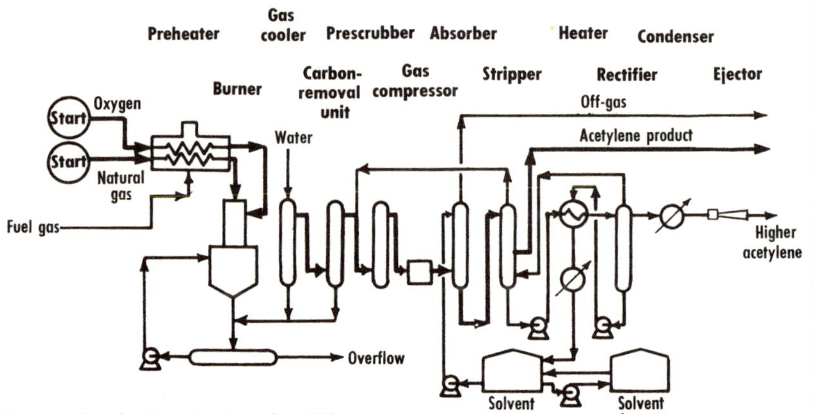

Source: Hydrocarbon Processing, November, 1969

FIG. 9.1 Acetylene by partial oxidation of methane

Thermal cracking

Acetylene can be produced by pyrolysis of ethane, propane, naphtha, or higher hydrocarbons. The Wulff process and the Farbwerke Hoechst high-temperature processes (HTP) are examples of this type of technology.

In the HTP process, a high temperature, high-energy system (2,000°C.) is created by complete combustion of fuel with oxygen. Vaporized naphtha is contacted with this system for 0.002 to 0.003 seconds, and is then quenched by contact with a spray of quench oil. The exit temperature of the gases is about 1,000°C. Ethylene is coproduced in the process and can be obtained in ethylene to acetylene weight ratios of 1 to 1 to 1.5 to 1, depending upon reaction temperature as shown in Fig. 9.2.

Coproduction of ethylene affords a distinct advantage over other acetylene processes as can be discerned from the data in Table 9.4. Nevertheless, acetylene production cost is still quite high, and implies a transfer price of at least 7–8¢/lb.

The acetylene to ethylene ratio can be similarly varied in a Wulff plant. An alternating cycle is used wherein the hydrocarbons are heated by a hot tile checkerwork at 1,100°C. to produce acetylene. The furnace operates on a four-minute cycle in which the checkerwork is heated for one minute and the hydrocarbons are pyrolyzed for one minute. The same sequence is repeated in the reverse direc-

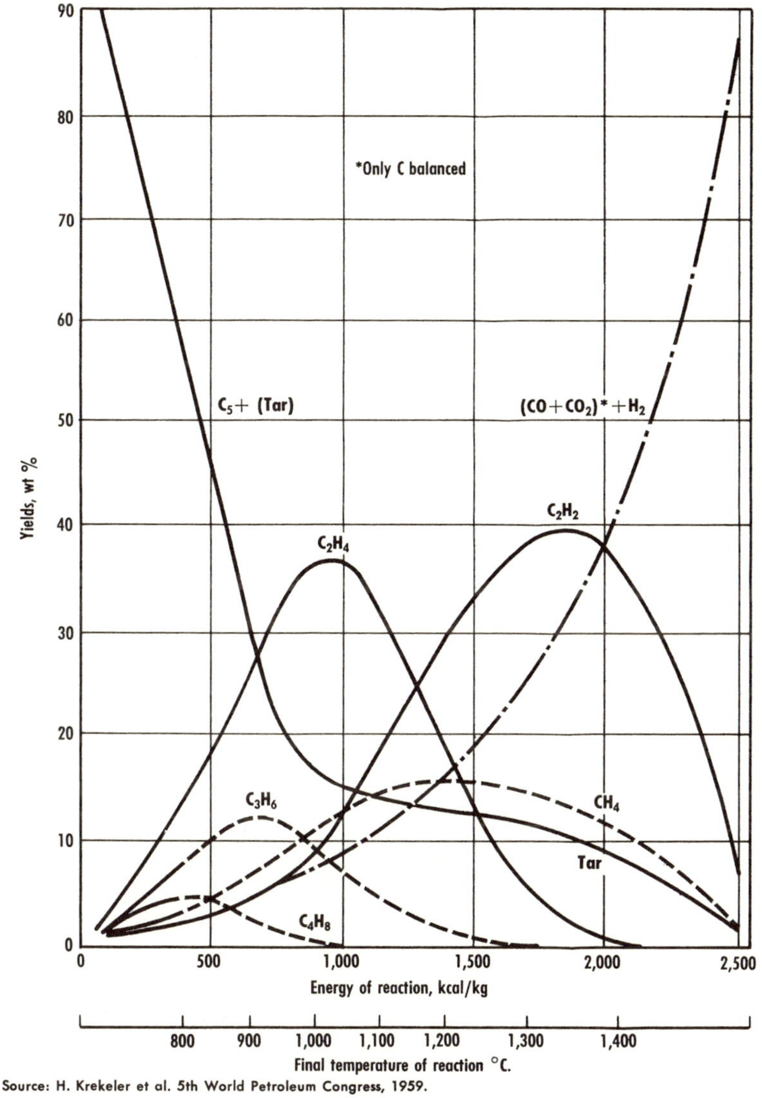

Source: H. Krekeler et al. 5th World Petroleum Congress, 1959.

FIG. 9.2 Product distribution for Hoechst HTP process

TABLE 9.4
Estimated economics for thermal process acetylene plant
(100-MM lb; 1972 construction)
Total fixed capital = $27.6 MM

	Production cost ¢/lb
Raw materials[1]	4.70
Labor	0.29
Utilities	1.00
Maintenance (6% ISBL + 3% OSBL)	1.33
Overhead (45% maintenance + labor)	0.73
Taxes and insurance (1.5% of investment)	0.42
Depreciation (10 years)	2.76
Total	11.23
Ethylene byproduct credit[2]	−4.88
Net production cost	6.35

[1] 4.83 lb naphtha/lb at 0.8¢/lb and other chemicals at 0.8¢/lb.
[2] 1.5 lb ethylene/lb of acetylene at 3.25¢/lb.

tion. The off gases from the acetylene recovery system are used as fuel for heating the combustion chamber.

Pyrolysis gases in the HTP and Wulff processes are contacted initially with DMF to remove water, diacetylene, and other products. This is followed by absorption of acetylene in DMF, from which it is finally recovered by stripping.

Recovery from ethylene plants

Ethylene plants produce a small amount of acetylene that must be reduced by hydrogenation below 5 ppm if the ethylene is intended for polymerization. In naphtha-based ethylene plants, acetylene production is about 2 wt % of the ethylene produced. In the large 1.2 billion-lb ethylene plants now being built, sufficient acetylene is produced (24 MM lb) to justify the expense for its recovery.

Acetylene can be separated from ethylene by a DMF or n-methylpyrrolidone wash. The gas-washing steps produce an acetylene-rich gas containing 50 to 90 vol % of acetylene. It is plausible that such acetylene may be cheaper than that made in acetylene plants per se, but not so cheap as ethylene and propylene against which acetylene must compete as a building block. Moreover, volumes of 24 MM lb annually are more conducive to specialty chemical production than for large-volume commodities.

MARKET DEMAND

Since 1962, acetylene use has been essentially static with demand, attaining a maximum 1.15 billion lb in 1965–66 (Table 9.5). Since that time, acetylene production has been declining steadily, and will probably continue to do so for the foreseeable future.

TABLE 9.5
Acetylene end use pattern
(MM lb)

Year	*Vinyl chloride*[1]	*Vinyl acetate*[2]	*Chloroprene*[3]	*Acrylonitrile*[4]	*Trichloroethylene*[5]	*Acrylics*[6]	*Other*[7]	*Total*
1962	315	105	186	136	74	16	103	935
1963	315	134	186	159	64	19	133	1,010
1964	320	145	201	171	69	21	173	1,100
1965	352	170	204	190	77	27	130	1,150
1966	385	165	208	86	79	36	191	1,150
1967	428	172	214	65	69	43	136	1,127
1968	376	208	221	80	74	43	130	1,132
1969	332	187	227	83	79	62	125	1,095
1970	270	151	240	69	74	52	130	986
1971	210	102	180	—	66	60	130	748

[1]0.44 lb/lb.
[2]0.32 lb/lb.
[3]0.65 lb/lb.
[4]0.60 lb/lb.
[5]0.22 lb/lb.
[6]0.42 lb/lb.
[7]Includes non-chemical uses and inventory.

The basis for the decline in acetylene demand is the development of new technology which makes it possible to replace acetylene with other, less expensive hydrocarbons. Thus, ethylene at 3–3.5¢/lb is more attractive for production of vinyl chloride and vinyl acetate than is acetylene at 7–8¢/lb. Similarly, propylene at 3¢/lb is more attractive for acrylonitrile manufacture than is acetylene.

Consumption of acetylene for non-chemical purposes (oxyacetylene welding) is about 100 MM lb/yr. Here too, natural gas and LP-gas have replaced acetylene for some applications. The advent of electric arc welding has also made inroads into this area.

Decline in acetylene production has been so great, that at the beginning of the 1970's, there is nearly sufficient hydrocarbon-based capacity to accommodate the entire market. Considerable shutdown of less economical carbide processes appears to be in the offing.

ACETYLENE MARKETS

The primary markets for acetylene are the manufacture of vinyl chloride, vinyl acetate, chloroprene, acrylonitrile, trichloroethylene, and acrylic acid and its esters. The total decline in acetylene demand is, of course, proportional to its decline for these end uses. Some uses have felt the impact of new technology sooner and more severely than others; notably, vinyl chloride and acrylonitrile. The decreasing contribution of these to total acetylene demand is:

% of total demand

Year	*Vinyl chloride*	*Vinyl acetate*	*Chloro-prene*	*Acrylo-nitrile*	*Trichloro-ethylene*	*Acrylics*	*Others*
1962	34	11	20	14	8	2	11
1971	28	14	24	0	9	8	17

Vinyl chloride

The facile addition of HCl to triple bonds prompted the original selection of acetylene for the manufacture of vinyl chloride. There not being any competitive processes at that time, this was the unquestioned route of choice.

$$H{-}C{\equiv}C{-}H + HCl \xrightarrow[250°C.]{HgCl_2} CH_2{=}CH{-}Cl$$

The reaction proceeds in the vapor phase with high-purity acetylene and anhydrous HCl. The mercuric chloride catalyst is supported on carbon, and yields of vinyl chloride are about 80 to 85%.

The preeminent position of acetylene was successfully challenged by ethylene with introduction of the oxychlorination process. This process proceeds:

$$(1)\quad CH_2{=}CH_2 + Cl_2 \rightarrow Cl{-}CH_2CH_2{-}Cl$$

$$(2)\quad 2ClCH_2CH_2Cl \xrightarrow{400°C.} 2CH_2{=}CH{-}CL + 2HCl$$

$$(3)\quad CH_2{=}CH_2 + 2HCl + \tfrac{1}{2}O_2 \xrightarrow[250°C.]{CuCl_2} ClCH_2CH_2Cl + H_2O$$

$$\text{NET}\quad 2CH_2{=}CH_2 + Cl_2 + \tfrac{1}{2}O_2 \rightarrow 2CH_2{=}CH{-}Cl + H_2O$$

The overall process is balanced, so that only vinyl chloride is obtained, and there is no net production of HCl. Selectivity is about 90 mole % based on ethylene. The process is further illustrated by the flow diagram in Fig. 9.3.

As shown in Table 9.6, ethylene-based technology is far more economical than acetylene. The principal cost advantage is the saving of nearly 2¢/lb in raw materials. This economic advantage of ethylene has forced the closing of many acetylene-based vinyl chloride units (Table 9.7). Active acetylene-based capacity in 1972 is only 10% of the total, as opposed to 40% in 1967.

TABLE 9.6
Estimated vinyl chloride economics
(500-MM-lb plant; 1972 construction)

Total fixed capital

Process	*$ MM*
Ethylene oxychlorination	17.9
Acetylene	18.9
Ethane (transcat)	18.0

Production cost

	¢/lb		
Process:	*Ethylene*	*Acetylene*	*Ethane*
Raw materials			
Ethane (0.59 lb/lb at 0.9¢/lb)	—	—	0.53
Ethylene (0.49 lb/lb at 3.0¢/lb)	1.46	—	—
Chlorine (0.67 lb/lb at 2.5¢/lb)	1.68	—	1.45[1]
Acetylene (0.44 lb/lb at 8.0¢/lb)	—	3.52	—
HCl (0.60 lb/lb at 2.5¢/lb)	—	1.49	—
Subtotal	3.14	5.01	1.98
Labor	0.09	0.06	0.09
Utilities	0.22	0.08	0.22
Maintenance (6% ISBL + 3% OSBL)	0.17	0.18	0.17
Overhead (45% maint. + labor)	0.12	0.11	0.12
Taxes & ins. (1.5% of investment)	0.05	0.06	0.05
Depreciation (10 years)	0.36	0.38	0.36
Total	4.15	5.88	2.99

[1]0.58 lb/lb at 2.5¢/lb.

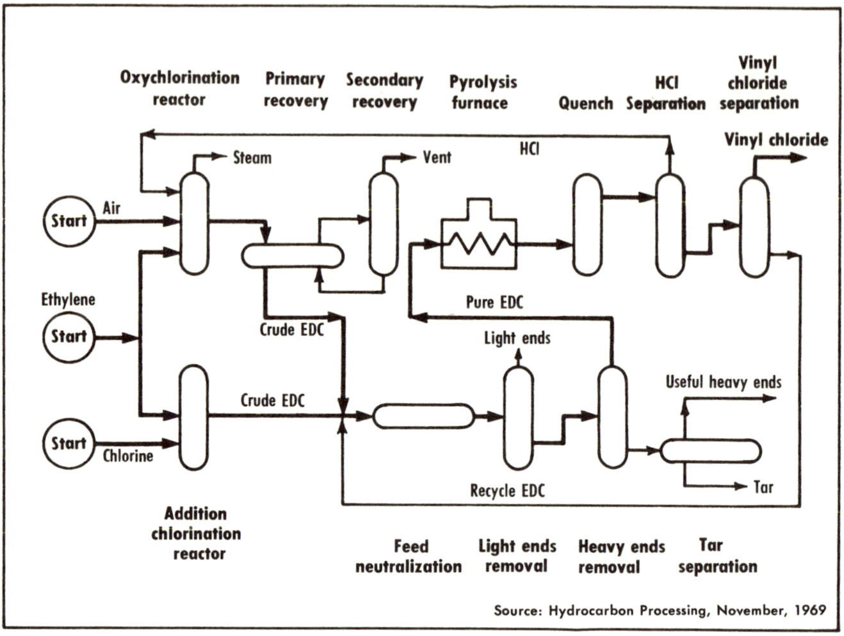

FIG. 9.3 Vinyl chloride oxychlorination process

The advantage of ethylene technology is further illustrated by comparison of Tables 9.8 and 9.5. Thus, although vinyl chloride production grew 14% annually from 1962–71, acetylene demand for this market fell a mean 5% annually.

The newest entry in vinyl chloride technology is an ethane process developed jointly by the Lummus Co. and Armstrong Cork. This technology may affect ethylene in this decade in the same manner as ethylene caused the demise of acetylene during the 1960's. The process is called "Transcat" and possesses a significant advantage in raw material costs as shown in Table 9.6.

The Transcat process comprises one step synthesis of vinyl chloride by oxychlorination of ethane. Vinyl chloride is readily produced by passing ethane, HCl and air (1 to 1 to 2.5 vol) through a molten catalyst of KCl and $CuCl/CuCl_2$ at 875°F.

Some acetylene-based processes still operating are of the balanced variety, employing both ethylene and acetylene. Instead of oxychlorination, the HCl produced by pyrolysis of ethylene dichloride is added to acetylene for additional vinyl chloride production.

TABLE 9.7
Vinyl chloride capacity[1]
(MM lb)

Company	*1967*	*1969*	*1972*	*Process*
Allied Chemical (Moundsville, W.Va.)	100	–	–	Acetylene
(Geismar, La.)	–	300	550	Ethylene
American Chemical (Long Beach, Calif.)	170	170	170	Ethylene
Continental Oil (Lake Charles, La.)	–	600	600	Ethylene
Cumberland Chemical (Calvert City, Ky.)	60	–	–	Acetylene
Diamond Shamrock (Deer Park, Tx.)	100	100	–	Acetylene
Dow Chemical (Freeport, Tx.)	200	200	525	Ethylene
(Plaquemine, La.)	250	300	575	Ethylene
Ethyl Corp. (Baton Rouge, La.)	270	270	270	Ethylene
(Houston, Tx.)	150	150	150	Ethylene
General Tire (Ashtabula, Ohio)	75	–	–	Acetylene
B. F. Goodrich (Calvert City, Ky.)	400	400	400	Ethylene
(Niagara Falls, N.Y.)	40	–	–	Acetylene
Goodyear (Niagara Falls, N.Y.)	70	–	–	Acetylene
Monochem (Geismar, La.)	250	250	250	Acetylene
PPG (Lake Charles, La.)	–	300	300	Ethylene
PPG-Corco (Puerto Rico)	–	–	500	Ethylene
Shell (Deer Park, Tx.)	–	–	700	Ethylene
Tenneco (Houston, Tx.)	200	200	200	Acetylene
Union Carbide (S. Charleston, W.Va.)	120	120	120	Ethylene &
(Texas City, Tx.)	230	230	–	Acetylene
Totals	2,685	3,590	5,310	

[1]Based on Oil, Paint & Drug Reporter, March 17, 1969.

TABLE 9.8
Vinyl chloride production[1]

Year	*MM lb*	*Unit value, ¢/lb*
1962	1,311	7.5
1963	1,435	7.0
1964	1,615	6.4
1965	2,000	6.1
1966	2,500	5.9
1967	2,424	5.3
1968	2,969	4.6
1969	3,736	4.4
1970	4,040	4.8[2]
1971	4,130	4.8[2]

[1]US Tariff Commission Reports; 1971 based on 7 months' reports.

[2]Oil, Paint & Drug Reporter posted price (polymer grade in tanks, freight equalized).

Vinyl acetate

In a manner analogous to vinyl chloride, ethylene appears to have become the preferred raw material for vinyl acetate production. Acetylene-based capacity for vinyl acetate has rapidly fallen from 77% to 38% of the total since 1966 (Table 9.9). Although vinyl acetate production has nearly tripled in the 1962–71 period, 1971 acetylene demand for this use is essentially the same volume as it was nine years earlier (compare Tables 9.10 and 9.5).

The acetylene route to vinyl acetate proceeds by vapor phase addition of acetic acid to acetylene in the presence of a zinc acetate on carbon catalyst. Conversion is about 65%/pass with yields greater than 90% based on acetylene.

The vapor-phase ethylene route introduced in the early 1960's, employs as a catalyst, palladium on alumina or silica-alumina:

$$CH_2{=}CH_2 + CH_3COOH + \tfrac{1}{2}O_2 \xrightarrow[150°C.]{Cat.} CH_2{=}CH{-}O\overset{\overset{O}{\|}}{C}CH_3 + H_2O$$

Conversions of ethylene per pass are 10–15% with 90% selectivity. This process is under a legal cloud at present with both Bayer and National Distillers claiming the basic patent rights.

The economic advantage of the ethylene route is not as apparent as in the case of vinyl chloride. Although the estimated investment cost for the ethylene plant is 25% higher, it possesses a production cost advantage of about 1¢/lb (Table 9.11.). Judging from the trend in installed capacity, the future is in the direction of ethylene-based processes.

Originally, liquid-phase ethylene processes were employed for vinyl acetate. These possessed significant corrosion problems, and in 1972 none was operating either domestically or abroad. Supposedly, they were less economical than the vapor phase routes.

Acrylonitrile

Once the only raw material for acrylonitrile production, acetylene has given way completely to ammoxidation of propylene. The last active plant using acetylene in the US shut down at the end of 1970 (Table 9.12).

TABLE 9.9

Vinyl acetate capacity[1]

(*MM lb*)

Company	*1966*	*1967*	*1969*	*1970*	*1972*	*Process*
Air Products (Calvert City, Tx.)	90	95	95	95	–	Acetylene
Borden Chemical (Geismar, La.)	75	90	115	115	115	Acetylene
(Geismar, La.)	–	–	–	75	75	Acetylene
Celanese Chemical (Bay City, Tx.)	100	100	100	100	–	Ethylene
(Pampa, Tx.)	65	65	65	–	–	Acetaldehyde-acetic anhydride
(Clear Lake, Tx.)	–	–	–	200	200	Ethylene
DuPont Co. (Niagara Falls, N.Y.)	75	75	75	75	–	Acetylene
(La Porte, Tx.)	–	–	–	–	400	Ethylene
Monsanto Co. (Texas City, Tx.)	55	65	65	80	–	Acetylene
National Starch (Long Mott, Tx.)	50	50	50	60	60	Acetylene
Union Carbide (S. Charleston, W.Va.)	50	55	55	55	–	Acetylene
(Texas City, Tx.)	145	145	195	300	300	Acetylene
US Industrial Chem. (LaPorte, Tx.)	–	–	–	300	300	Ethylene
Total	705	740	815	1,455	1,450	
% acetylene	77	78	80	59	38	

[1]Oil, Paint & Drug Reporter Profile, Jan. 1, 1970 and other trade publications.

TABLE 9.10
Vinyl acetate[1]

Year	*Production (MM lb)*	*Sales (MM lb)*	*Unit value, (¢/lb)*
1962	318	189	14.0
1963	405	172	13.2
1964	440	217	11.4
1965	512	259	10.7
1966	606	254	10.7
1967	603	231	11.1
1968	718	248	10.7
1969	729	293	10.0
1970	799	na	10.0[2]
1971	835	na	8.8[2]

[1] US Tariff Commission; 1971 based on 7 months' reports.
[2] Oil, Paint & Drug Reporter posted price (tanks, delivered).

TABLE 9.11
Comparative vinyl acetate economics
(*300-MM-lb plants; 1972 construction period*)

Estimated total investment cost:
Acetylene process = $12.6 MM
Ethylene process (gas phase) = $17.3 MM

Estimated production costs

	¢/lb vinyl acetate	
	Acetylene route	*Ethylene route*
Raw materials		
Acetic acid (6.0¢/lb)	4.31	4.23
Ethylene (3.0¢/lb)	—	1.02
Acetylene (8.0¢/lb)	2.56	—
Catalyst & chemicals	0.32	0.29
Total materials	7.19	5.54
Labor	0.17	0.19
Utilities	0.29	0.70
Maint. (6% ISBL + 3% OSBL)	0.20	0.27
Overhead (45% of maint. + labor)	0.17	0.21
Taxes & ins. (1.5% of investment)	0.07	0.09
Depreciation (10 years)	0.42	0.57
Total	8.51	7.57

TABLE 9.12
Acrylonitrile capacity[1]
(*MM lb*)

Company	*1966*	*1968*	*1970*	*1972*	
American Cyanamid (Fortier, La.)	110	–	–	–	Acetylene–HCN
(Fortier, La.)	125	130	130	175	Propylene–NH_3
DuPont (Memphis, Tenn.)	150	180	180	180	Propylene–NH_3
(Beaumont, Tx.)	125	–	200[2]	200	Propylene–NO
B. F. Goodrich (Calvert City, Ky.)	45	45	45	45	Propylene–NH_3
Monsanto (Alvin, Texas)	280	320	370	370	Propylene–NH_3
(Texas City, Tx.)	130	130	130	–	Acetylene–HCN
Vistron (Lima, Ohio)	200	200	200	240	Propylene–NH_3
Total	1,165	1,005	1,255	1,210	

[1] Oil, Paint & Drug Reporter, Aug. 9, 1971.
[2] Propylene ammoxidation process.

The acetylene route which accounted for 72% of domestic capacity in 1960 is:

$$H{-}C{\equiv}C{-}H \;+\; HCN \xrightarrow[70^\circ C.]{aq\ Cu_2Cl_2} CH_2{=}CH{-}CN$$

The reaction takes place at atmospheric pressure, and acrylonitrile is recovered from the aqueous phase by steam distillation. Yields are about 80% based on acetylene.

The propylene ammoxidation route introduced by Sohio originally used a bismuth phosphomolybdate catalyst on silica. This was later supplanted by a more efficient antimony-uranium oxide system. The process proceeds in the vapor phase to give 60–70% selectivity on 90% propylene conversion:

$$CH_2{=}CH{-}CH_3 \;+\; NH_3 \xrightarrow[\substack{Catalyst\\425^\circ C.}]{Air} CH_2{=}CH{-}CN \;+\; HCN \;+\; CH_3CN \;+\; H_2O \;+\; CO_2$$

The advantage of the propylene ammoxidation process lies in greatly reduced raw material costs as shown in Table 9.13. Moreover, rather than be needed as a raw material, HCN is produced as a valuable by-product in the propylene route.

TABLE 9.13
Comparative acrylonitrile raw material costs

	¢/lb	
	Propylene	*Acetylene*
Propylene (1.16 lb/lb at 3.0¢/lb)	3.48	–
Acetylene (0.6 lb/lb at 8¢/lb)	–	4.80
Ammonia (0.37 lb/lb at 2.0¢/lb)	0.74	–
Hydrogen cyanide (0.6 lb/lb at 10¢/lb)	(0.05)*	6.00
Total	4.17	10.80
Acetonitrile by-product credit		
(0.05 lb/lb at 24¢/lb)	1.25	–
Net raw-material cost	2.92	10.80

*HCN produced as by-product (0.05 lb/lb at 10¢/lb.)

Further discussions of vinyl chloride, vinyl acetate and acrylonitrile appear in Chapters 10 and 11 dealing with ethylene and propylene.

Chloroprene

Although chloroprene consumption of acetylene has risen to 24% of total demand in the last decade, its continuance at this level is momentary. As is the case with other end uses, acetylene technology for chloroprene manufacture is rapidly yielding to more economical butadiene-based processes.

There are only two domestic producers of chloroprene, DuPont and Petro-Tex (Table 9.14). Until Petro-Tex started up its Houston plant in 1970, chloroprene output was based exclusively on acetylene. DuPont brought on stream its own butadiene-based plant in 1970, shortly after Petro-Tex. The swing to butadiene technology will be rapid. DuPont's Montague, Mich. plant was shut down in 1972, while at the same time its Louisville plant was being converted to butadiene. Hence, by year-end 1972, chloroprene will have joined acrylonitrile in oblivion as an end use for acetylene.

Chloroprene is used captively for polychloroprene (neoprene) production. The processes are contiguous, with the monomer being transferred to the polymerization system. Polychloroprene is a synthetic rubber which has excellent resistance to oxidative, solvent, and chemical attack.

Little market data is available on polychloroprene production. Non-government sources have offered the data in Table 9.15.

TABLE 9.14
Neoprene capacity[1]
(MM lb)

	1970	*1972*	*Process*
DuPont (La Place, La.)	70	70	Butadiene
DuPont (Louisville, Ky.)	300	300[2]	Acetylene
DuPont (Montague, Mich.)	70	—	Acetylene
Petro-Tex (Houston, Tx.)	45	45	Butadiene
Total	485	415	

[1] Based on Oil, Paint & Drug Reporter, Aug. 24, 1970.
[2] Converting to butadiene.

TABLE 9.15
Polychloroprene production[1]

Year	*MM lb*
1967	330
1968	340
1969	350
1970	370
1979	460[2]

[1] Oil, Paint & Drug Reporter, Aug. 24, 1970.
[2] International Institute of Synthetic Rubber Producers.

The butadiene-based technology is a vapor-phase process which proceeds:

$$CH_2{=}CH{-}CH{=}CH_2 + Cl_2 \xrightarrow{300°C.} CH_2{=}CH{-}\overset{\overset{Cl}{|}}{C}H{-}CH_2Cl$$
$$+ \; ClCH_2{-}CH{=}CH{-}CH_2Cl$$

Quenching reduces byproduct formation, and the reaction products are fractionated to recover the 3,4-dichlorobutene-1. The 1,4-dichlorobutene-2 is isomerized to the desired isomer by passage over a Cu-Cu_2Cl_2 catalyst.

$$ClCH_2{-}CH{=}CH{-}CH_2Cl \rightarrow CH_2{=}CH{-}\overset{\overset{CL}{|}}{C}H{-}CH_2Cl$$

Finally, heating with aqueous NaOH produces the desired chloroprene:

$$CH_2{=}CH{-}\underset{}{\overset{\displaystyle Cl}{\overset{|}{C}}}H{-}CH_2Cl + NaOH \rightarrow CH_2{=}CH{-}\overset{\displaystyle Cl}{\overset{|}{C}}{=}CH_2 + NaCl$$

The monomer is passed to storage and thence to a polymerization system for polychloroprene production. The flow diagram for chloroprene production is shown in Fig. 9.4.

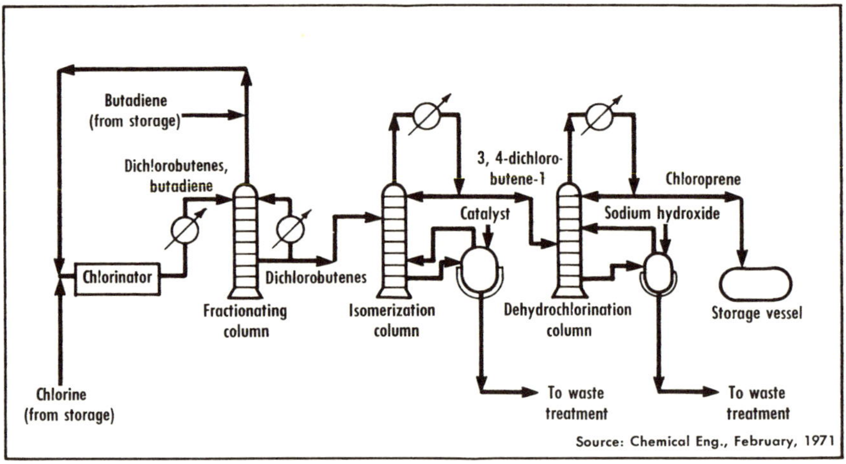

FIG. 9.4 Butadiene-based chloroprene process

The now-obsolete acetylene-based process proceeded:

$$2HC{\equiv}CH \xrightarrow[\substack{NH_4Cl\\80^\circ C.}]{Cu_2Cl_2} CH_2{=}CH{-}C{\equiv}CH$$

$$CH_2{=}CH{-}C{\equiv}CH + HCl \xrightarrow[\substack{NH_4Cl\\50^\circ C.}]{Cu_2Cl_2} CH_2{=}\overset{\displaystyle Cl}{\overset{|}{C}}{-}CH{=}CH_2$$

Trichloroethylene

This material is a chlorinated solvent which is primarily used for vapor-degreasing of fabricated metal parts. The market has grown steadily in the last decade, and may continue to grow at the same rate, since this material is not considered an offender for air pollution. Market and price data are summarized in Table 9.16.

TABLE 9.16
Trichloroethylene production[1]

Year	*MM lb*	*Unit value, ¢/lb*
1962	356	10.0
1963	325	9.0
1964	370	8.7
1965	435	8.4
1966	480	8.4
1967	489	8.2
1968	519	7.2
1969	597	7.0
1970	611	9.8[2]
1971	545	8.8[2]

[1] US Tariff Commission Reports; 1971 based on 7 months' data.
[2] Oil, Paint & Drug Reporter posted price (tanks, delivered).

Trichloroethylene and its homolog, perchloroethylene, are considered as one of the last bastions of acetylene technology. Trichloroethylene is produced almost entirely from tetrachloroethane which can be made from acetylene or ethylene.

$$HC{\equiv}CH \quad + \quad 2Cl_2 \rightarrow Cl_2CHCHCl_2 \leftarrow 4Cl_2 \quad + \quad CH_2{=}CH_2$$

Hydrogen chloride is then removed by passing the tetrachloroethane over a supported $BaCl_2$ catalyst at 300–500°C.

$$Cl_2CHCHCl_2 \xrightarrow{\Delta} ClCH{=}CCl_2 \quad + \quad HCl$$

The market is apparently shared evenly by ethylene and acetylene processes, as evidenced by Table 9.17.

Installed capacities are highly variable, since a variety of chlorinated hydrocarbons can be made in the same equipment, depending upon demand.

TABLE 9.17
Trichloroethylene capacity[1]
(MM lb)

Producer	*1968*	*1972*	*Process*
Detrex (Ashtabula, Ohio)	100	100	Acetylene
Diamond Shamrock (Deer Park, Tx.)	–	60	Ethylene
Dow (Freeport, La.)	120	200	Ethylene
(Plaquemine, La.)	35	35	Ethylene
DuPont (Niagara Falls, N.Y.)	275	275	Acetylene
Ethyl (Baton Rouge, La.)	40	40	Ethylene
Hooker (Niagara Falls, N.Y.)	60	60	Acetylene
(Tacoma, Wash.)	30	30	Acetylene
(Taft, La.)	40	40	Acetylene
PPG (Lake Charles, La.)	80	80	Ethylene
Total	780	920	

[1] Oil, Paint & Drug Reporter Profile (April 1, 1970).

Acrylics

This is a small volume market for acetylene with little prospects for growth for the latter. Nearly every producer seems to have his own process, and acetylene technology constitutes only 39% of total capacity (Table 9.18). Propylene-based technology appears to be predominating, however.

TABLE 9.18
US acrylate capacity[1]
(MM lb)

	Year end					
Company	*1965*	*1967*	*1968*	*1969*	*1972*	*Process*
Celanese (Pampa, Tx.)	35	45	45	45	45	β-propiolactone
Dow Badische (Freeport, Tx.)	15	40	40	40	40	Acetylene–CO
B. F. Goodrich (Calvert City, Ky.)	5	5	5	5	5	β-propiolactone
Minnesota Mining (St. Paul, Minn.)	5	5	5	5	5	Acrylonitrile
Rohm & Haas (Deer Park, Tx.)	80	120	120	200	250	Acetylene–CO
(Deer Park, Tx.)	–	–	–	–	150	Propylene
Union Carbide (Institute, W.Va.)	45	50	50	50	50	Ethylene oxide–HCN
(Taft, La.)	–	–	200	200	200	Propylene
Total	185	265	465	545	745	

[1] Based on Oil, Paint & Drug Reporter, Jan. 1, 1968.

Acrylic acid per se is not marketed in large volume, but is generally converted to one of its esters. The ester most commonly produced is ethyl acrylate as shown in Table 9.19.

TABLE 9.19
US acrylate production[1]
(MM lb)

Year	*Acrylic acid*	*Ethyl acrylate*	*2-Ethyl-hexylacrylate*	*Butyl acrylate*	*Methyl acrylate*	*Total*
1962	4	72	12	5	11	104
1963	5	80	15	6	13	119
1964	4	91	23	7	15	140
1965	6	117	25	8	17	173
1966	11	130	32	9	19	201
1967	14	141	32	34	21	242
1968	16	165	33	66	23	303
1969	13	200	33	65	25	336
1970E	15	210	35	63	25	348
1971E	16	220	35	65	27	363

[1] 1962–1969 data based on US Tariff Commission; 1970 and 1971 are estimates.

The market for acrylics has grown considerably primarily because polyacrylate resins made from them are employed. in water-base paints. These products possess good scrub resistance, do not yellow on aging, dry rapidly, and can be applied over a variety of surfaces. The market for acrylics is delineated in Table 9.20.

TABLE 9.20
Market pattern for acrylics[1]

Market	%
Paint	48
Textiles	15
Specialties	7
Fiber	6
Exports	10
Miscellaneous	14
	100

[1] Oil, Paint & Drug Reporter, Jan. 1, 1968.

Acrylates are readily produced from acetylene at mild temperatures and pressures by using carbon monoxide and nickel carbonyl as a catalyst:

$$HC{\equiv}CH + CO + ROH \xrightarrow[45°C.\ MEK]{Ni(CO)_4} \underset{85\%}{CH_2{=}CH{-}COOR}$$

Esterification and carbonylation can be a simultaneous process or successive. If water is used in place of alcohol, acrylic acid is produced.

The propylene route employs a vapor-phase reaction in which a mixture of propylene and air are passed over a transition metal catalyst, such as MoO_3:

$$CH_2{=}CH{-}CH_3 + O_2 \xrightarrow{300\text{–}350°C.} CH_2{=}CH{-}\overset{\overset{\displaystyle O}{\|}}{C}H \rightarrow CH_2{=}CH{-}COOH$$

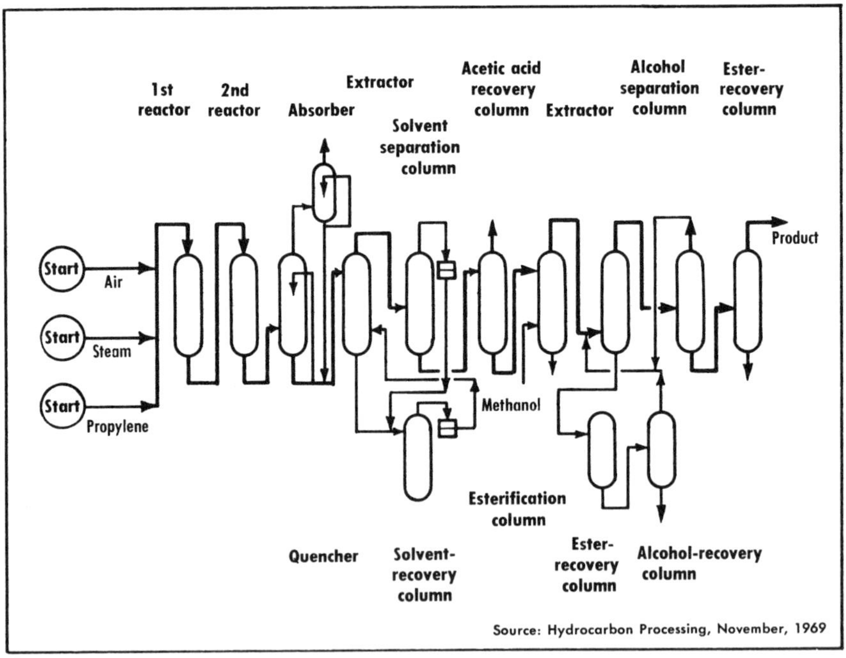

FIG. 9.5 Acrylates via propylene oxidation

Acrolein is an intermediate product, which necessitates a two stage oxidation system as shown in Fig. 9.5.

The acrylic acid is esterified with methanol or other alcohol at 80°C. in the presence of H_2SO_4. Acrylic acid yield based on propylene is about 67%.

An economic comparison of the acetylene and propylene processes for acrylic acid is given in Table 9.21. The propylene route has a much lower manufacturing cost, which is largely offset by a much higher investment requirement.

TABLE 9.21
Estimated acrylic acid economics
(150-MM lb plant; 1972 Construction)

Total investment cost

Process	*$ MM*
Acetylene	10.0
Propylene	16.9

Production cost

	¢/lb	
Route:	*Acetylene*	*Propylene*
Raw materials	6.85[1]	3.24[2]
Utilities	0.80	1.12
Labor	0.27	0.33
Maintenance (6% ISBL + 3% OSBL)	0.32	0.54
Overhead (45% maint. + labor)	0.27	0.39
Taxes & insurance (1.5% of invest.)	0.10	0.17
Depreciation	0.67	1.14
Total	9.28	6.93

[1] 0.42 lb/lb at 8.0¢/lb.
[2] 0.88 lb/lb at 3.0¢/lb.

10

ARTHUR M. BROWNSTEIN

Ethylene

ETHYLENE is a colorless, flammable gas which boils at −104°C. After ammonia, it ranks as the largest volume petrochemical with 1971 production at 18 billion lb.

Ethylene manufacture was extensively described in Chapter 5 and will not be repeated here. However, some discussion is warranted regarding changes in manufacturing which will occur during this decade. Through 1971, production was principally based on natural-gas liquids, i.e., ethane, propane and butane. These are commonly referred to as LP-gas (liquified petroleum gas), and are primarily obtained from natural gas. With an expected shortage of natural gas commencing about 1975, these natural-gas liquids will become costly and more difficult to obtain. As a consequence, it has been projected that heavier feedstocks (C_5^+), such as catalytic gas oil and naphtha will be employed in greater amounts for ethylene production as shown in Table 10.1.

With a shift to heavier feedstocks, there will be a substantial increase in byproduct output. The comparative product distribution between propane and naphtha is illustrated in Table 10.2.

The use of heavier liquid feedstocks will mean considerable quantities of higher olefins et al. The revenues from the sale of by-

TABLE 10.1
US ethylene production by type of feedstock[1]
(Percent)

	1971	*1975*	*1980*
Ethane, propane and butane	82	60	57
Naphtha and cat gas oil	18	40	43
	100	100	100

[1] N. J. Lewis (Pace Co.); 35th API Refining Meeting, May 13, 1970.

TABLE 10.2
Comparative analysis of propane and naphtha cracking[1]
(Wt %)

	Propane	*Naphtha*
Methane	24.3	17.8
Acetylene	0.4	0.2
Ethylene	29.7	28.3
Ethane	3.3	4.5
Propylene	16.8	13.4
Propane	17.5	0.4
C_4^+	5.6	34.5
Others[2]	2.4	0.9
	100.0	100.0

[1] M. R. Kitzen, Petrochem. Eng., Sept., 1970.
[2] H_2, N_2, O_2, CO and CO_2.

products will most likely be higher than the ethylene product revenue. Thus, ethylene and propylene prices will become more sensitive to byproduct values. This anticipated trend is shown in Table 10.3.

Another consequence of the use of heavier feedstocks, is that co-product production in the C_4^+ fraction will largely preempt synthetic production of some of these materials. For example, in 1970, over 50% of the world's butadiene production came from ethylene plants. In 1980, this is expected to increase to 90%, thus discouraging sizable introduction of synthetic butadiene.

Ethylene production has been undergoing another significant change, a continuing increase in plant size. As recently as 1966, most units were less than 400 MM lb in size and had an average capacity

TABLE 10.3
Ethylene revenues as a function of feedstock[1]

	% of total revenues		
Feedstock	*Ethylene*	*Propylene*	*Others*
Ethane	92	3	5
Propane	45	20	35
Naphtha	38	24	38
Gas oil	28	21	51

[1] B. Struth (Chem Systems) in the Oil & Gas Journal, August 2, 1971, p. 70.

of 345 MM lb. At that time, there was only one unit with 1 billion lb of capacity. In 1972, eight plants are at least one billion lb in size, and average capacity has increased 57% to 540 MM lb (Table 10.4).

TABLE 10.4
Ethylene units—size and number

Ethylene plant size (MM lb)	*1966*	*1969*	*1972*
to 199	8	6	6
200–299	3	2	3
300–399	3	2	2
400–499	4	7	6
500–999	8	13	15
1,000+	1	5	8
Total capacity (MM lb)	12,100	19,865	21,630
Average capacity (MM lb)	345	460	540

The rapid increase in plant size is not likely to continue. The 1.2–1.5 billion-lb plant may be the limit for a single train ethylene plant. This assumption is based on the combination of economics and reliability of equipment, especially compressors, furnaces and transfer-line exchangers.

Ethylene is not easily shipped over long distances, and the tendency has been to connect producer and consumer by pipelines. As a consequence, an elaborate pipeline system has evolved on the Gulf Coast, connecting ethylene producers and manufacturers of first derivatives, such as styrene and polyethylene. A simplified example of one such pipeline for the Houston area is shown in Fig. 10.1. Most ethylene plants are located on the Gulf Coast because of proximity to feedstocks (Table 10.5).

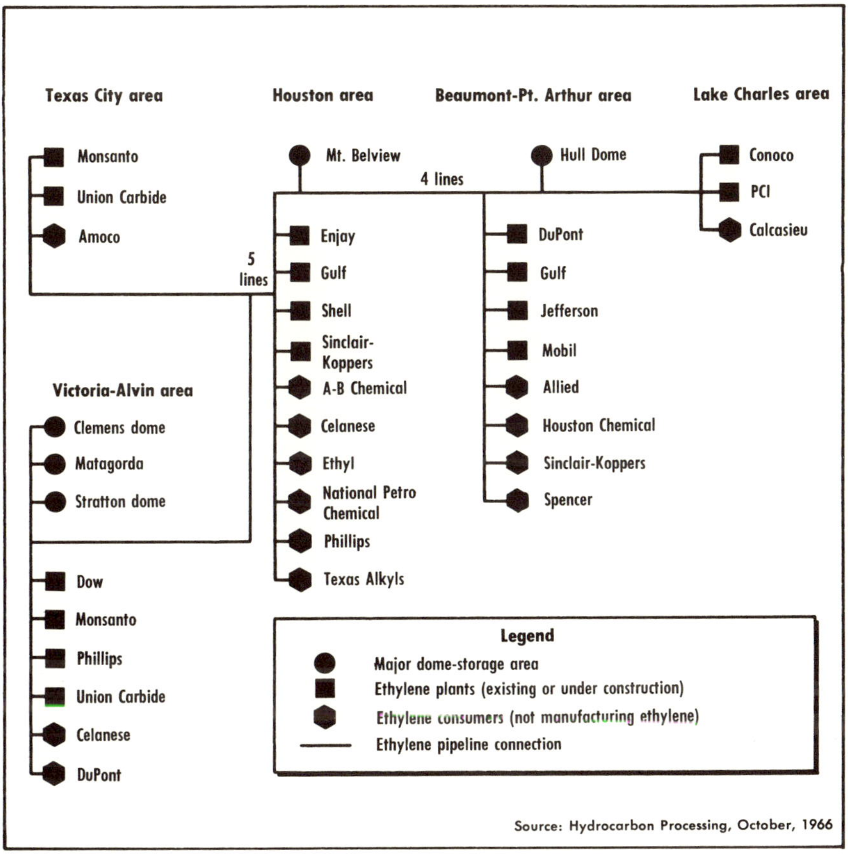

FIG. 10.1 Ethylene pipeline network for greater Houston area

TABLE 10.5
1972 ethylene capacity[1]

Company	*Feedstock*	*MM lb*
Allied Chemical (Geismar, La.)	ethane-propane	500
Arco (Wilmington, Calif.)	refinery gas	100
Chemplex (Clinton, Ia.)	ethane-propane	500
Cities Service (Lake Charles, La.)	propane	1,000
Conoco (Lake Charles, La.)	ethane-propane	500
Corco (Penuelas, P.R.)	naphtha	1,000
Dow (Bay City, Michigan)	naphtha	170
(Freeport, Tx.)	ethane-propane	1,400
(Plaquemine, La.)	ethane-propane	600

TABLE 10.5
1972 ethylene capacity[1]
(Cont'd)

DuPont (Orange, Tx.)	ethane-propane	750
Eastman (Longview, Tx.)	ethane-propane	450
El Paso (Odessa, Tx.)	ethane-propane	400
Enjay (Baton Rouge, La.)	ethane, gas oil & naphtha	1,000
(Baytown, Tx.)	refinery gas	85
(Bayway, N.J.)	refinery gas	175
Goodrich (Calvert City, Ky.)	propane	250
Gulf (Cedar Bayou, Tx.)	propane	400
(Port Arthur, Tx.)	refinery gas & propane	1,000
Jefferson Chemical (Port Neches, Tx.)	refinery gas, ethane & propane	500
Mobil (Beaumont, Tx.)	refinery gas & naphtha	500
Monsanto (Alvin, Tx.)	refinery gas	600
(Texas City, Tx.)	refinery gas	100
National Distillers (Tuscola, Ill.)	ethane-propane	350
Northern Petrochemicals (Joliet, Ill.)	ethane-propane	800
Olin (Brandenberg, Ky.)	ethane	90
Phillips (Sweeny, Tx.)	ethane-propane & refinery gas	600
Phillips-Houston (Sweeny, Tx.)	ethane-propane	500
Shell (Deer Park, Tx.)	propane & refinery gas	1,200
(Norco, La.)	ethane-propane & refinery gas	500
(Torrance, Calif.)	propane	70
Sinclair-Koppers (Houston, Tx.)	ethane-propane & refinery gas	500
Sun Olin (Claymont, Delaware)	refinery gas	220
Union Carbide (Institute, W.Va.)	ethane-propane	350
(Ponce, P.R.)	refinery gas & naphtha	1,000
(Seadrift, Tx.)	ethane-propane	900
(S. Charleston, W.Va.)	ethane-propane	400
(Taft, La.)	ethane-propane & naphtha	1,000
(Texas City, Tx.)	ethane-propane	750
(Torrance, Calif.)	refinery gas	150
(Whiting, Ind.)	refinery gas	270
		21,630

[1] Based on Informations Chimie, May, 1970, p. 157.

MARKET STRUCTURE

Ethylene production has tripled in the last decade from 5.9 to 17.8 billion lb. As production has grown with increasingly larger, more economical plants, the price for ethylene has declined to about 3.3¢/lb (delivered). These data are elaborated in Table 10.6, where it is also apparent that the ethylene market has become increasingly captive. Sales remained steadily at a mean 3.5 billion lb as production grew. In 1969, sales of ethylene only constituted 24% of production. It is now believed by many that the maximum economy of scale for ethylene has been achieved, and that as investment costs increase, so will ethylene prices. These have been forecast at 4–5¢/lb in the 1975–80 period.

TABLE 10.6
Ethylene production and sales[1]
(MM lb)

Year	*Production*	*Sales*	*Unit value, ¢/lb*
1961	5,856	3,361	5.0
1962	6,283	3,979	4.7
1963	7,518	2,208	4.5
1964	9,570	2,377	4.7
1965	8,641	2,714	4.0
1966	11,241	3,277	4.1
1967	11,851	3,353	4.0
1968	13,151	3,367	3.4
1969	16,436	3,877	3.3
1970	17,000	na	3.3[2]
1971	17,800	na	3.3[2]

[1]US Tariff Commission; 1971 based on 7 months' data; 1970 is estimated by Chemical Week.

[2]Oil, Paint & Drug Reporter posted price (contract; delivered).

An estimated distribution of ethylene use is illustrated in Table 10.7. Unlike acetylene, ethylene has not forfeited any of its markets to another raw material. On the contrary, it is possible to see where ethylene has been the beneficiary of acetylene's demise in vinyl chloride and vinyl acetate. Ethylene consumption for ethylene dichloride (the vinyl chloride intermediate) grew five-fold in the last decade, 70% faster than total ethylene. In the same vein, usage for vinyl

TABLE 10.7

Ethylene end use pattern[9]

(MM lb)

Year	Ethylene oxide[1]	Polyethylene Low density[2]	Polyethylene High density[3]	Styrene[4]	Ethylene dichloride[5]	Vinyl acetate[6]	Ethanol[7]	Acetal-dehyde[8]	Misc.[10]	Total[9]
1961	1,356	1,320	300	506	430	–	1,230	na	514	5,656
1962	1,518	1,592	445	560	560	–	1,230	na	378	6,283
1963	1,889	1,754	542	620	565	–	1,420	na	728	7,518
1964	2,163	1,955	691	745	695	–	1,510	160	1,651	9,570
1965	2,190	2,263	824	825	775	64	1,500	226	1,260	8,641
1966	2,327	2,648	955	900	1,140	70	1,380	280	1,541	11,241
1967	2,308	2,716	1,140	945	1,250	70	1,400	310	1,716	11,855
1968	2,625	3,306	1,330	1,100	1,510	70	1,560	380	1,270	13,151
1969	3,408	3,831	1,690	1,320	1,930	70	1,730	385	2,072	16,436
1970	3,671	4,168	1,790	1,254	2,060	130	1,660	545	1,722	17,000[11]
1971	3,500	4,500	1,900	1,166	2,100	177	1,700	595	1,862	17,800

[1] 1.0 lb/lb.
[2] 1.0 lb/lb.
[3] 1.0 lb/lb.
[4] 0.32 lb/lb; 90% of capacity based on synthetic ethylbenzene.
[5] 0.32 lb/lb.
[6] 0.70 lb/lb (liq. phase); 0.34 lb/lb (vapor phase).
[7] 0.77 lb/lb.
[8] 0.68 lb/lb.
[9] Based on US Tariff Commission Reports; 1971 based on 7 months' data.
[10] Includes ethylene dibromide, ethylene propylene elastomers, alpha-olefins and ethyl chloride.
[11] Chemical Week, May 6, 1970.

acetate grew from nothing in 1964 to nearly 200 MM lbs in 1971. A percentage distribution of ethylene use over the last decade is shown in Table 10.8.

TABLE 10.8
Percent distribution of ethylene consumption
(%)

	Ethylene oxide	*Polyethylene*		*Styrene*	*Ethylene dichloride*	*Vinyl acetate*	*Ethanol*	*Other*
		LD	*HD*					
1961	7.9	23.2	5.3	9.0	7.6	0	21.7	25.3
1971	19.7	25.4	10.7	6.5	11.8	1.0	9.6	15.3

POLYETHYLENE

This resin has long constituted the principal use for ethylene and has been a major factor in its growth. Polyethylene accounted for more than 36% of ethylene consumption in 1971. It is produced as two varieties, low and high-density polyethylene.

Low-density polyethylene is a partially crystalline (50–60%) solid melting at 115°C., and has a density of 0.91–0.94 gm/cc. It is the older of the two polyethylenes, and is produced by high pressure polymerization of ethylene. In contrast, high density polyethylene (0.95–0.97 gm/cc) is more than 90% crystalline and is produced via low-pressure polymerization of ethylene. In addition, high-density polyethylene is linear, while low-density polyethylene is branched in structure. Hence, low-density polyethylene is often referred to as high-pressure or branched polyethylene, while high-density polyethylene is frequently described as low-pressure or linear polyethylene. These differences in structure affect the mechanical properties of these resins and therefore their market applications.

Low-density polyethylene was first produced by ICI before World War II in a fortuitous experiment in which an effort was being made to combine ethylene and benzene under high pressure (1400 atm). Traces of oxygen caused polymerization to occur. The current commercial processes employ high-purity ethylene (99+%) at 1,000–2,500 atmospheres and about 200°C. In place of oxygen, such peroxides as t-butyl peroxide are used as initiators. Impurities, such as acetylene, act as chain-transfer agents, and must be removed if high-molecular-weight polymers are to be obtained.

Using, t-butyl peroxide as the initiator, the polymerization proceeds:

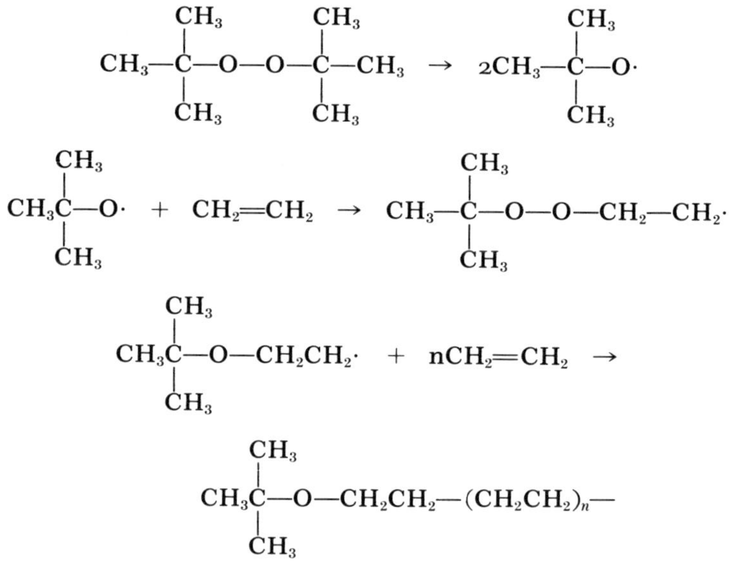

The process scheme is illustrated by Fig. 10.2.

Short-chain branching (not shown in these equations) has a predominant effect on the degree of crystallinity and therefore on the density of high pressure polyethylene. Properties dependent on crystallinity, such as stiffness, tear strength, hardness, and chemical resistance tend to increase with increasing density or decreasing branching. Thus, these properties are greater in polyethylene produced under low rather than high-pressure conditions.

As shown in Table 10.9, about one-half the low-density polyethylene is used in film and sheeting. Few competitive materials have low-density polyethylene's combination of flexibility, resilience and chemical resistance. As illustrated in Table 10.10, packaging accounts for nearly all the film and sheet market.

Injection molding, coatings and wire, and cable comprise secondary applications for low-density polyethylene. Production of housewares and toys accounts for a substantial degree of injection molded products. The non-polar nature of low-density polyethylene has made it ideal for use as wire and cable insulation. In this respect, polyethylene has filled a long-standing need for a material which could effectively insulate cables without introducing electrical losses at high frequencies.

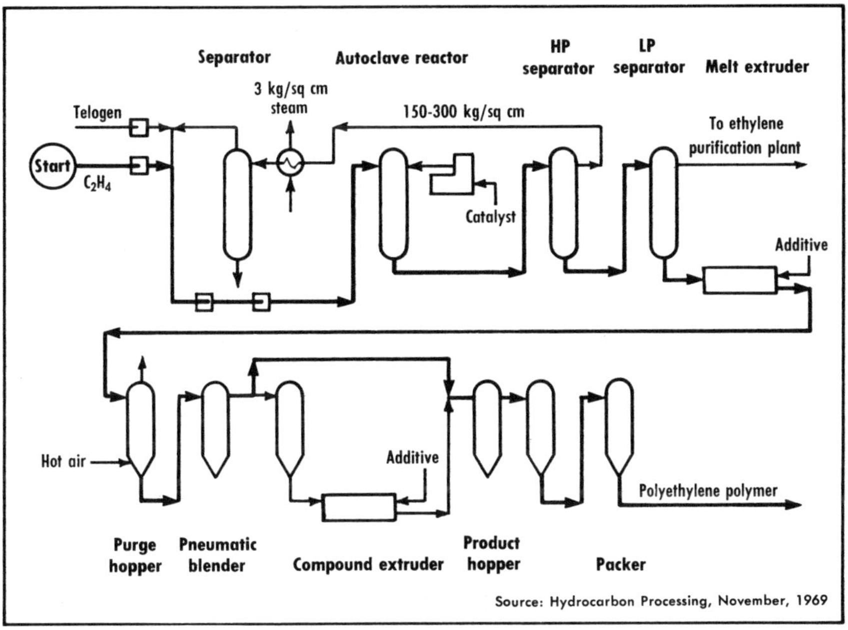

FIG. 10.2 Low-density polyethylene flow scheme

High-density polyethylene can be produced in several ways, one of the more common being polymerization of ethylene with supported metal oxide catalysts. In this technique, high-purity ethylene is contacted with Cr_2O_3 on silica-alumina at 100–500 psig and 100°–150°C. Cyclohexane is a commonly used solvent. Polymer recovery is about 98% on 100% ethylene conversion. The polymer is essentially linear and highly crystalline.

$$n\ CH_2{=}CH_2 \rightarrow -(CH_2CH_2)_n-$$

The melting point of low-pressure polyethylene is 135°C. vs. about 115°C. for high-pressure material. This linear material is decidedly stiffer than the branched resin. The process scheme for low density polyethylene is set forth in Fig. 10.3.

High-density polyethylene is a more recent industrial entry than its low-density counterpart. Hence, it is produced in smaller, but more rapidly growing volume than the low-density material. As is apparent from Table 10.11, high-density polyethylene has grown six-fold in the last decade to 1,810 MM lb. In this same period, the low-density resin grew a respectable three-and-a-half fold to an enormous 4.5 billion lb.

TABLE 10.9
Low-density polyethylene end use pattern[1]

Year	*Injection molding*	*Blow molding*	*Film & sheet*	*Extrusion coating on paper*	*Wire & cable*	*Exports*	*Misc.*	*Total*[2]
1963	187	35	617	209	172	297	205	1,752
1964	257	39	745	255	194	224	239	2,100
1965	291	40	938	283	216	198	210	2,282
1966	326	42	1,097	314	276	203	216	2,527
1967	353	45	1,094	307	265	257	251	2,573
1968	412	52	1,419	348	285	282	306	3,195
1969	496	50	1,631	398	358	447	316	3,696
1970	491	42	2,049	381	404	360	320	4,047
1971[3]	476	86	1,800	405	405	273	468	3,913

[1]US Tariff Commission.
[2]Represents consumption rather than production.
[3]Based on 7 months' data.

TABLE 10.10
Low-density polyethylene film markets (1970)[1]

Market	*MM lb*
Food packaging	
Baked goods	178
Candy	24
Frozen food	37
Meat and poultry	35
Produce	230
Misc.	198
Non-food packaging	
Garment bags	92
Industrial liners, shipping bags	204
Rack and counter	81
Soft goods	96
Misc.	266
Non-packaging	
Agriculture	70
Construction	92
Household wrap and bag	200
Rubber industry	32
Misc.	214
Total	2,049

[1]Based on Modern Plastics, Jan., 1971.

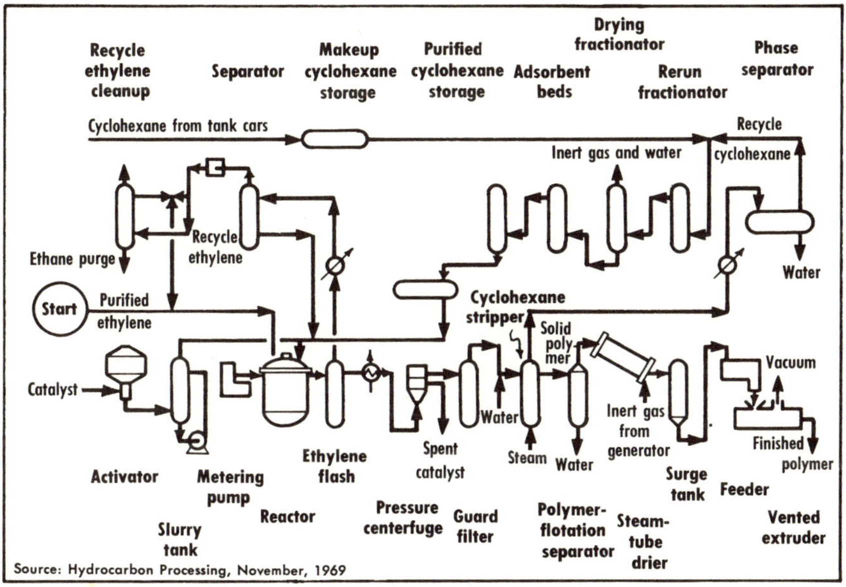

FIG. 10.3 High-density polyethylene flow diagram

TABLE 10.11
Polyethylene production[1]
(MM lb)

Year	*High density*	*Low density*	*Total*
1961	286	1,320	1,606
1962	424	1,592	2,016
1963	516	1,754	2,270
1964	658	1,955	2,613
1965	784	2,263	3,047
1966	910	2,648	3,558
1967	1,082	2,716	3,798
1968	1,261	3,306	4,567
1969	1,610	3,831	5,441
1970	1,704	4,168	5,872
1971	1,810	4,500	6,310

[1]US Tariff Commission; 1971 based on 7 months' data.

Injection and blow-molding applications constitute by far the major uses for high-density polyethylene (Table 10.12). The higher stiffness and heat resistance of the linear material have led to its replacement of branched polyethylene in applications where these properties are important.

TABLE 10.12
High-density polyethylene end use pattern[1]
(MM lb)

Year	*Injection molding*	*Blow molding*	*Film & sheet*	*Wire & cable*	*Pipe & conduit*	*Exports*	*Misc.*	*Total*[2]
1963	72	196	20	13	17	44	93	454
1964	100	239	25	20	21	64	137	660
1965	130	213	38	23	30	88	159	740
1966	179	348	44	30	38	82	168	930
1967	217	403	38	35	38	88	157	976
1968	236	489	49	39	44	153	213	1,224
1969	289	539	55	33	46	165	283	1,410
1970	326	592	63	31	60	229	345	1,646
1971[3]	354	615	74	33	85	256	419	1,836

[1]US Tariff Commission.
[2]Represents consumption rather than production.
[3]Based on 7 months' data.

TABLE 10.13
1972 High-density polyethylene capacity[1]

Company		*MM lb*
Allied	(Orange, Tx.)	225
Amoco	(Chocolate Bayou, Tx.)	100
Celanese	(Pasadena, Tx.)	255
Chemplex	(Clinton, Ia.)	125
Dow	(Freeport, Tx.)	100
	(Plaquemine, La.)	100
DuPont	(Orange, Tx.)	150
Gulf	(Orange, Tx.)	100
Hercules	(Parlin, N.J.)	80
	(Lake Charles, La.)	90
Koppers-Sinclair	(Port Arthur, Tx.)	200
Monsanto	(Texas City, Tx.)	180
National Petrochemicals	(Houston, Tx.)	190
Phillips	(Pasadena, Tx.)	240
	(Guayama, P.R.)	90
Union Carbide	(Seadrift, Tx.)	250
	Total	2,475

[1]Oil, Paint & Drug Reporter, Jan. 18, 1971.

The production of bottles for detergents and bleach are common applications for blow molding. The ability to resist environmental stress cracking was a large factor in the growth of high-density polyethylene for these uses.

Producers of high and low density polyethylene are listed in Tables 10.13 and 10.14. These facilities are based near ethylene plants for easy access to raw materials. Consequently, most polyethylene plants are located on the Gulf Coast although their markets are in the north.

TABLE 10.14
1972 Low-density polyethylene capacity[1]

		MM lb
Amoco	(Alvin, Tx.)	28
Chemplex	(Clinton, Ia.)	180
Columbian Carbon	(Lake Charles, La.)	220
Dow	(Freeport, Tx.)	300
	(Plaquemine, La.)	200
DuPont	(Orange, Tx.)	520
	(Victoria, Tx.)	150
Eastman	(Longview, Tx.)	250
El Paso	(Odessa, Tx.)	300
Enjay	(Baton Rouge, La.)	200
Gulf	(Cedar Bayou, Tx.)	200
	(Orange, Tx.)	200
Monsanto	(Texas City, Tx.)	130
National Distillers	(Deer Park, Tx.)	300
	(Tuscola, Ill.)	150
Northern Petrochemical	(Texas City, Tx.)	135
	(Clinton, Ia.)	580
Phillips	(Pasadena, Tx.)	94
Rexene Polymers	(El Paso, Tx.)	310
Sinclair-Koppers	(Port Arthur, Tx.)	175
Union Carbide	(Seadrift, Tx.)	340
	(S. Charleston, W.Va.)	120
	(Texas City, Tx.)	225
	(Torrance, Calif.)	120
	(Whiting, Ind.)	240
	(Taft, La.)	500
		5,657

[1]Oil, Paint & Drug Reporter, Jan. 11, 1971.

An economic comparison of low and high density polyethylene appears in Table 10.15. Production costs for low-density material are about 30% lower than high density. Although this is largely traceable to much larger units for low density polyethylene, it nevertheless appears that raw material and utilities cost may be somewhat higher for high density polyethylene. These factors are, of course, not capital dependent and would remain higher regardless of plant size.

TABLE 10.15
Estimated economics for polyethylene (1972 construction)

Total capital investment

Resin	*$ MM*
Low density[1]	44.0
High density[2]	27.7

[1] 300-MM-lb plant.
[2] 150-MM-lb plant; Phillips Process.

Production costs

	¢/lb	
	Low density[3]	*High density*[4]
Raw materials	3.57[5]	4.19[6]
Utilities	0.81	2.13
Labor	0.19	0.35
Maintenance (6% ISBL + 3% OSBL)	0.70	0.88
Overhead (45% maint. + labor)	0.40	0.55
Taxes & insurance (1.5% of invest.)	0.22	0.27
Depreciation (10%/yr)	1.46	1.85
Total	7.35	10.22

[3] 300-MM-lb plant.
[4] 150-MM-lb plant.
[5] Includes 1.0 lb/lb ethylene at 3.3¢/lb.
[6] Includes 1.0 lb/lb ethylene at 3.3¢/lb.

ETHYLENE OXIDE

Other than low-density polyethylene, this constitutes the principal use for ethylene (Table 10.7). Ethylene oxide was at one time exclusively produced by the chlorohydrin process. This is a two step reaction wherein ethylene, chlorine and water are passed into a packed reactor to form ethylene chlorohydrin. This occurs via in situ formation of hypochlorous acid from chlorine hydrolysis:

$$Cl_2 + H_2O \rightarrow HOCl + HCl$$

$$HOCl + CH_2{=}CH_2 \xrightarrow{50°C.} HO{-}CH_2CH_2{-}Cl$$

In a second step, alkali such as hydrated lime, converts the chlorohydrin to ethylene oxide:

$$2HOCH_2CH_2Cl + Ca(OH)_2 \xrightarrow{100°C.} 2\overset{\;\;O}{CH_2{-}CH_2} + CaCl_2 + 2H_2O$$

Overall yield based on ethylene is 80 mole %. Ethylene dichloride is produced as a byproduct (0.1 lb/lb ethylene oxide) as is β,β'-dichloroethyl ether (0.09 lb/lb).

TABLE 10.16
Ethylene oxide capacity[1]
(MM lb)

Company		*1970*	*1972*
Calcasieu Chemical	(Lake Charles, La.)	150	150
Celanese	(Clear Lake, Tx.)	300	300
Dow	(Freeport, Tx.)	425	425
	(Placequemine, La.)	150	150
Eastman	(Longview, Tx.)	60	60
GAF	(Linden, N.J.)	65	–
Houston Chemical	(Beaumont, Tx.)	80	80
Jefferson Chemical	(Port Neches, Tx.)	500	500
Matador Chemical	(Orange, Tx.)	45	45
Northern Natural Gas	(Joliet, Illinois)	240	240
Olin Mathieson	(Brandenberg, Ky.)	100	100
Shell	(Geismar, La.)	125	125
Sun Olin	(Claymont, Del.)	80	80
Union Carbide	(Institute, W.Va.)	220	220
	(Seadrift, Tx.)	330	430
	(S. Charleston, W.Va.)	60	60
	(Texas City, Tx.)	700	500[2]
	(Torrance, Calif.)	50	50
	(Whiting, Ind.)	150	150
	(Ponce, P.R.)	100	100
	(Taft, La.)	350	450
	Total	4,280	4,215

[1]Based on Oil, Paint & Drug Reporter, Oct. 1, 1969.
[2]One unit shut down at this site.

Introduction of direct, vapor phase oxidation of ethylene has almost completely eliminated the use of chlorohydrin technology. Of nearly 4.2 billion lb. of installed capacity (Table 10.16), not more than 10% is still believed to employ chlorohydrin processes.

The vapor-phase route comprises passing ethylene and oxygen or air over a silver oxide catalyst supported on alumina:

$$CH_2{=}CH_2 \;+\; \tfrac{1}{2}O_2 \;\rightarrow\; \overset{O}{\overbrace{CH_2{-}CH_2}} \;+\; CO_2 \;+\; H_2O$$

Reaction conditions are usually 270°C. and one atmosphere pressure. Yields are about 60 mole %, with substantial CO_2 and H_2O produced as byproducts. As shown in Fig. 10.4, the reaction gases are passed to a scrubber where the ethylene oxide is absorbed in a dilute aqueous solution.

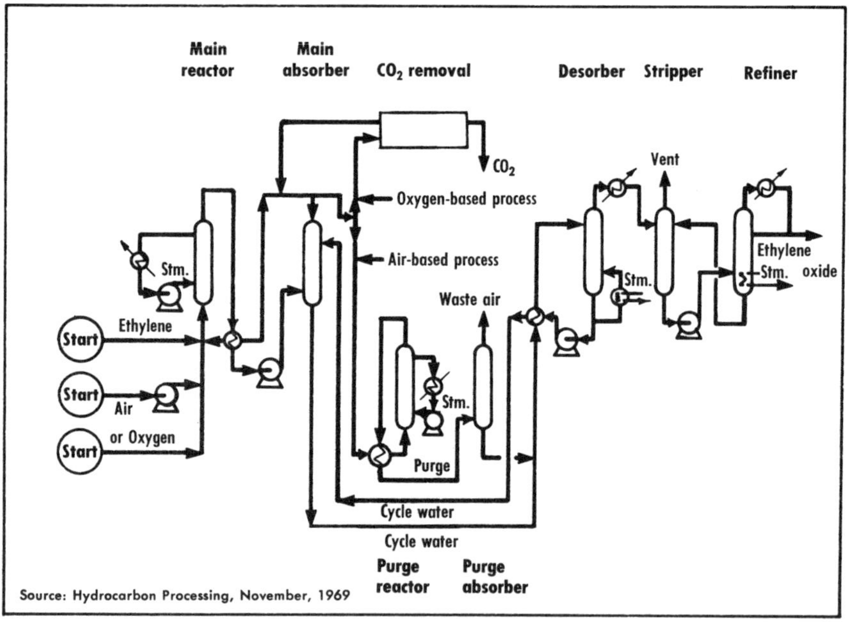

FIG. 10.4 Scientific Design ethylene oxide process

The economic superiority of the catalytic air oxidation technology is evident from Table 10.17. Excluding depreciation, the production cost for this technology is less than 50% that of the chlorohydrin

TABLE 10.17
Estimated ethylene oxide economics
(300-MM-lb plant; 1972 construction)

Total fixed investment cost

Process	*$ MM*
Chlorohydrin	15.20
Catalytic air oxidation	38.60

Production cost

	¢/lb	
	Chlorohydrin	*Air oxidation*
Raw materials	9.61[1]	3.30[2]
Utilities	0.78	0.28
Labor	0.20	0.14
Maintenance (6% ISBL + 3% OSBL)	0.24	0.64
Overhead (45% maint. + labor)	0.20	0.35
Taxes & insurance (1.5% of invest.)	0.08	0.20
Depreciation (10 years)	0.50	1.32
Total	11.61	6.23

[1] Ethylene at 0.75 lb/lb and 3.3¢/lb; and chlorine at 1.8 lb/lb and 3.25¢/lb.
[2] Ethylene at 1.0 lb/lb and 3.3¢/lb.

route. It has been some time since a new chlorohydrin-based ethylene oxide plant has been built in the US. Many existing plants have been converted for propylene oxide production. There is no commercial route for direct vapor-phase oxidation of propylene.

Ethylene oxide production has grown nearly three-fold in the last decade (Table 10.18). Accompanying this growth has been a continuing increase in plant size which has led to a corresponding decline in sales price. A factor here, of course, is utilization of economical, direct oxidation technology. 1971 posted prices are considerably below the mill cost for a new chlorohydrin plant.

The data in Table 10.17 further indicate that most ethylene oxide production is captive; i.e., sales in 1969 only accounted for 13% of production. In a sense, this may be correct. However, ethylene glycol and polyethylene glycol manufacture account for two-thirds of ethylene oxide consumption (Table 10.19). Catalytic hydration to these glycols is readily achieved by the oxide producer:

$$\underset{CH_2-CH_2}{\overset{O}{\diagup\ \diagdown}} + H_2O \rightarrow HOCH_2CH_2OH +$$
$$HO-CH_2CH_2-(O-CH_2CH_2)_n-OCH_2CH_2OH$$

TABLE 10.18

Ethylene oxide production and sales[1]

(MM lb)

Year	*Production*	*Sales*	*Unit price ¢/lb*
1961	1,356	128	13
1962	1,518	157	12
1963	1,889	170	11
1964	2,163	198	10
1965	2,190	256	10
1966	2,327	304	10
1967	2,308	302	9
1968	2,625	394	8
1969	3,408	438	8
1970	3,671	na	8[3]
1971[2]	3,500	na	7[3]

[1] US Tariff Commission.

[2] Based on 8 months' data.

[3] Oil, Paint & Drug Reporter posted price (FOB, tanks).

In turn, ethylene glycol is almost entirely a merchant product as can be noted from Table 10.20. Hence, ethylene oxide production is

TABLE 10.19

Ethylene oxide end use pattern (1971)[1]

Use	*MM lb*
Ethylene glycol	1,960
Surfactants	595
Ethanolamines	280
Polyethylene glycols	245
Miscellaneous	420
Total	3,500

[1] Based on Oil, Paint & Drug Reporter, October 1, 1969.

essentially merchant, but as the glycol and its corresponding ethers. It can further be seen from Table 10.20 that the merchant price for ethylene glycol has been the same or less than that of ethylene oxide. This implies that the internal transfer price for ethylene oxide is substantially lower than the market price.

TABLE 10.20
Ethylene glycol production and sales
(MM lb)

Year	*Production*	*Sales*	*Unit price ¢/lb*
1961	1,183	818	11
1962	1,434	964	9
1963	1,660	1,068	9
1964	1,815	1,203	9
1965	1,798	1,198	9
1966	2,081	1,250	8
1967	1,989	1,305	8
1968	2,043	1,627	7
1969	2,571	1,936	6
1970E	2,640	na	8[2]
1971E	2,760	na	8[2]

[1]US Tariff Commission.
[2]Oil, Paint & Drug Reporter posted price in tanks (delivered).

Use as antifreeze is the principal use for ethylene glycol. This is summarized in Table 10.21. Two significant, but far smaller applications, are manufacture of polyester fiber and film and latex paints.

TABLE 10.21
Estimated 1971 ethylene glycol end-use pattern[1]

Use	*MM lb*
Anti-freeze	1,740
Polyester fibers & film	470
Latex paints & Unsaturated polyesters	470
Other[2]	80
Total	2,760

[1]Based on Oil, Paint & Drug Reporter, Oct. 1, 1969.
[2]Includes exports.

ETHYLENE DICHLORIDE

Ethylene dichloride is largely employed for vinyl chloride manufacture. Adoption of oxychlorination technology (Chapter 9) has aided the five-fold growth of ethylene for this use in the last decade. In contrast to ranking fifth among ethylene uses in 1961, ethylene dichloride is now third. Other than vinyl chloride manufacture, production of chlorinated solvents, such as trichloro- and perchloroethylene, are the major applications for ethylene dichloride (Table 10.22). Use as a lead scavenger in motor gasoline requires about 200 MM lb of ethylene dichloride. This demand will largely vanish by 1975 as lead-free gasoline comes into general use.

TABLE 10.22
Estimated 1971 ethylene dichloride end use pattern[1]

End use	*MM lb*
Vinyl chloride	4,910
Chlorinated-solvent intermediate	590
Lead scavenger	198
Amines	262
Misc.	590
Total	6,550

[1] Based on Oil, Paint & Drug Reporter, Sept. 20, 1971.

TABLE 10.23
Ethylene dichloride production[1]
(MM lb)

Year	*Production*	*Sales*	*Unit value ¢/lb*
1961	1,368	443	4.75
1962	1,774	295	4.65
1963	1,793	326	4.55
1964	2,199	446	4.65
1965	2,456	309	4.50
1966	3,617	291	4.23
1967	3,971	324	3.89
1968	4,799	492	3.46
1969	6,037	1,227	2.96
1970	6,450	na	9.0[2]
1971	6,550	na	9.0[2]

[1] US Tariff Commission.
[2] Oil, Paint & Drug Reporter posted price (tanks, delivered).

The captive consumption of ethylene dichloride for vinyl chloride is evident from the data in Table 10.23. Of more than 6 billion lb of production, only 1.2 billion lb, or 20%, finds its way into merchant sales. Moreover, the posted price of 9¢/lb is not at all in keeping with the unit value, especially for vinyl chloride manufacture.

Comparison of the manufacturers of ethylene dichloride in Table 10.24 and of vinyl chloride in Table 9.7 of Chapter 9 further illustrate the captive use of this chemical for vinyl chloride.

TABLE 10.24
Ethylene dichloride capacity (1972)[1]

		MM lb
Allied	(Baton Rouge, La.)	500
American Chemical	(Long Beach, Calif.)	325
Continental Oil	(Lake Charles, La.)	1,000
Diamond Shamrock	(Deer Park, Tx.)	110
Dow	(Freeport, Tx.)	2,100
	(Plaquemine, La.)	1,100
Ethyl Corp.	(Baton Rouge, La.)	500
	(Houston, Tx.)	250
B. F. Goodrich	(Calvert City, Ky.)	800
PPG	(Lake Charles, La.)	500
	(Guayanilla, P.R.)	835
Shell	(Deer Park, Tx.)	1,700
Union Carbide	(S. Charleston, W.Va.)	150
	(Texas City, Tx.)	150
Vulcan	(Geismar, La.)	240
	Total	10,260

[1] Based on Oil, Paint & Drug Reporter, Sept. 20, 1971.

Vinyl chloride itself is exclusively employed for the manufacture of polyvinylchloride (PVC) homo- and copolymers. This market has become more merchant in character in the last decade, moving from sales at 39% of production in 1961 to 63% in 1969 (Table 10.25). Whereas vinyl chloride is produced close by ethylene plants, PVC itself is produced mostly near its markets in northeastern United States (Table 10.26). This is a reflection of the easier transportation of vinyl chloride versus ethylene.

Because of the low boiling point of vinyl chloride (−14°C.) and the limited stability of PVC, polymerization is carried out under conditions giving rapid reaction at low temperature. Four basic processes are used for commercial PVC production. These are bulk, emulsion,

TABLE 10.25
Vinyl chloride production[1]
(MM lb)

Year	*Production*	*Sales*	*Unit value, ¢/lb*
1962	1,311	516	7.5
1963	1,435	501	7.0
1964	1,615	597	6.4
1965	2,000	688	6.1
1966	2,500	836	5.9
1967	2,424	952	5.3
1968	2,969	1,463	4.6
1969	3,736	2,359	4.4
1970	4,040	na	4.8[2]
1971	4,130	na	4.8[2]

[1] US Tariff Commission Reports; 1971 based on 7 months' data.
[2] Oil, Paint & Drug Reporter posted price in tanks (works, freight equalized).

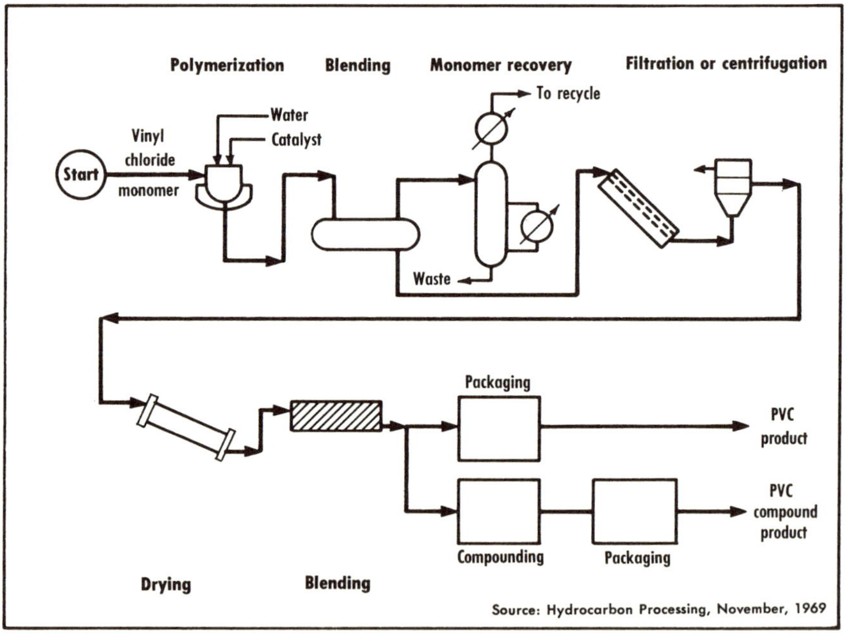

FIG. 10.5 PVC process flow diagram

TABLE 10.26
PVC capacity (1972)[1]

Company		MM lb
Air Products	(Cumberland, Va.)	120
Allied Chemical	(Painesville, Ohio)	200
American Chemical	(Long Beach, Calif.)	200
Borden	(Illiopolis, Ill.) (Leominster, Mass.)	260
Conoco Plastics	(Aberdeen, Miss.)	150
	(Oklahoma City, Okla.)	75
Diamond Shamrock	(Deer Park, Tx.)	125
	(Delaware City, Delaware)	125
Escambia	(Pensacola, Fla.)	50
Ethyl Corp.	(Baton Rouge, La.)	125
Firestone	(Pottstown, Pa.)	150
	(Perryville, Md.)	100
General Tire	(Ashtabula, Ohio)	100
B. F. Goodrich	(Avon Lake, Ohio) (Henry, Illinois) (Long Beach, Calif.) (Louisville, Ky.) (Niagara Falls, N.Y.)	600
Goodyear	(Niagara Falls, N.Y.)	180
Great American Chem.	(Fitchburg, Mass.)	40
Hooker	(Hicksville, N.Y.)	80
	(Burlington, N.J.)	175
Keysor	(Saugus, Calif.)	60
Monsanto	(Springfield, Mass.)	150
Pantasote	(Passaic, N.J.)	75
	(Pt. Pleasant, W.Va.)	75
Stauffer	(Delaware City, Del.)	175
Tenneco	(Burlington, N.J.) (East Brunswick, N.J.) (Flemington, N.J.)	220
Thompson Apex	(Aberdeen, Miss.)	150
	(Assonet, Mass.)	200
Union Carbide	(S. Charleston, W.Va.) (Texas City, Tx.)	350
Uniroyal	(Fairport, Ohio) (Painesville, Ohio)	125
	Total	4,435

[1] Based on Modern Plastics, March 2, 1971.

solution, and suspension polymerization. Of these, suspension polymerization is the most common. In a typical process, vinyl chloride is contacted with a free-radical initiator, such as diisopropyl peroxydicarbonate, in an aqueous medium at 50°C.:

$$n\ CH_2{=}CH{-}Cl \rightarrow {-}(CH_2{-}\underset{\displaystyle Cl}{\underset{|}{CH}})_n{-}$$

The yield of PVC exceeds 96 wt % of the monomer charged. A diagram depicting the process scheme is shown in Fig. 10.5.

The principal end uses for PVC resins are summarized in Table 10.27. The largest use, calendering, accounts for 16% of the total. This is used for film and sheeting. Familiar applications are rain wear, shower curtains, and handbags. Such other principal applications as flooring, wire and cable coating, and sound records are self-explanatory. The export market accounts for a sizable portion of vinyl chloride resin demand.

TABLE 10.27
Vinyl chloride homo- and copolymer end use pattern[1]
(MM lb)

Year	*Calendering*	*Flooring*	*Wire & cable*	*Sound records*	*Exports*	*Other*	*Total*[2]
1965	334	262	217	90	68	885	1,856
1966	439	359	226	114	69	919	2,126
1967	395	317	196	105	71	1,028	2,112
1968	406	301	289	122	114	1,318	2,550
1969	472	331	371	132	151	1,211	2,668
1970	458	322	409	141	188	1,467	2,985
1971[3]	493	360	286	114	174	1,687	3,114

[1] US Tariff Commission.
[2] Sales and use.
[3] Based on 7 months' data.

STYRENE

Ethylene is the other member of the chemical pair needed to produce ethylbenzene, the precursor of styrene. Technology and economics of ethylbenzene-styrene manufacture are discussed in Chapter 6.

As in the case of vinyl chloride, styrene production is located on the Gulf Coast whereas polystyrene manufacture is placed chiefly near its markets in northeastern US. Comparison of Table 6.16 in Chapter 6 and Table 10.28 readily indicate this. One reason for the different placement of monomer and polymer capacities is that styrene can be more readily shipped than can ethylene. A second probable factor is the high steam cost for styrene production, which is supposedly less expensive on the Gulf Coast than in the Northeast.

TABLE 10.28
Polystyrene capacity (1972)[1]

Company	*MM lb*
A & E Plastik Pak[2]	20
Amoco (Willow Springs, Ill.)	265
BASF (Jamesburg, N.J.)	80
Cosden (Big Spring, Tx.)	300
Dart (Joliet, Ill.)	140
Diamond Plastics[2]	50
Dow (Allyn's Point, Conn.)	700
Foster Grant (Peru, Ill.)	320
Hammond Plastics (Oxford, Mass.)	35
Johns Manville[2]	10
Monsanto (Springfield, Mass.)	375
National Distillers[2]	25
Petrochemical Investment Corp. (Houston, Tx.)	60
Richardson (West Haven, Conn.)	50
Rexene Polymers[2]	140
Shell (Wallingford, Conn.)	130
Sinclair-Koppers (Kobuta, Penn.)	300
Solar Chemical (Houston, Tx.)	125
Union Carbide (Bound Brook, N.J.)	280
USS Chemicals[2]	200
Total	3,605

[1] Oil, Paint & Drug Reporter, May 10, 1971. Does not include SBR elastomers.

[2] Location not given.

Styrene is employed for the manufacture of a variety of homo- and copolymers. Principal among these are straight polystyrene, rubber-modified polystyrene, styrene-butadiene elastomers (SBR), acrylonitrile-butadiene-styrene (ABS), and styrene-acrylonitrile (SAN) resins. Estimated consumption of styrene for each of these uses is presented in Chapter 6. Comparable data in terms of actual resin production are summarized in Table 10.29.

TABLE 10.29
Styrene resins production[1]
(MM lb)

Year	*ABS + SAN*	*Straight polystyrene*	*Rubber-modified polystyrene*	*SBR*	*Copolymers*
1969	559	1,057	1,080	3,090	405
1970	513	1,112	1,166	2,930	380
1971	718	1,160	1,230	na	411

[1] US Tariff Commission; 1971 based on 8 months' data.

It is evident from Table 10.29 that SBR is the largest volume resin produced from styrene, and is followed distantly by straight and rubber-modified polystyrene.

Suspension and bulk polymerization techniques are commonly used for polystyrene production. Bulk polymerization readily occurs at 50–120°C., with or without a peroxide catalyst:

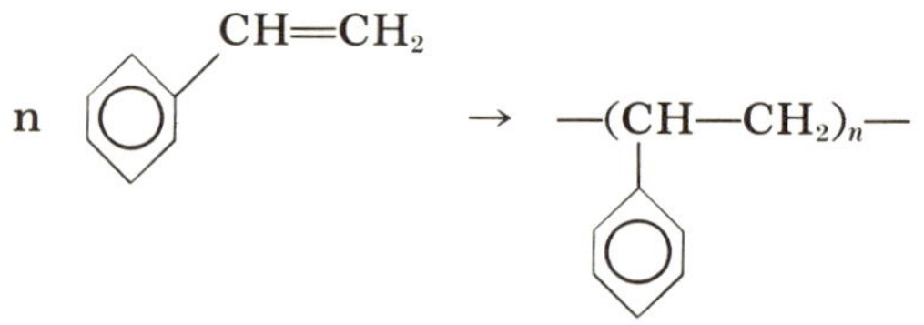

Commercial polystyrene is offered in many different grades. Since bulk polymerization is the most widely used industrial method, plants using this technique produce a range of products. The commercial product is atactic in structure and therefore amorphous. Isotactic polystyrene can also be produced, but offers little advantage in properties except between the glass-transition temperature (+80°C.) and its crystalline melting point range (ca. 230°C.), where it is much like other crystalline polymers. Like most polyolefins, polystyrene is quite resistant to alkali and oxidizing and reducing agents.

Polystyrene is very easy to process, especially by injection molding techniques. It has found great application in packaging, housewares, toys, and furniture. A summary of its applications is given in Table 10.30.

Rubber is incorporated into polystyrene mainly to improve the impact strength of the polymer. The mechanical properties of the product vary according to the type and amount of rubber incorporated. Although rubber-modified polystyrene has greater impact strength

TABLE 10.30
Major polystyrene molding and extrusion markets (1971)[1]

Use	MM lb
Foamed sheet	126
Furniture	86
Housewares	237
Packaging	545
Radio and TV	85
Refrigerators	81
Toys	257
Others	543
Total	1,960

[1] Based on Modern Plastics (Jan., 1971).

than ordinary polystyrene, it also has poorer tensile strength, hardness, and clarity.

ABS resins are rigid and tough with high tensile strength and thermal stability. These are basically premium plastics, but are versatile and capable of being processed into a broad range of grades that compete with high-priced engineering plastics as well as cheaper materials, such as PVC in pipe. Typical copolymer compositions are 20–30% acrylonitrile, 20–30% butadiene, and 40–60% styrene. A summary of ABS markets is given in Table 10.31.

TABLE 10.31
ABS markets (1971)[1]

Use	MM lb
Automotive	92
Business machines, phones	54
Major appliances	89
Pipe and fittings	120
Recreational	78
Exports	28
Miscellaneous	249
Total	710

[1] Based on Modern Plastics (Jan., 1971).

SBR is by far the largest volume synthetic rubber produced in the US and will probably remain so for the next decade (Table 10.32).

TABLE 10.32
US consumption of rubber by type[1]
(M metric tons)

	1969	*1979*
SBR	1,332	1,620
Butyl	96	110
Nitrile	69	110
Polybutadiene	267	590
Polyisoprene	87	300
Ethylene-propylene	46	320
Neoprene	160	210
Total	2,057	3,620

[1]R. Lamberson, NPRA Meeting, March 21, 1971.

SBR is a random copolymer containing about 25 wt % styrene. Tire tread stocks account for 55% of SBR usage. At one time, it was thought that ethylene-propylene rubber would severely encroach on this market, but this has not occurred.

VINYL ACETATE

The apparent trend toward ethylene-based processes for vinyl acetate portends significant ethylene consumption for this chemical in the future. Ethylene-based processes may have significant economic advantages over acetylene routes as described in Chapter 6.

The only end use for vinyl acetate is polyvinylacetate. As is the case with other vinyl resins, polyvinylacetate can be prepared by bulk, emulsion, or suspension systems, in the presence of peroxide initiators:

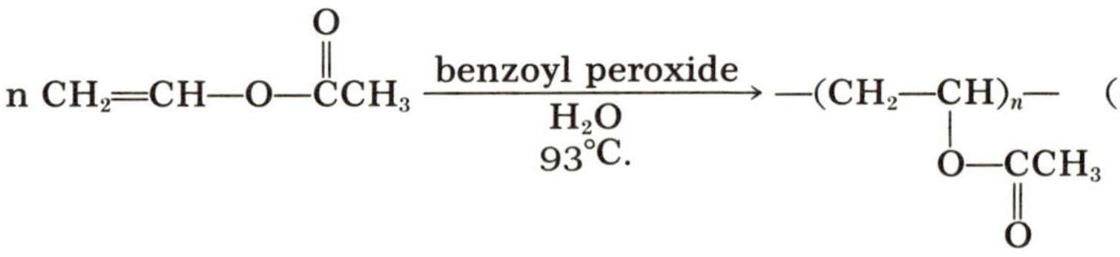

(99.5%)

The resin produced is atactic and hence amorphous. Its glass transition temperature is only slightly above room temperature (28°C.). Consequently, although polyvinyl acetate is tough and has dimensional stability, it becomes sticky and exhibits severe cold flow at only slightly elevated temperatures.

Polyvinyl acetate is principally used for water-base paints and adhesives, both the emulsion and hot melt type. Data which describe these and other polyvinyl acetate uses are found in Table 10.33.

TABLE 10.33
Polyvinylacetate market applications[1]

Year	*Paints*	*Adhesives*	*Paper treating*	*Textile treating*	*Other*	*Exports*	*Total*[2]
1965	104	111	20	na	45	3	283
1966	110	105	26	10	52	2	305
1967	95	117	25	8	63	2	310
1968	121	134	29	12	61	3	358
1969	120	112	29	15	63	6	338
1970	178	146	28	22	76	4	448
1971[3]	139	136	28	18	89	6	416

[1] US Tariff Commission.
[2] Sales and use.
[3] Based on 8 months' data.

Alcoholysis of polyvinyl acetate readily yields polyvinyl alcohol:

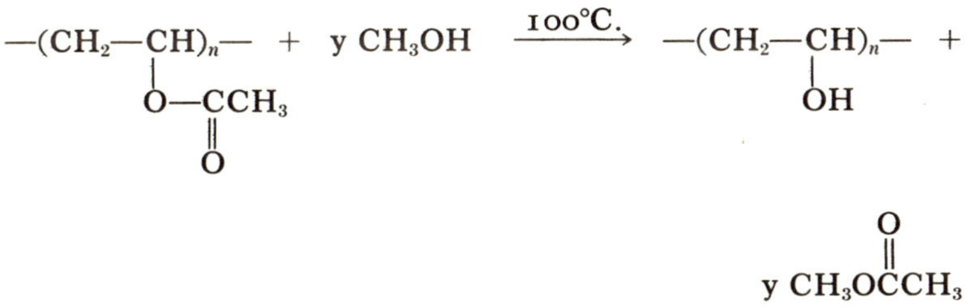

Ethanol may be used in place of methanol to yield ethyl acetate as a coproduct. Optionally, one can employ alkaline hydrolysis. Production of polyvinyl alcohol is relatively small as noted in Table 10.34.

Polyvinyl alcohol is water-soluble. It dissolves slowly in cold water and much more rapidly at higher temperatures, especially above 90°C. It is primarily used in wet-strength adhesives. About a decade ago, polyvinyl alcohol films were being touted as the big breakthrough in water-soluble film. Packaged food that would dissolve when thrown into water excited marketing men, but never caught on with the consumer in the US.

One primary use for polyvinyl alcohol is conversion to polyvinyl

TABLE 10.34
Vinyl acetate principal end uses[1]
(As MM lb of resin)

Year	*Polyvinylacetate*	*Polyvinylalcohol*
1964	282	na
1965	313	na
1966	335	38
1967	342	43
1968	384	45
1969	386	56
1970	403	49
1971[2]	432	31

[1] US Tariff Commission.
[2] Based on 8 months' data.

butyral. Prepared by reaction of polyvinyl alcohol and butanal, these resins are useful as laminates for safety glass.

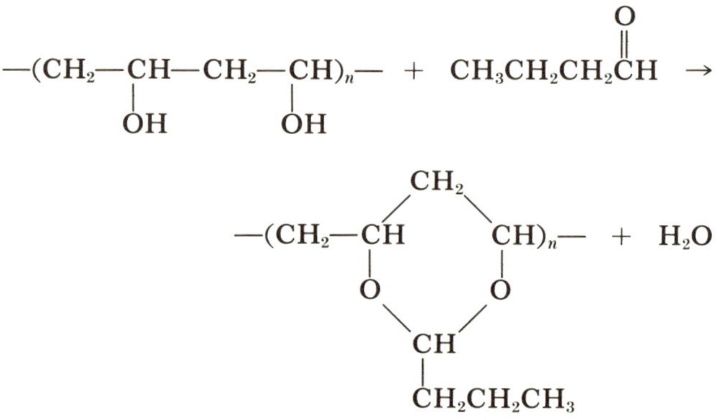

ETHANOL

Ethylene consumption for ethanol production is confined to industrial ethanol and is not used for ethanol in alcoholic beverages. The latter is made by fermentation and is beyond the scope of this discussion. Some fermentation alcohol is used for industrial purposes and is about 20% of total capacity (Table 10.35).

The indirect hydration process was the original synthetic route to ethanol. This comprised absorption of ethylene in concentrated

TABLE 10.35
1972 ethanol capacity[1]

		MM lb
Enjay	(Baton Rouge, La.)	380
Eastman	(Longview, Tx.)	160
Publicker	(Philadelphia, Pa.)	160
Shell	(Houston, Tx.)	255
Union Carbide	(S. Charleston, W.Va.)	380
	(Texas City, Tx.)	635
USI	(Tuscola, Ill.)	415
Georgia Pacific	(Bellingham, Wash.)[2]	
Hercules	(Hopewell, Va.)	130
Publicker	(Philadelphia, Pa.)	
	Total	2,515

[1] 6.3 lb/gal.
[2] Produced by fermentation and waste liquor recovery.

sulfuric acid to form the mono- and diethylsulfates followed by hydrolysis:

$$CH_2{=}CH_2 + H_2SO_4 \xrightarrow[200\ \text{psig}]{75°\text{C.}} CH_3CH_2OSO_3H + CH_3CH_2OSO_2OCH_2CH_3$$

$$CH_3CH_2OSO_3H + CH_3CH_2OSO_2OCH_2CH_3 \xrightarrow[H_2SO_4,\ 70°\text{C.}]{H_2O} CH_3CH_2OH + H_2SO_4$$

Combined molar yields of mono- and diethylsulfate are greater than 90%. The hydrolysis step proceeds in about 95% yield. Diethyl ether is the only significant byproduct, and is probably formed by reaction of ethanol and unhydrolyzed diethylsulfate.

Effective recovery of H_2SO_4 for reuse is essential to the economics of the process. This was achieved by concentration of the dilute, aqueous acid, and posed a serious corrosion problem to the recovery vessels.

Development of the direct hydration process for ethanol has

largely supplanted the indirect hydration technique. All new plants are based on this technology, and some of the older indirect hydration units are also being converted.

The direct hydration process was originally developed by Shell Development Co. In this process, a mixture of water and ethylene (0.6 mole ratio) are passed over a supported phosphoric acid catalyst at 325°C. and 1,000 psig. Conversion per pass is 4.2%, with greater than 97% selectivity:

$$CH_2{=}CH_2 \xrightarrow[\text{catalyst}]{H_2O} CH_3CH_2OH$$

As in the indirect hydration route, diethyl ether is again a minor byproduct. The process scheme is summarized in Fig. 10.6.

The estimated economics for this process are noted in Table 10.36. Material costs are the largest single factor in the cost of production.

TABLE 10.36
Estimated ethanol economics

Fixed investment cost
(60 MM gal; 1972 construction)

Process	*$ MM*
Catalytic hydration	16.3

Production cost

	¢/lb ethanol[1]
Raw materials	2.54[2]
Utilities	0.85
Labor	0.12
Maintenance (6% ISBL + 3% OSBL)	0.21
Overhead (45% of maintenance & labor)	0.15
Taxes and insurance (1.5% of investment)	0.07
Depreciation (10 years)	0.43
Total	4.36

[1] 6.3 lb/gal for 95% ethanol.
[2] 0.77 lb ethylene at 3.3¢/lb.

Ethanol's share of ethylene consumption has fallen from 21.7 to 9.6% of the total in the last decade (Table 10.8). This decline is not due to more economical routes for ethanol production, but to a decline in

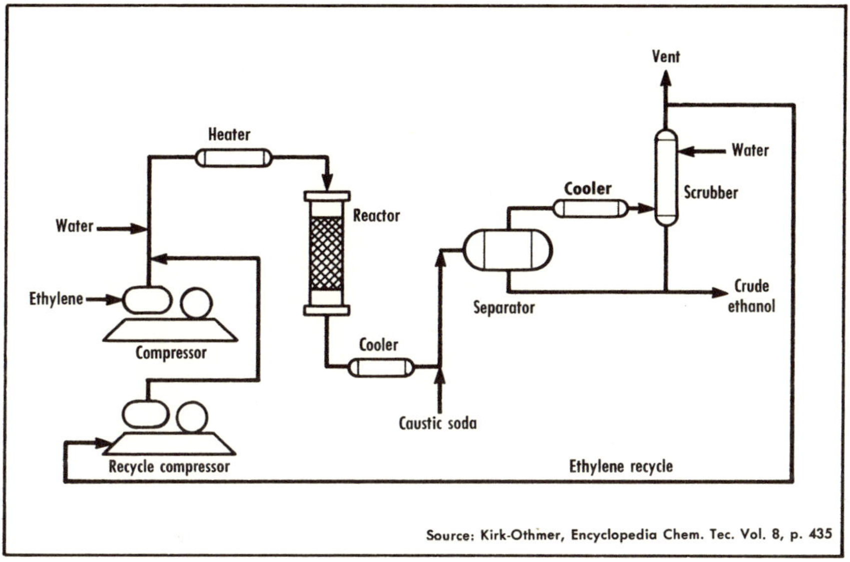

FIG. 10.6 Direct hydration of ethylene

TABLE 10.37
Synthetic ethanol statistics[1]
(*MM lb*)

	Production	*Sales*	*Unit price ¢/lb*
1961	1,685	935	6
1962	1,680	989	6
1963	1,951	1,103	6
1964	2,071	1,163	6
1965	2,039	1,315	6
1966	1,881	1,176	6
1967	1,919	1,135	6
1968	2,127	1,247	6
1969	2,361	1,215	6
1970[2]	2,150	na	8.3[3]
1971[2]	2,200	na	8.6[3]

[1] US Tariff Commission.

[2] Estimate based on Oil, Paint & Drug Reporter, March 1, 1971.

[3] Oil, Paint & Drug Reporter posted price (190 proof, USP, tanks, delivered).

demand for this chemical per se. Demand for ethanol has remained at about 2.0 to 2.4 billion lb since 1963 (Table 10.37). This static situation is caused by competing routes to ethanol's principal uses. This competitive situation is noted in Fig. 10.7. Of the products shown, acetic acid and acetaldehyde account for the bulk of ethanol uses.

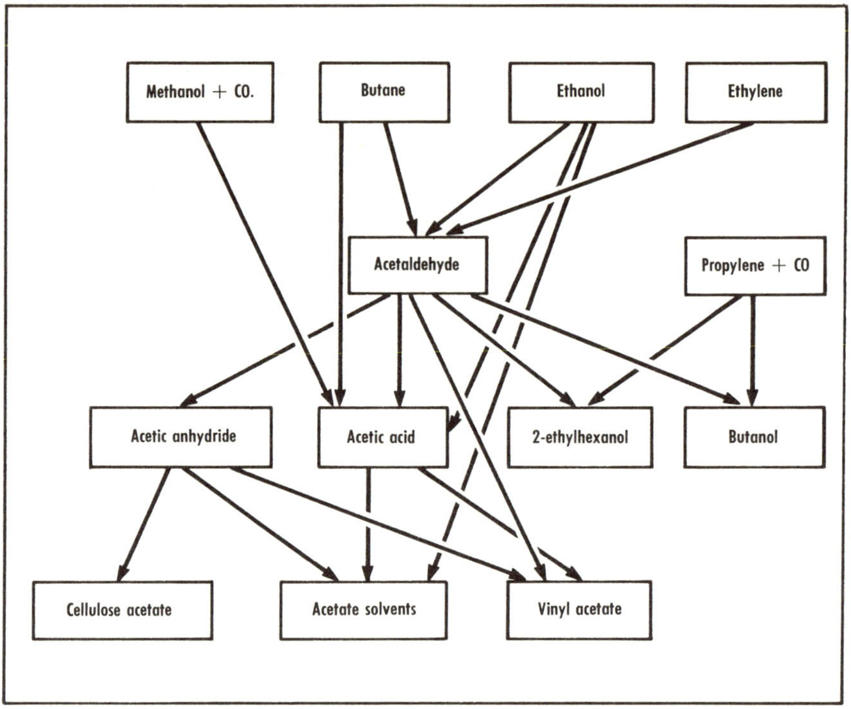

FIG. 10.7 Ethanol competitive end-use pattern

ACETALDEHYDE

Acetaldehyde, once far and away the chief use for ethanol, has slipped considerably because of other routes to this product as shown in Figure 10.7. Acetaldehyde is produced by a variety of processes as illustrated in Table 10.38. The ethanol route of acetaldehyde comprises vapor phase oxidation over a silver catalyst:

$$CH_3CH_2OH \xrightarrow[\substack{450^\circ C. \\ \text{air}}]{Ag} CH_3CHO + H_2O$$

Ethanol at 6¢/lb is not particularly competitive with LP-gas and

TABLE 10.38
Acetaldehyde capacity
(MM lb)

Company		*1967*	*1972*	*Process*
Celanese	(Bay City, Tx.)	210	210	Ethylene
	(Bishop, Tx.)	200	200	LP-gas
	(Clear Lake, Tx.)	175	375	Ethylene
	(Pampa, Tx.)	10	10	Byproduct petroleum gas
Commercial Solvents	(Agnew, Calif.)	1	1	Ethanol
DuPont Company	(Louisville, Ky.)	10	10	Byproduct petroleum gas
Eastman	(Kingsport, Tenn.)	200	–	Ethanol
	(Longview, Tx.)	250	500	Ethylene
Goodrich	(Calvert City, Ky.)	1	1	Byproduct petroleum gas
Hercules	(Parlin, N.J.)	35	–	Ethanol
Monsanto	(Texas City, Tx.)	5	5	Byproduct petroleum gas
Publicker	(Philadelphia, Penn.)	80	80	Ethanol
Shell	(Norco, La.)	5	5	Byproduct petroleum gas
Union Carbide	(Institute, W.Va.) (S. Charleston, W.Va.) (Texas City, Tx.)	650	650	Ethanol
	Totals	1,832	2,047	

byproduct refinery gases valued at about 1¢/lb. The economics of acetaldehyde production by oxidation of these materials is difficult to establish because of the variety of products that are obtained.

Vapor phase oxidation of LP-gas proceeds:

$$C_3H_8 + C_4H_{10} \xrightarrow[\substack{700°F. \\ 100\ psi}]{air} CH_3CHO + HCHO + CH_3OH + \text{higher alcohols, ketones, acids and glycols}$$

The most attractive route to acetaldehyde is the Wacker process, which is catalytic hydration of ethylene. This is a liquid phase process employing a $PdCl_2$–$CuCl_2$ catalytic system:

$$CH_2{=}CH_2 + H_2O + 2CuCl_2 \xrightarrow{PdCl_2} CH_3\overset{\overset{\displaystyle O}{\|}}{C}H + 2HCl + 2CuCl$$

$$2CuCl + 2HCl + \tfrac{1}{2}O_2 \rightarrow 2CuCl_2 + H_2O$$

During the reaction, palladium chloride is reduced to elemental palladium which is reoxidized by cupric chloride. Oxygen converts the reduced cuprous chloride back to cupric chloride.

The reaction proceeds at room temperature with yields of 93 to 95%. The process flow diagram is illustrated in Fig. 10.8. Estimated economics for a 200-MM-lb plant built in 1972 are summarized in Table 10.39.

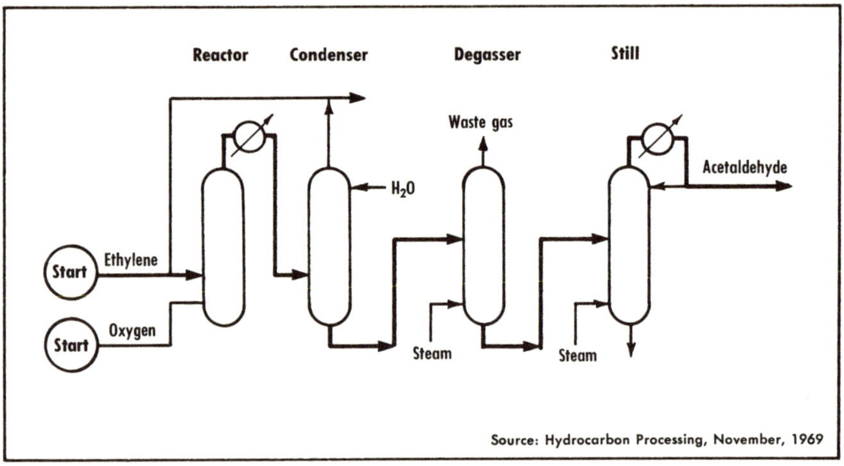

FIG. 10.8 Acetaldehyde by the Wacker process

Acetaldehyde production and sales are summarized in Table 10.40. As can be discerned from these data, the bulk of US production is probably used captively. Acetic acid and acetic anhydride command the major share of the acetaldehyde market (Table 10.41).

Use of acetaldehyde for 2-ethylhexanol and n-butanol will decline considerably as producers of these chemicals switch to propylene-based technology. This is discussed in detail in Chapter 11.

Moreover, non-acetaldehyde routes for acetic acid have been adopted by a number of firms (Table 10.42). Oxidation of byproduct refinery gases is the process used in greatest quantity. This is reminiscent of the acetaldehyde process of the same nature, and may be superior to the acetaldehyde oxidation route. This is especially likely since acetic acid at 6¢/lb is probably the same price as acetaldehyde (Table 10.43). Refinery gases at about 1¢/lb look quite attractive in comparison as feedstocks. Again, because of the number of products that are produced, process economics are difficult to discern.

TABLE 10.39

Estimated acetaldehyde economics
(200 MM-lb plant; 1972 construction)

Fixed capital investment

Process	*$ MM*
Ethylene	14.80

Production cost

	Ethylene, ¢/lb
Raw materials[1]	2.45
Utilities	0.82
Labor	0.24
Maintenance (6% ISBL + 3% OSBL)	0.35
Overhead (45% labor & maintenance)	0.27
Taxes and insurance (1.5% of investment)	0.11
Depreciation (10 years)	0.75
Total	4.99

[1] Includes 0.68 lb ethylene/lb at 3.3¢/lb.

TABLE 10.40

Acetaldehyde production and sales[1]
(MM lb)

Year	*Production*	*Sales*	*Unit value, ¢/lb*
1961	na	78	8
1962	na	88	8
1963	na	111	8
1964	1,058	100	7
1965	1,230	102	6
1966	1,300	242	5
1967	1,409	na	9[2]
1968	1,585	na	9[2]
1969	1,652	na	9[2]
1970[3]	1,600	na	9[2]
1971[3]	1,650	na	9[2]

[1] US Tariff Commission.

[2] Oil, Paint & Drug Reporter posted price in tanks (99%); freight allowed.

[3] Based on Oil, Paint & Drug Reporter, Jan. 1, 1971.

TABLE 10.41
Estimated 1971 end use pattern for acetaldehyde

Use	MM lb
Acetic acid and acetic anhydride	750
2-ethylhexanol	280
n-butanol	315
miscellaneous	305
Total	1,650

$$C_3H_8 + C_4H_{10} \xrightarrow[\substack{800 \text{ psig} \\ 300^\circ\text{–}475^\circ\text{F.}}]{\text{air}} CH_3COOH + HCOOH +$$

2-4 carbon alcohols, ketones, and acids

The process comprises liquid-phase oxidation using cobalt, manganese, or chromium acetates in acetic acid solution.

TABLE 10.42
Acetic acid capacity (1972)

		MM lb	Process
Borden	(Geismar, La.)	100	CO-methanol
Celanese	(Bishop, Tx.)	200	Petroleum gases
	(Pampa, Tx.)	600	Petroleum gases
	(Clear Lake, Tx.)	300	Acetaldehyde
Eastman	(Kingsport, Tenn.)	325	Acetaldehyde-ethanol
FMC	(Bayport, Tx.)	45	Acetaldehyde
Hercules	(Parlin, N.J.)	20	Acetaldehyde
Monsanto	(Texas City, Tx.)	300	CO-methanol
Publicker	(Philadelphia, Pa.)	80	Acetaldehyde-ethanol
Union Carbide	(Brownsville, Tx.)	400	Petroleum gases
	(Texas City, Tx.)	100	Petroleum gases
	(S. Charleston, W.Va.)	140	Petroleum gases
	(Taft, La.)	90	Acetaldehyde
Others		100	
	Total	2,800	

The most publicized of the newer acetic acid processes is carbonylation of methanol. The catalyst system is CoI_2 or an iodide-promoted rhodium complex. Molar selectivity can be virtually quantitative even when the CO feed contains 50% hydrogen.

TABLE 10.43
Synthetic acetic acid production[1]

Year	MM lb	Unit value, ¢/lb
1961	764	8
1962	969	8
1963	1,028	7
1964	1,100	7
1965	1,347	7
1966	1,409	7
1967	1,560	7
1968	1,738	6
1969	1,770	6
1970[2]	1,800	9[3]
1971E	1,800	9[3]

[1] US Tariff Commission.
[2] Hydrocarbon Processing, Jan., 1970, p. 145.
[3] Oil, Paint & Drug Reporter posted price (glacial, tech., tanks, delivered).

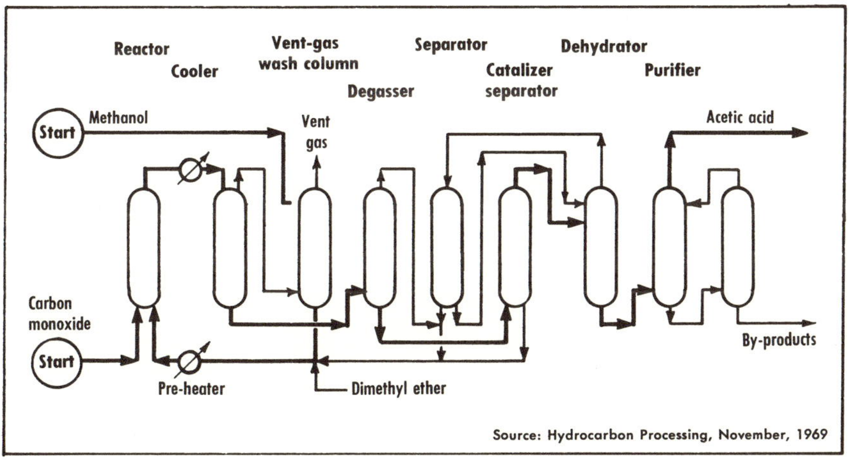

FIG. 10.9 Acetic acid by methanol carbonylation

The flow scheme for this process is illustrated in Fig. 10.9.

Acetic acid is used mainly for cellulose acetate and vinyl acetate manufacture (Table 10.44).

TABLE 10.44
Acetic acid use pattern (1971)[1]

Use	*MM lb*
Cellulose acetate	810
Vinyl acetate	485
Acetate esters	270
Chloroacetic acid	55
Miscellaneous	180
Total	1,800

[1] Based on Oil, Paint & Drug Reporter, Oct. 1, 1968.

11

ARTHUR M. BROWNSTEIN

Propylene

PROPYLENE differs from ethylene, inasmuch as there is no primary process for its manufacture. In this respect, propylene is unique as a large volume, chemical building block. Although there have been several proposed processes for its production, none of these have been sufficiently economical to gain widespread acceptance. Hence, propylene is nearly exclusively available only as a byproduct from ethylene plants and refinery catalytic cracking (Chapters 5 and 10). Consequently, growth in propylene supply is a function of gasoline and ethylene demand.

Propylene differs from ethylene in yet another fundamental manner. Unlike ethylene, nearly two-thirds of propylene demand is for gasoline manufacture. This is illustrated in Figure 11.1. Thus, both supply and demand of propylene are overwhelmingly governed by the refinery industry. A quantitative description of this situation is given in Table 11.1. Because of the intimate relationships of propylene with refining, exact information on supply and demand are difficult to ascertain. The data of Table 11.1, therefore, should be considered as reasonably good approximations.

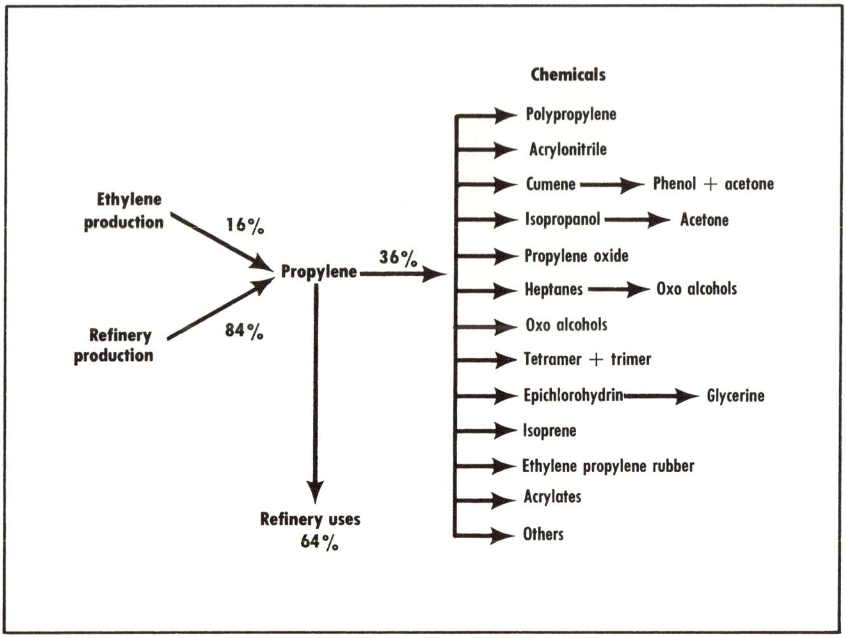

FIG. 11.1 US propylene production-use profile

TABLE 11.1
US propylene supply-demand[1]
1966–70
(*M metric tons/year*)

	1966	*1968*	*1970*
Demand			
Chemicals[2]	2,380	3,035	3,575
Refinery	6,350	6,150	6,400
Total demand	8,730	9,185	9,975
Supply			
Refinery production	8,280	7,900–8,750	8,410–9,250
Ethylene units	1,000	1,230	1,590
Total available	9,280	9,130–9,980	10,000–10,840

[1] Based on A. M. Brownstein, et al., European Petrochemical Assoc. Meeting, October, 1969.

[2] Excludes polymer gasoline products such as heptene, trimer, and tetramer which are produced in refinery polymerization units.

The difficulty in establishing propylene demand for chemicals is additionally compounded because some intermediates, such as heptenes, trimer, and tetramer, are obtained as byproducts of refinery operations. In Table 11.1, propylene consumption for their manufacture has been arbitrarily consigned to refinery demand. This is not the case in Table 11.5, however, where these compounds have been included for convenience with other chemical uses.

A partial listing of US propylene producers is illustrated in Table 11.2. These data represent those refiners and operators of ethylene plants who have made propylene available for chemical uses. Refiners who retain their propylene for gasoline production, and who could, if they wanted, sell to the chemical market, are not included here.

TABLE 11.2
US chemical propylene capacity (1972)[1]
(MM lb)

Company	*Refinery*	*Ethylene plant*	*Total*
Amerada-Hess (Port Reading, N.J.)	130	–	130
Ashland (Ashland, Ky.)	135	–	135
Arco (Wilmington, Calif.)	240	–	240
British Petroleum (Marcus Hook, Pa. and Houston, Tx.)	400	–	400
Chemplex (Clinton, Ia.)	–	130	130
Chevron (El Segundo, Calif.)	200	–	200
Cities Service (Lake Charles, La.)	300	550	850
Clark Oil (Blue Island, Ill.)	70	–	70
Continental Oil (Lake Charles, La. and Ponca City, Okla.)	120	–	120
Corco-PPG (Penuelas, P.R.)	–	760	760
Dart-El Paso (Odessa, Tx.)	–	150	150
Dow (Bay City & Midland, Mich., Freeport, Tex. & Plaquemine, La.)	–	500	500
DuPont (Orange, Tx.)	–	200	200
Eastman (Longview, Tx.)	–	225	225
Enjay (Baton Rouge, La. & Baytown, Tx.)	1,900	–	1,900
Goodrich (Calvert City, Ky.)	–	100	100
Gulf (Cedar Bayon & Pt. Arthur, Tx.; Philadelphia, Pa.)	600	770	1,370
Jefferson (Port Neches, Tx.)	200	–	200
Marathon (Detroit, Mich. & Texas City, Tx.)	150	–	150
Mobil (Beaumont, Tx.)	250	250	500

TABLE 11.2 (cont'd)

Monsanto (Alvin, Sweeny & Texas City, Tx.)	—	350	350
Northern Petrochemical (Joliet, Ill.)	—	200	200
Phillips (Sweeny, Tx.)	250	150	400
Shell (Deer Park, Tx. & Norco, La.)	875	700	1,575
Signal (Houston, Tx.)	80	—	80
Sinclair-Koppers (Houston, Tx.)	—	100	100
Skelly (El Dorado, Kansas)	90	—	90
Sohio (Lima, Ohio)	300	—	300
Standard of Indiana (Whiting, Ind. and Wood River, Ill.)	1,000	—	1,000
Sun Oil (Marcus Hook, Pa. and Corpus Christi, Tx.)	450	—	450
Texaco (Westville, N.J.)	200	—	200
Texas City Refining (Texas City, Tx.)	100	—	100
Texas-US Chemical (Pt. Neches, Tx.)	—	30	30
Tidewater (Delaware City, Del.)	250	—	250
Union Oil (Wilmington, Calif.)	200	—	200
Union Carbide (Institute & S. Charleston, W.Va.; Taft, La.; Texas City & Seadrift, Tx.; Whiting, Ind.; Penuelas, P.R.)	—	1,000	1,000
Union Texas	—	120	120
Total	8,490	6,285	14,775

[1] Based on *Chemical Week,* May 20, 1970.

ALKYLATION AND POLYMERIZATION

These two processes comprise the refinery demand for propylene. Polymerization is the older of the two operations, and is steadily being phased out in favor of alkylation. As can be noted from Table 11.3, polymerization processes currently consist of less than 10% of refinery propylene consumption.

Alkylate is a better quality gasoline than that obtained from polymerization units (poly plants).

The alkylation process reacts propylene and/or butenes with isobutane to produce branched hydrocarbons. Refinery grade propylene (50 to 70% propylene; 30 to 50% propane) is used in conjunction with an H_2SO_4 or HF catalyst. The propane passes unchanged through the reactor and is recovered for sale as LP-gas. With propylene as the feed, the principal reaction products are 2,3- and 2,4-dimethylpentane.

TABLE 11.3
US capacity of refinery processes consuming propylene[1]
(1,000 bbl/stream day)

	Polymerization	*Alkylation*
1960	148	407
1965	126	530
1966	118	560
1967	96	595
1968	96	646
1970	80	749
1972	60	808

[1] Based on annual refining summaries of the Oil & Gas Journal.

These arise through a series of skeletal rearrangements arising from carbonium ion intermediates:

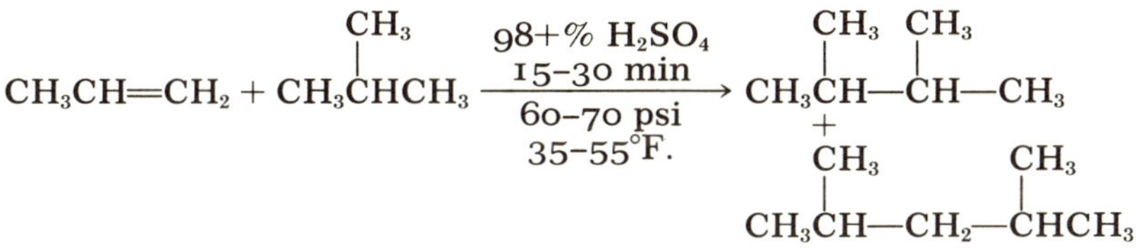

A schematic flow diagram of the process is shown in Fig. 11.2.

Poly gasoline is a mixture of oligomers obtained via the acid-catalyzed polymerization of propylene and/or butenes. Reaction conditions commonly comprise a phosphoric acid catalyst supported on kieselguhr with which the olefins are contacted at 200°C. and 500 to 1,000 psig. Nonenes (trimer) and dodecenes (tetramer) are the principal products if propylene alone is employed. If the reaction feed is a mixture of propylene and butenes, heptenes are the principal product. A description of typical product mixtures appears in Table 11.4.

The greater value of tetramer, trimer, and heptenes for chemical use, as opposed to gasoline value, has prompted the recovery of these materials for chemical sales. With the decline of poly gasoline, production of trimer, tetramer, and heptenes have taken on greater significance. It is likely that many poly plants now in operation are being run for this purpose rather than for gasoline.

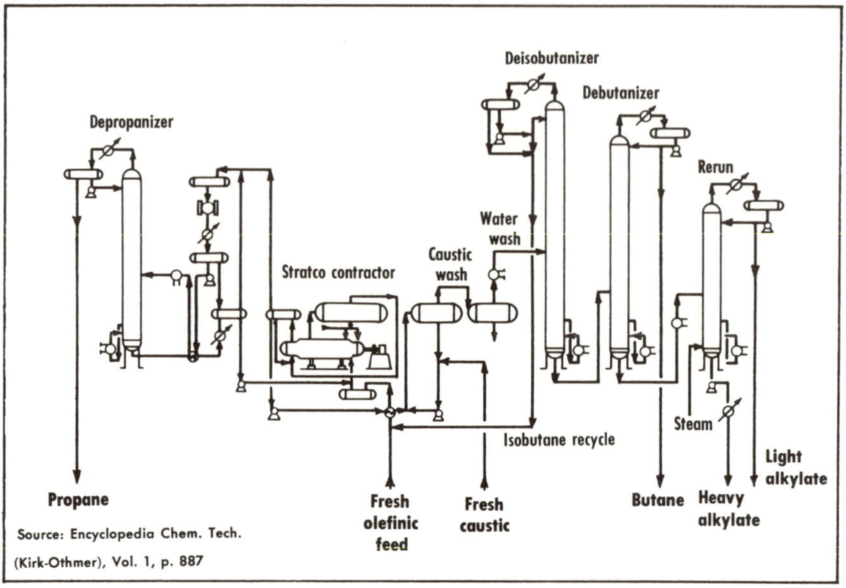

FIG. 11.2 Sulfuric acid alkylation unit

TABLE 11.4
Oligomers from the acid-catalyzed polymerization of propylene

	Polymer, mole %						
Feed	C_6	C_7	C_8	C_9	$C_{10/11}$	C_{12}	*Heavier*
Propylene (49% propylene, 51% saturates)	16	5	5	53	6	20	
Propylene-butylene mix (19% propylene, 30% C_4 olefins, 51% saturates)	4	33	17	9	20	12	4

*Encyclo. Chem. Tech. (Kirk-Othmer); Vol. 16, p. 581. Second edition © 1968 by John Wiley & Sons, Inc.

TRIMER AND TETRAMER

Not only has the growth of trimer and tetramer been hampered by changes in refining practices in favor of alkylate, but these products have also been casualties in the battle against pollution.

Demand for tetramer in 1971 was about the same level as in 1965 (Table 11.6). Tetramer is principally used for dodecylbenzene produc-

TABLE 11.5

US propylene end use pattern for chemicals

(MM lb)

Year	*Cumene*[1]	*Isopropanol*[2]	*Poly-propylene*[3]	*Propylene oxide*[4]	*Trimer & tetramer*[5]	*Acrylo-nitrile*[6]	*Oxo*[7] *alcohols*	*Acrylic acid*[8]	*Other*[9]	*Total*[10]
1961	100	1,000	110	370	720	280	na	–	95	2,675
1962	135	1,130	155	435	710	420	na	–	155	3,140
1963	150	1,250	210	500	675	530	665	–	210	4,190
1964	195	1,280	290	560	740	690	970	–	275	5,000
1965	230	1,310	400	595	875	895	1,055	–	310	5,670
1966	310	1,460	590	695	925	830	1,130	–	350	6,290
1967	400	1,760	705	800	890	780	1,140	–	370	6,845
1968	470	1,770	935	910	819	1,180	1,145	–	415	7,644
1969	590	1,710	1,130	1,120	915	1,340	1,080	90	450	8,425
1970	600	1,820	1,075	1,065	940	1,210	1,030	125	485	8,350
1971	630	1,950	1,240	1,040	955	1,120	1,070	150	500	8,655

[1] 0.35 lb/lb.
[2] 0.85 lb/lb.
[3] 1.04 lb/lb.
[4] 0.98 lb/lb through 1968, then a mean 0.95 lb/lb.
[5] 1.27 lb/lb tetramer and 1.21 lb/lb trimer.
[6] 1.16 lb/lb.
[7] 1.1 lb propylene/lb 2-ethylhexanol and 2.6 lb propylene/lb heptene.
[8] 0.88 lb/lb.
[9] Glycerol, epichlorohydrin, ethylene-propylene terpolymer, butanols, isoprene, linear detergent alkylate, etc.
[10] Reflects demand rather than reported production.

tion, which in turn is employed for detergent manufacture. Dodecylbenzene, often called branched detergent alkylate, is prepared by reacting tetramer with benzene to give a highly branched product:

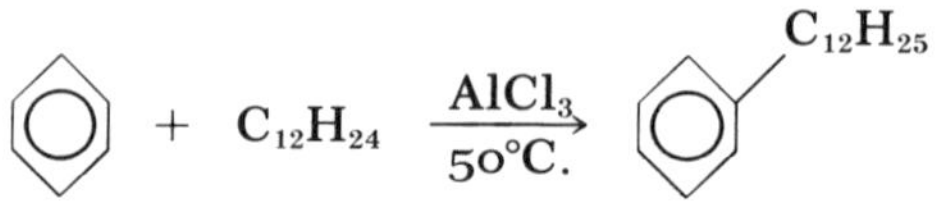

This in turn is converted to the detergent by reaction with oleum to yield the sulfonic acid.

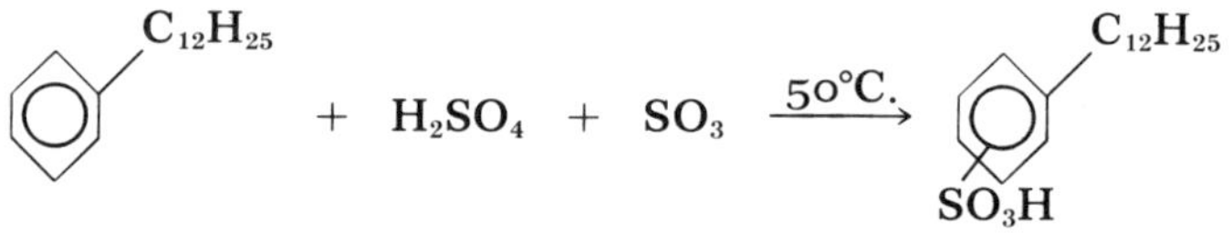

Lack of growth in tetramer demand dates back to 1964 when the US detergent industry switched from branched to linear detergent alkylate. The former, which was based on tetramer, was not biodegradable and was regarded as a pollutant. Linear detergent alkylate is based on ethylene, and the side chain on the benzene ring is linear. This material is biodegradable and is now extensively used throughout the US.

Tetramer produced in the US is exported per se, or converted to branched dodecylbenzene, which is exported. Like the US, Western

TABLE 11.6
Propylene tetramer[1]
(MM lb)

Year	*Production*	*Sales*	*Unit value, ¢/lb*
1961	416	209	3.3
1962	439	264	3.1
1963	400	335	3.2
1964	432	344	2.8
1965	462	359	2.5
1966	457	306	2.7
1967	434	252	3.3
1968	384	221	4.2
1969	459	393	5.9
1970[2]	455	na	na
1971[2]	460	na	na

[1] US Tariff Commission.
[2] Estimates.

Europe no longer uses non-biodegradable detergents. Export sales are to those parts of the world not now concerned about ecology.

Trimer has a more diverse market use than tetramer. Demand falls into two major areas, isodecyl alcohol and nonylphenol (Table 11.7).

TABLE 11.7
Propylene trimer end use pattern[1]
(MM lb)

Year	*Isodecyl alcohol*[2]	*Nonylphenol*[3]	*Misc.*[4]	*Total*	*Unit value ¢/lb*
1961	74	37	54	165	3.6
1962	62	40	28	130	2.7
1963	65	33	46	144	3.0
1964	72	43	47	162	3.0
1965	117	44	83	244	3.0
1966	134	46	109	289	3.1
1967	151	47	88	286	3.4
1968	168	45	60	273	3.6
1969	178	48	78	304	3.3
1970[5]	na	na	na	300	na
1971[5]	na	na	na	305	na

[1] US Tariff Commission.
[2] 1.10 lb/lb isodecanol.
[3] 0.72 lb/lb nonylphenol.
[4] Largely exports. Includes nonylbenzene.
[5] Estimates.

Nonylphenol is obtained by reaction of trimer with phenol in the presence of an acid catalyst, such as BF_3. The resulting phenol is readily condensed with ethylene oxide to yield the corresponding polyether. This ethoxylated nonylphenol is an active, nonionic detergent. It does not form salts with metal ions, and is therefore equally effective in hard and soft water. This sequence of reactions can be summarized:

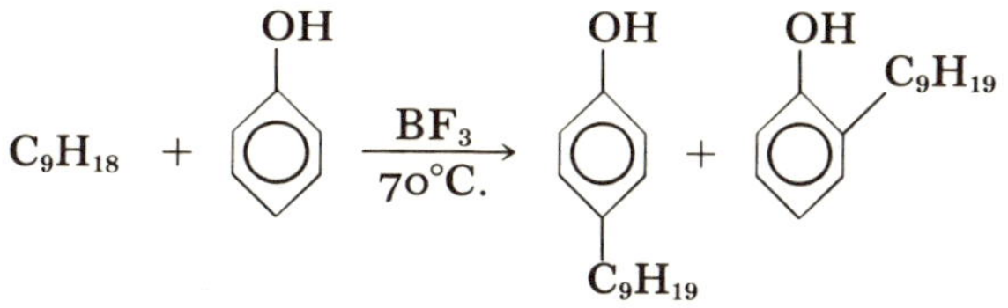

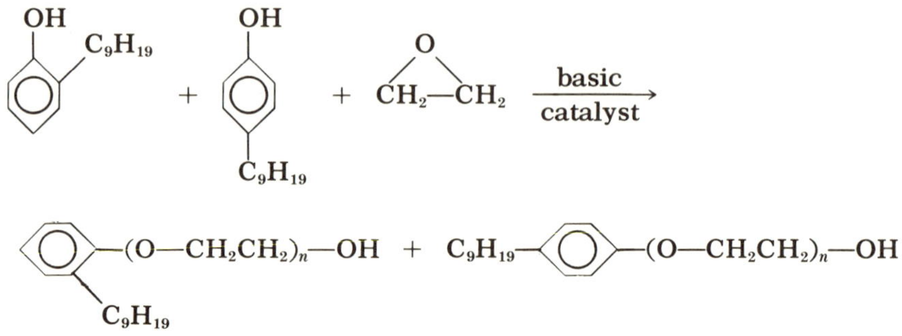

As described in Table 11.7, nonylphenol growth has been slight in recent years. As in the case of dodecylbenzene, nonylphenol ethoxylates are non-biodegradable. The loss of household detergent markets to biodegradable detergents has been offset by nonylphenol through increased use in industrial applications, such as surfactants, oil additives, and rubber chemicals.

The major use of trimer is isodecyl alcohol. This highly branched alcohol is made by carbonylation of the olefin in the manner described in Chapter 3. Substantially all of this material is used for the manufacture of phthalate and adipate plasticizers for PVC resins.

Installed capacity for trimer and tetramer is summarized in Table 11.8. Capacities from these units are flexible, since trimer can

TABLE 11.8
1972 tetramer and trimer capacity[1]

Company	*MM lb*	*Products*
Air Products (Delaware City, Del.)	55	trimer & tetramer
Arco (Wilmington, Calif.)	105	trimer & tetramer
Chevron (Richmond, Calif.)	200	tetramer
Conoco (Lake Charles, La. and Ponca City, Okla.)	60	tetramer
Enjay (Baytown, Tex.; Baton Rouge, La.; Bayway, N.J.)	655	trimer & tetramer
Pure Oil (Nederland, Tx.)	10	tetramer
Sun Oil (Marcus Hook, Pa.; Toledo, O.; Corpus Christi, Tx.)	120	trimer & tetramer
Texaco (Port Arthur, Tx.)	40	tetramer
Union Oil (Nederland, Tx.)	35	trimer
Total	1,280	

[1] Based on the Oil & Gas Journal annual refinery statistics, US Tariff Commission Reports, and Chemical Week, May 20, 1970.

be converted into additional tetramer and conversely. Hence, increased production of one leads to a decrease of the other.

OXO ALCOHOLS

As explained in Chapter 3, oxo alcohols are produced by hydrocarbonylation of olefins, primarily propylene and heptene. There are two principal oxo alcohols, 2-ethylhexanol obtained indirectly via propylene carbonylation, and isooctanol via carbonylation of heptene.

Isooctanol is synthesized in a manner similar to 2-ethylhexanol except for the elimination of aldol condensation of intermediate aldehydes (Chapter 3). The process can be summarized as:

$$CH_3CH_2CH_2CH_2CH_2CH{=}CH_2 \xrightarrow[\substack{150°C.\\250\ atm}]{H_2/CO(1:1)} CH_3CH_2CH_2CH_2CH_2\overset{\overset{\displaystyle CH_3}{|}}{C}H{-}\overset{\overset{\displaystyle O}{\|}}{C}H$$

$$\downarrow H_2$$

$$CH_3CH_2CH_2CH_2CH_2\overset{\overset{\displaystyle CH_3}{|}}{C}H{-}CH_2OH$$

This reaction is idealized. The heptene precursor is an isomeric mixture obtained from poly plants. Hence, isooctanol is not a pure compound, but an isomeric mixture having a distribution similar to:

Isomer	*Wt %*
3,4-dimethyl-1-hexanol	20
3,5-dimethyl-1-hexanol	30
4,5-dimethyl-1-hexanol	30
Isomeric heptanols	20

Phthalate plasticizers account for the principal market for both isooctanol and 2-ethylhexanol. The phthalates are substantially identical in both properties and price (Table 11.9), as are the alcohols themselves (Tables 11.10 and 11.11). Nevertheless, 2-ethylhexanol commands a much larger market than isooctanol and shows steady growth, while isooctanol appears to be in decline. This situation appears puzzling especially in view of the estimated manufacturing costs of the two products as shown in Table 11.12. Although the estimated production costs of isooctanol and 2-ethylhexanol are about the same, there appears to be a distinctly lower capital investment required for the former.

TABLE 11.9
Phthalate plasticizer production[1]
(MM lb)

Year	*Isooctyl* (¢/lb)		*2-Ethylhexyl* (¢/lb)	
1961	43	(22)	138	(24)
1962	160	(15)	175	(15)
1963	64	(20)	171	(18)
1964	135	(14)	189	(14)
1965	122	(14)	212	(13)
1966	103	(15)	253	(14)
1967	98	(16)	293	(16)
1968	94	(16)	330	(15)
1969	83	(14)	355	(13)
1970	85	(12)	350	(11)
1971	na		na	

[1]US Tariff Commission; 0.70 lb 2-EH and isooctanol/lb phthalate.

TABLE 11.10
2-Ethylhexanol[1]
(MM lb)

Year	*Production*	*Sales*	*Unit value, ¢/lb*
1963	230	na	na
1964	280	na	na
1965	293	149	11
1966	319	149	11
1967	352	157	13
1968	387	176	12
1969	405	201	10
1970	440[2]	na	11[3]
1971	475[2]	na	11[3]

[1]US Tariff Commission.
[2]Based on Oil, Paint & Drug Reporter, Jan. 1, 1971.
[3]Oil, Paint & Drug Reporter posted price (tanks, delivered).

TABLE 11.11
Isooctanol[1]
(MM lb)

Year	*Production*	*Sales*	*Unit value, ¢/lb*
1964	142	128	12
1965	127	113	11
1966	100	99	12
1967	122	109	12
1968	132	80	12
1969	109	83	10
1970	95[3]	na	12[2]
1971	95[3]	na	12[2]

[1] US Tariff Commission.
[2] Oil, Paint & Drug Reporter posted price (tanks, delivered).
[3] Estimates.

TABLE 11.12
Estimated oxo alcohol exonomics
(125-MM-lb plant; 1972 construction)

Total fixed capital

Process	*$ MM*
Isooctanol	7.70
2-Ethylhexanol (2-EH)	16.30

Production cost

	¢/lb	
	2-EH	*Isooctanol*
Raw materials	4.50[1]	6.02[2]
Utilities	0.67	0.45
Labor	0.23	0.23
Maintenance (6% ISBL + 3% OSBL)	0.64	0.30
Overhead (45% maint. + labor)	0.39	0.24
Taxes & insurance (1.5% of invest.)	0.19	0.09
Depreciation (10 years)	1.31	0.62
Total	7.93	7.95

[1] Includes 1.1 lb propylene/lb at 3.0¢/lb.
[2] Includes 1.0 lb heptene/lb at 4.2¢/lb.

The solution to this situation is traceable to the heptene raw material for isooctanol. There does not yet appear to be an economically attractive route for heptene per se, and its availability is contingent upon poly plant operation. As active poly plants continue to dwindle in number, so does US heptene supply, and hence isooctanol production (Table 11.13). Heptene production in 1971 was at its lowest level since 1964 (Table 11.14). A substantial portion of US heptene production is exported for foreign isooctanol manufacture.

TABLE 11.13
US heptene capacity
(MM lb)

Company	*1968*	*1972*
Amoco (Yorktown, Va.)	30	30
B. P. (Marcus Hook, Pa.)	50	50
Clark Oil (Wood River, Ill.)	25	25
Cosden (Big Spring, Tx.)	30	30
Enjay* (Baytown, Tx.)	80	80
Gulf* (Port Arthur, Tx.)	40	—
Tidewater Air Products* (Delaware City, Del.)	85	85
Total	340	300

*Isooctanol producers. Others are USS Chemicals and Union Carbide.

TABLE 11.14
Heptenes[1]
(MM lb)

Year	*Production*	*Sales*	*Unit value, ¢/lb*
1961	na	na	na
1962	86	20	4.6
1963	158	71	3.4
1964	254	141	3.8
1965	280	173	3.8
1966	301	232	3.7
1967	288	196	3.9
1968	275	178	4.0
1969	244	135	4.2
1970	210	na	na
1971	210	na	na

[1]US Tariff Commission; 1970 and 1971 are estimates.

An alternate route to 2-ethylhexanol is based on ethylene, not propylene, as raw material. The ethylene route involves acetaldehyde production, which is in turn converted to n-butyraldehyde. The latter is self-condensed as in the propylene route to yield 2-ethylhexanol.

$$2CH_2{=}CH_2 \xrightarrow[PdCl_2]{H_2O} 2CH_3\overset{\overset{\displaystyle O}{\|}}{C}H \xrightarrow{NaOH} CH_3CH{=}CH{-}\overset{\overset{\displaystyle O}{\|}}{C}H$$

$$2CH_2CH{=}CH\overset{\overset{\displaystyle O}{\|}}{C}H \xrightarrow{H_2} 2CH_3CH_2CH_2\overset{\overset{\displaystyle O}{\|}}{C}H \xrightarrow{base}$$

$$CH_3CH_2CH_2CH_2\overset{\overset{\displaystyle C_2H_5}{|}}{C}HCHO \xrightarrow{H_2} CH_3CH_2CH_2CH_2\overset{\overset{\displaystyle C_2H_5}{|}}{C}HCH_2OH$$

Although the process does not produce isobutanal and isobutanol as undesirable products as in the propylene route, it is apparently more costly. Only one ethylene-based plant in the US is currently operating. Apparently, co-production of acetic acid and acetates as practiced by Celanese is a deciding factor in the process economics (Table 11.15).

TABLE 11.15
US 2-ethylhexanol capacity (1972)[1]

Company	*MM lb*	*Process*
Celanese (Bay City, Tx.)	80	ethylene
Dow Badische (Freeport, Tx.)	40	propylene
Eastman (Longview, Tx.)	75	propylene
Oxochem (Penuelas, P.R.)	170	propylene
Shell (Houston, Texas)	80	propylene
Union Carbide (S. Charleston, W.Va.)	(75)[2]	ethylene
(Texas City, Tx.)	150	propylene
(Ponce, P.R.)	135	propylene
Total	730	

[1] Based on Oil, Paint & Drug Reporter, Jan. 1, 1971.
[2] Shut down.

CUMENE

Cumene production is exclusively tied to phenol demand and was discussed in some detail previously with regard to benzene in Chapter 6. Propylene demand for cumene has increased more than six-fold in the last decade, and has mirrored the transition to cumene-based processes for phenol manufacture (Table 11.5).

The growth in cumene for phenol has prompted many would-be manufacturers to consider existing facilities which could be inexpensively converted to cumene production. One popular choice has been to convert less profitable poly plants for this purpose. An impetus for this decision has been the ability to use the same catalyst as for polymer gasoline. Moreover, one can also use refinery grade propylene and still produce LP-gas as previously.

ISOPROPANOL

This chemical is the patriarch of propylene demand (Table 11.5). Its premier position is owed to classical chemistry which recognized the ease of propylene hydration.

$$CH_3CH{=}CH_2 + H_2SO_4 \xrightarrow{50°C.} CH_3\overset{\displaystyle OSO_3H}{\overset{|}{C}}HCH_3 \xrightarrow{H_2O} CH_3\overset{\displaystyle OH}{\overset{|}{C}}HCH_3 + H_2SO_4$$

Although the above indirect hydration technique is effective, the preferred process is now direct catalytic hydration. The first company to employ this technique was ICI, who used the following process.

$$CH_3CH{=}CH_2 + H_2O \xrightarrow[\substack{225\ atm \\ 250°C.}]{WO_2} CH_3\overset{\displaystyle OH}{\overset{|}{C}}HCH_3$$

Chemical-grade propylene (90 to 95% propylene; 5 to 10% propane) is employed, with yields of 97% of theory. The flow sheet for isopropanol is quite analogous to ethanol as illustrated in Fig. 10.6 of Chapter 10. The direct catalytic hydration process eliminates the corrosive sulfuric acid system previously encountered, and also obviates the need for H_2SO_4 recovery. The economics for the direct hydration process are summarized in Table 11.16.

It should be noted that propylene is manufactured in three grades. Refinery grade, the cheapest, has already been described and is used for gasoline production and for some chemicals, such as cumene. Chemical grade (90 to 95% propylene) is employed for production of most intermediates, such as isopropanol and acrylonitrile. Polymer-grade propylene is the highest purity and most expensive grade. Minimum purity for this is 99.5%.

Although isopropanol continues to be the major market for propyl-

TABLE 11.16
Estimated isopropanol economics
(500-MM-lb[1] plant; 1972 construction)

Total capital investment	$11.30 MM
Production cost	
Item	*¢/lb*
Raw materials[2]	2.34
Utilities	0.47
Labor	0.06
Maintenance (6% ISBL + 3% OSBL)	0.11
Overhead (45% labor + maintenance)	0.08
Taxes & insurance (1.5% of total investment)	0.03
Depreciation (10%/yr)	0.22
Total	3.31

[1] 6.55 lb/gal.
[2] Includes 0.72 lb propylene at 3.0¢/lb.

ene (nearly 2 billion lb in 1971), its recent growth rate has been sluggish. In 1961, isopropanol accounted for 37% of chemical demand for propylene, 14% more than a decade later (Table 11.5). Production increased 10% annually from 1961 to 1967, but has limped along since then at barely 2%/yr (Table 11.17).

TABLE 11.17
Isopropanol production and sales[1]

Year	*MM lb production*	*Sales*	*Unit value, ¢/lb*
1961	1,189	444	6
1962	1,326	471	6
1963	1,466	542	6
1964	1,504	539	6
1965	1,538	582	6
1966	1,714	695	6
1967	2,069	739	6
1968	2,074	795	6
1969	2,013	851	6
1970[2]	2,140	na	7.3[3]
1971[2]	2,290	na	7.5[3]

[1] US Tariff Commission.
[2] Based on *Chemical Week*, Oct. 14, 1970.
[3] Oil, Paint & Drug Reporter posted price (anhydrous, tanks, delivered.

The downturn in isopropanol growth in recent years is largely attributable to an alternate process for its principal end use, acetone (Table 11.18). This accounts for nearly 50% of isopropanol use and can be inexpensively produced by vapor phase oxidation:

$$CH_3CH(OH)CH_3 \xrightarrow[\text{40–50 psi} \atop \text{Air}]{450\text{–}550^{\circ}\text{C.}} CH_3C(=O)CH_3 + H_2O$$

The reaction proceeds readily in 98% selectivity at 90% conversion/pass. Silver and copper catalysts are commonly used.

TABLE 11.18
Isopropanol end use pattern (1971)[1]

Item	*MM lb*
Acetone	1,030
Other chemicals	230
Coatings solvent	140
Drugs and cosmetics	110
Misc. (including exports)	780
Total	2,290

[1]Based on Oil, Paint & Drug Reporter, Jan. 5, 1970.

TABLE 11.19
US acetone production[1]
(MM lb)

Year	*Isopropanol based*	*Cumene based*	*Total*	*Unit value, ¢/lb*
1961	591	158	750	6.2
1962	668	181	849	5.6
1963	718	226	943	4.9
1964	761	293	1,054	4.8
1965	747	377	1,124	4.6
1966	881	449	1,330	4.8
1967	792	492	1,284	5.2
1968	799	562	1,360	4.9
1969	929	589	1,518	4.8
1970	na	na	1,598	5.9[2]
1971[2]	na	na	1,580	5.9[2]

[1]US Tariff Commission; 1971 based on 8 months' data.
[2]Oil, Paint & Drug Reporter posted price (tanks, delivered).

Although the isopropanol dehydrogenation process is seemingly quite effective, it has met a strong competitor in acetone byproduct from cumene-based phenol plants.

In Table 11.19, it is apparent that byproduct acetone from phenol has steadily increased to account for 39% of total production in 1969, compared to only 21% in 1961. This situation is destined to become more aggravating to the isopropanol-based producers of acetone as the swing to phenol-based cumene capacity continues (Table 11.20).

This subject was presented in some detail in Chapter 6. It is ap-

TABLE 11.20
US acetone capacity[1]
(MM lb)

Company	*Location*	*1967*	*1968*	*1970*	*1972*	*Process*
Amoco	Texas City, Tx.	–	–	–	240	Isopropanol
Allied Chemical	Philadelphia, Pa.	150	190	190	300	Cumene
Celanese	Bishop, Tx.	35	35	35	35	Propane
Chevron	Richmond, Calif.	35	35	35	35	Cumene
Clark Oil	Chicago, Ill.	35	35	35	35	Cumene
Dow	Freeport, Tx.	–	–	–	240	Cumene
Eastman	Kingsport, Tenn.	90	90	90	90	Isopropanol
Enjay	Bayway, N.J.	110	110	110	110	Isopropanol
Georgia Pacific	Plaquemine, La.	–	–	120	120	Cumene
Hercules	Gibbstown, N.J.	30	30	30	–	Cumene
Monsanto	Alvin, Tx.	35	80	80	225	Cumene
Shell Chemical	Houston, Tx.	30	30	30	30	Cumene
	Dominguez, Calif.	150	150	150	150	Isopropanol
	Norco, La.	100	100	100	100	Isopropanol
	Houston, Tx.	180	180	245	400	Isopropanol
Skelly Oil	El Dorado, Kansas	30	30	30	30	Cumene
Union Carbide	Marietta, Ohio	–	–	–	175	Cumene
	Bound Brook, N.J.	87	87	87	87	Cumene
	Institute, W.Va.	120	120	120	120	Isopropanol
	Texas City, Tx.	130	130	130	130	Isopropanol
	Whiting, Ind.	120	120	120	120	Isopropanol
	Ponce, P.R.	–	–	–	120	Cumene
USS Chemicals	Haverhill, Ohio	–	–	120	120	Isopropanol
	Total	1,467	1,552	1,857	3,012	

	1967	*1968*	*1970*	*1972*
% cumene based	29.4	33.4	34.3	46.4
% isopropanol based	68.3	64.3	63.8	52.5
% propane based	2.3	2.3	1.9	1.1

[1] Based on Oil, Paint & Drug Reporter, Oct. 4, 1970.

parent that the supply of acetone from cumene-based phenol plants is governed by the vagaries of the phenol market, and is indifferent to acetone supply and demand.

For each pound of phenol produced by the cumene route, 0.6 lb of acetone are obtained in accordance with this equation.

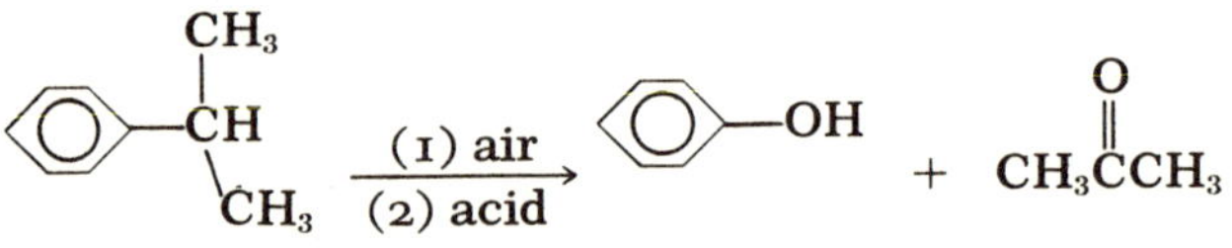

It has been claimed that acetone can be sold from such phenol plants for as little as 3½–4¢/lb, and that the phenol producer would still break even. Such prices are substantially below the unit values of acetone in Table 11.19.

In view of the isopropanol-acetone market situation, it is unlikely that substantial isopropanol capacity increases are in the offing (Table 11.21).

TABLE 11.21

Isopropanol capacity (1972)[1]

Company		*MM lb*
Enjay	(Baton Rouge, La.)	680
Shell	(Dominguez, Calif.)	275
	(Houston, Tx.)	530
Union Carbide	(Texas City, Tx.)	550
	(Whiting, Ind.)	475
	Total	2,510

[1] Based on Oil, Paint & Drug Reporter, Jan. 5, 1970.

The principal market for acetone is methylisobutyl ketone (MIBK) closely followed by methylmethacrylate as shown in Table 11.22. MIBK is readily obtained in 70 to 80% yield by aldol condensation of acetone.

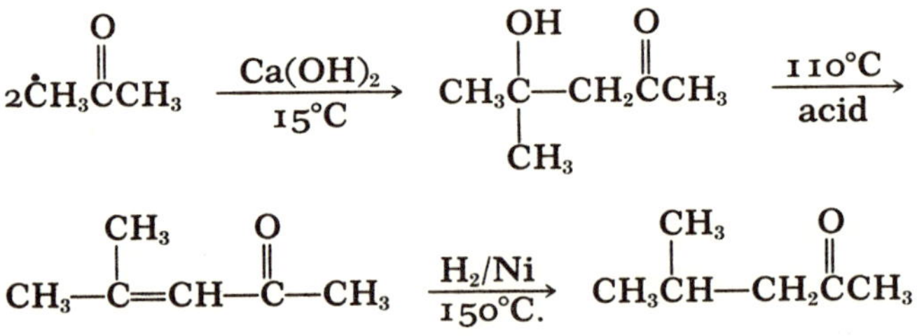

TABLE 11.22
Acetone end use pattern (1971)[1]

Use	MM lb
Methylisobutyl ketone	300
Methyl isobutyl carbinol	80
Methyl methacrylate	285
Cellulose Acetate	220
Bisphenol – A	55
Solvent for paint, varnish, lacquers, and drugs	205
Misc. including exports	435
Total	1,580

[1] Based on Oil, Paint & Drug Reporter, Oct. 4, 1971.

MIBK is commonly used for solvent casting of coatings and film. It has run afoul of anti-pollution criteria in Los Angeles, and processes in that area have been formulated away from MIBK. If the Environmental Protective Agency establishes these same criteria nationally, it would spell the demise of MIBK and significantly hinder acetone growth.

Methylmethacrylate has a brighter outlook than MIBK. It is principally produced by Rohm and Haas via the cyanohydrin process:

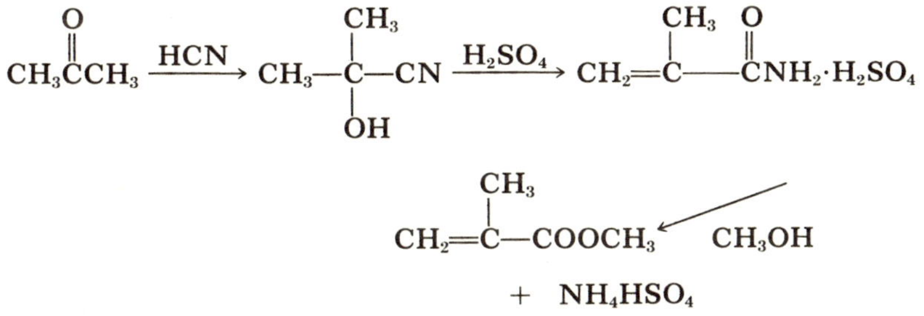

POLYPROPYLENE

This resin has grown rapidly in the last decade to the point where it ranks second only to isopropanol in demand for propylene (Table 11.5). Other than poly 4-methyl-1-pentene, polypropylene is the lightest of all commercially available plastics and has a density of 0.905 gm/cc.

Polypropylene exists in three stereoisomeric forms, atactic, syndiotactic, and isotactic. The last of these is the most crystalline and is of greatest commercial importance. The high crystallinity of isotactic polypropylene (60%+) imparts to it high tensile strength, stiffness, and hardness. Hence, this polymer has a high strength-to-weight ratio that is advantageous in many applications.

To make isotactic polypropylene, polymer-grade propylene is polymerized in the liquid phase at 50 to 100°C. and 1 to 10 atm pressure in the presence of a $TiCl_3$–$Al(C_2H_5)_3$ catalyst system.

Paraffins, such as cyclohexane and heptane are commonly used solvents:

$$CH_3CH{=}CH_2 \xrightarrow{TiCl_3 + Al(C_2H_5)_3} -(CH_2-\underset{\displaystyle CH_3}{\underset{|}{CH}})_n-$$

Propylene of high purity (99.5%+) is fed to the reactor under constant pressure as shown in Fig. 11.3. The crystalline polymer is insoluble and precipitates as a finely, granular solid enveloping the catalyst particles. The polymer slurry is transferred to a stripping tank

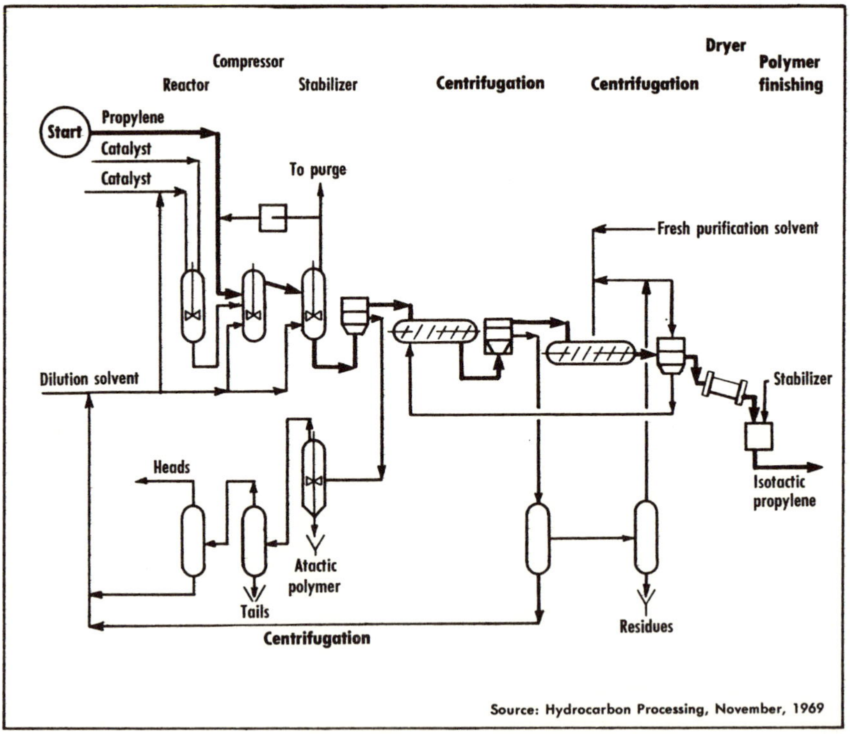

FIG. 11.3 Isotactic polypropylene

where unreacted propylene is flashed off and recycled. The catalyst is solubilized and removed from the polymer by addition of methanol.

Depending on the stereospecificity of the catalyst and the reaction conditions, variable amounts of atactic and syndiotactic polymer are formed as byproducts. The estimated manufacturing cost for isotactic polypropylene is given in Table 11.23.

TABLE 11.23
Estimated polypropylene economics
(200-MM-lb plant; 1972 construction)

Total capital investment	$28.3 MM
Production cost	
	¢/lb
Raw materials[1]	5.10
Utilities	0.49
Labor	0.22
Maintenance (6% ISBL + 3% OSBL)	0.68
Overhead (45% labor + maintenance)	0.41
Taxes and insurance (1.5% of investment)	0.21
Depreciation (10 yr)	1.42
Total	8.53

[1] Includes 1.04 lb propylene at 3.5¢/lb (polymer grade).

TABLE 11.24
Polypropylene production and sales[1]
(MM lb)

Year	*Production*	*Sales*	*Unit value, ¢/lb*
1961	97	57	38
1962	145	107	na
1963	197	146	29
1964	270	226	24
1965	374	301	22
1966	554	372	23
1967	662	599	21
1968	878	925	21
1969	1,090	972	22
1970	1,038	863	na
1971[2]	1,190	1,080	na

[1] US Tariff Commission.
[2] Based on 8 months' data.

Polypropylene is substantially a merchant material, not captive, as can be discerned from Table 11.24. By far, polypropylene fiber and filament are the largest end uses for this polymer (Table 11.25). Biaxially oriented film having high strength and low water-vapor transmission, has high potential for packaging applications. Principal producers of polypropylene are summarized in Table 11.26.

TABLE 11.25
1971 polypropylene end use pattern[1]

Market	*MM lb*
Appliances	60
Filament and fiber	320
Film and sheet (packaging)	184
Housewares	110
Transportation	133
Exports[2]	130
Other[3]	253
Total	1,190

[1] Based on *Modern Plastics,* Jan., 1971.
[2] US Census Bureau; based on 7 months' data.
[3] Appliances, wire and cable, luggage, pipe, toys, etc.

TABLE 11.26
1972 polypropylene capacity[1]
(*MM lb*)

Company		*1970*	*1972*
Avisun	(New Castle, Del.)	250	250
	(Chocolate Bayou, Tx.)	–	150
Rexene Polymers	(Odessa, Tx.)	50	130
Diamond Shamrock	(Houston, Tx.)	70	–
Enjay	(Baytown, Tx.)	300	300
Hercules	(Lake Charles, La.)	370	370
	(Parlin, N.J.)	100	–
Novamont	(Neal, W.Va.)	70	70
Shell	(Woodbury, N.J.)	250	250
Texas-Eastman	(Longview, Tx.)	95	95
	Total	1,555	1,615

[1] Based on *Modern Plastics,* Jan., 1971.

Polypropylene has excellent electrical properties and is completely free from environmental stress cracking. However, it is inherently less stable than is polyethylene to thermal, light, and oxidative attack.

This is presumably due to the tertiary hydrogen atoms present in polypropylene, but lacking in polyethylene. Hence, antioxidants and uv stabilizers are necessary for satisfactory processing and weathering.

PROPYLENE OXIDE

Unlike ethylene oxide, propylene oxide cannot be produced in any economical yield by direct, vapor phase oxidation. Thus, the trend has been to produce propylene oxide by the chlorohydrin technology once employed extensively for ethylene oxide (Chapter 10):

$$CH_3CH{=}CH_2 + HOCl \rightarrow CH_3\overset{\displaystyle OH}{\overset{|}{C}}HCH_2Cl \xrightarrow{Ca(OH)_2} CH_3\underset{\diagdown O \diagup}{CH{-}CH_2} + CaCl_2$$

To accomplish this end, propylene oxide has been largely produced in former chlorohydrin-based ethylene oxide plants.

Addition of propylene and CO_2 to aqueous calcium hypochlorite, readily yields hypochlorous acid which then adds to the olefin to yield the chlorohydrin. The CO_2 is necessary to create the acidic conditions for HOCl production. The chlorohydrin process is usually operated to produce a chlorohydrin concentration of only 3 to 4% to minimize bis(chloroisopropyl)ether as a byproduct. Under the proper operating conditions, this could become the major product. Propylene chloride is also a byproduct, and operation in dilute solutions is necessary to avoid a separate oil phase of this chemical. This would preferentially dissolve chlorine and propylene and increase propylene chloride yield. By contrast, ethylene oxide production employs a chlorohydrin solution concentration of 7%.

The dilute propylene chlorohydrin solution is contacted with a 10% aqueous slurry of $Ca(OH)_2$ to yield propylene oxide. Yield of propylene oxide is about 67 mole %. With the exception of Celanese and Oxirane, the chlorohydrin process is used by every US producer (Table 11.27).

Propylene oxide process technology has undergone a dramatic change in recent years, due to the introduction of the Oxirane process. This is a joint development by Halcon International and Arco Chemical.

The Oxirane process can be summarized by these equations.

TABLE 11.27
US propylene oxide capacity (1971)[1]
(MM lb)

Company	*1970*	*1972*	*Process*
Celanese (Bishop, Tx.)	10	10	propane butane oxidation
Dow (Freeport, Tx.)	650	650	chlorohydrin
(Midland, Mich.)	25	–	chlorohydrin
(Plaquemine, La.)	200	200	chlorohydrin
Jefferson (Port Neches, Tx.)	120	120	chlorohydrin
Olin (Brandenberg, Ky.)	115	115	chlorohydrin
Oxirane (Bayport, Tx.)	160	340	hydroperoxide
Union Carbide (S. Charleston, W.Va.)	200	200	chlorohydrin
Wyandotte (Wyandotte, Mich.)	320	160[2]	chlorohydrin
Total	1,800	1,795	

[1] Based on Oil, Paint & Drug Reporter, April 1, 1971.
[2] Shut down of older unit.

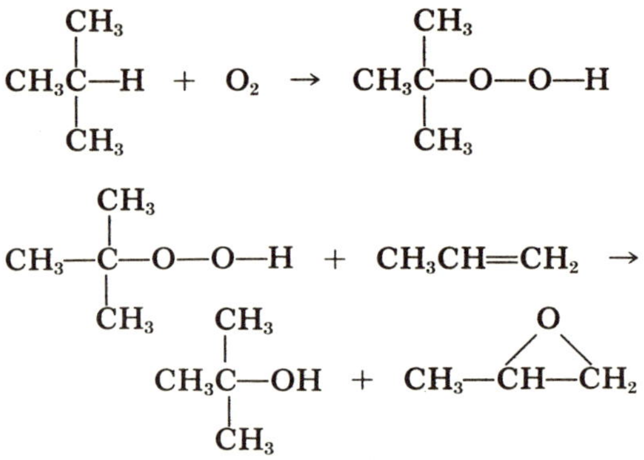

The isobutane (obtained from refinery gases or natural gas) can be oxidized in the liquid phase at 125 to 150°C. and 500 psig to obtain the hydroperoxide. Tert-butyl alcohol is the principal byproduct and can also function as a solvent. Epoxidation also occurs in the liquid phase via a tungsten, vanadium, or molybdenum catalyst system, such as $Mo(CO)_6$.

The reaction proceeds at about 110°C. and 550 psig in 82% selectivity on 22% propylene conversion. The t-butanol coproduct is recovered, easily dehydrated, and sold as isobutylene. Byproduct credit for the latter is essential to the economics of the process as illustrated in Table 11.28. Production costs are substantially lower than the

TABLE 11.28
Estimated economics for propylene oxide
(200-MM-lb plant; 1972 construction)

Total fixed investment

Process	*$ MM*
Chlorohydrin[1]	2.12
Oxirane	26.6

[1] Modified ethylene oxide unit.

Production cost

	¢/lb	
	Chlorohydrin	*Oxirane*
Raw materials	9.00[1]	5.47[3]
Utilities	0.55	1.15
Labor	0.20	0.25
Maintenance (6% ISBL + 3% OSBL)	0.15[2]	0.64
Overhead (45% labor + maintenance)	0.16	0.40
Taxes and insurance (1.5% of investment)	0.05[2]	0.20
Depreciation (10 years)	0.11	1.33
Total	10.22	9.44
Byproduct credit	—	5.00[4]
Net	10.22	4.44

[1] Includes propylene (0.98 lb/lb at 3.0¢/lb) and chlorine (1.43 lb/lb at 3.25¢/lb).
[2] Based on original ethylene oxide investment.
[3] Includes propylene (0.80 lb/lb at 3.0¢/lb) and isobutane (2.5 lb/lb at 1.1¢/lb).
[4] 2.0 lb isobutylene/lb propylene oxide at 2.5¢/lb.

chlorohydrin route. Abroad, this process has been commercialized using ethylbenzene in place of isobutane. In that instance, styrene would be produced as a byproduct in place of isobutylene. A schematic diagram which uses ethylbenzene hydroperoxide as a coreactant is given in Fig. 11.4.

In view of the disparate economics of Table 11.28, it is difficult to find justification for new chlorohydrin-based plants for propylene oxide.

Another process of interest is that developed by Burmah Oil. This is also an epoxidation route and proceeds: (1) isopropanol is oxidized by air to yield H_2O_2 and acetone; (2) unreacted isopropanol and acetone are removed, and the H_2O_2 is reacted with acetic acid to yield peracetic

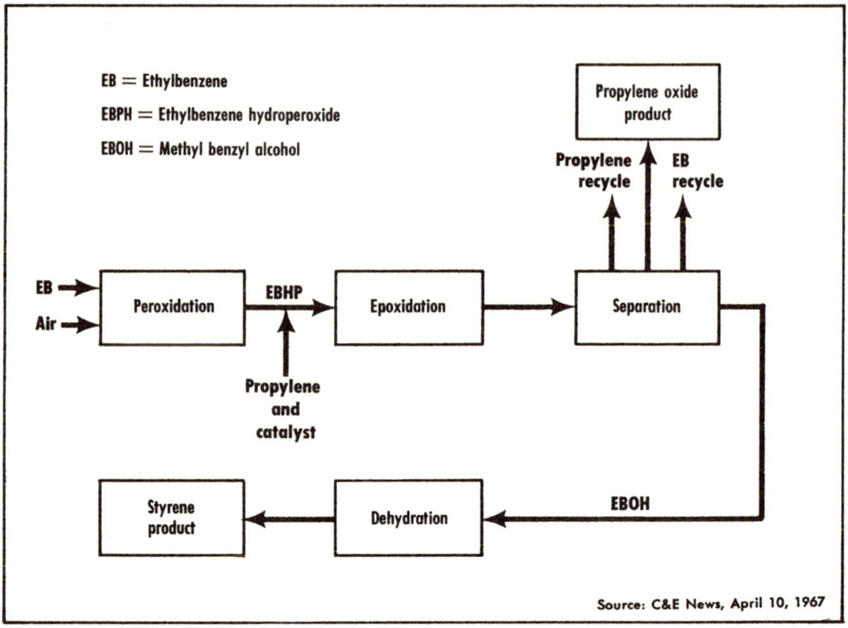

FIG. 11.4 Oxirane process for propylene oxide

acid; (3) propylene is contacted with the peracetic acid to form propylene oxide and again, acetic acid. Yields foı each step are reported to be 90%. This route has not yet been commercialized.

Propylene oxide production just topped 1 billion lb in 1971, with about 90% of the chemical being used captively (Table 11.29). Prices have dwindled 40% in the last decade to about 8¢/lb.

Propylene glycol, obtained by hydrolysis of propylene oxide, is the principal use for the latter (Table 11.30). This is accomplished commercially in much the same manner that ethylene glycol is obtained from ethylene oxide (Chapter 10). Although all propylene glycol is presently produced in this manner, there may be some significant changes in the near future. PPG, who is not now a propylene glycol producer (Table 11.31), has licensed a new process from Kuraray, Ltd. (Japan). The process produces the glycol directly from propylene via oxidation using a palladium nitrate catalyst:

$$CH_3CH{=}CH_2 + O_2 \xrightarrow[\text{acetic acid}]{Pd(NO_3)_2} CH_3\underset{|}{\overset{OH}{C}}H\underset{|}{\overset{OH}{C}}H_2$$

TABLE 11.29
Propylene oxide[1]

Year	Production	Sales	Unit value, ¢/lb
1961	374	36	13
1962	446	60	12
1963	497	52	12
1964	569	60	13
1965	605	69	12
1966	710	83	10
1967	814	76	10
1968	958	90	9
1969	1,177	151	8
1970	1,121	na	15[2]
1971	1,090	na	8.5[3]

[1] US Tariff Commission; 1971 based on 8 months' data.
[2] Oil, Paint & Drug Reporter posted price (tanks, delivered).
[3] Oil, Paint & Drug Reporter posted price (FOB plant; freight equalized).

Production of unsaturated polyester resins comprises the principal market for mono- and dipropylene glycol (Table 11.32).

The other major markets for propylene oxide – polypropylene glycol and polypropoxyethers – are in the manufacture of flexible and rigid polyurethane foams.

TABLE 11.30
Major propylene oxide end uses[1]
(MM lb)

Year	Polypropoxy ethers	Propylene glycol	Polypropylene glycol	Di-propylene glycol
1962	117	184	87	23
1963	135	200	79	26
1964	173	236	96	27
1965	221	213	94	34
1966	271	259	101	37
1967	329	303	135	31
1968	367	353	196	39
1969	369	461	226	48
1970	na	450[2]	na	na
1971	na	460[2]	na	na

[1] US Tariff Commission.
[2] Oil, Paint & Drug Reporter, September 6, 1971.

TABLE 11.31
US propylene glycol capacity (1972)[1]

Company		*MM lb*
Celanese	(Bishop, Tx.)	10
Dow	(Freeport, Tx.)	150
	(Plaquemine, La.)	100
Jefferson	(Port Neches, Tx.)	80
Olin	(Brandenberg, Ky.)	45
Oxirane	(Bayport, Tx.)	180
Union Carbide	(S. Charleston, W.Va.)	80
	Total	645

[1] Oil, Paint & Drug Reporter, Sept. 6, 1971.

TABLE 11.32
1971 propylene glycol end use pattern[1]

Use	*MM lb*
Polyester resins	193
Cellophane	45
Tobacco	37
Polymeric plasticizers	32
Brake fluids and other functional fluids	18
Exports[2]	84
Misc.	51
Total	460

[1] Based on Oil, Paint & Drug Reporter, Sept. 6, 1971.
[2] Based on 7 months' data from US Census Bureau.

Polypropylene glycol is obtained by the base-catalyzed addition of propylene oxide to propylene glycol:

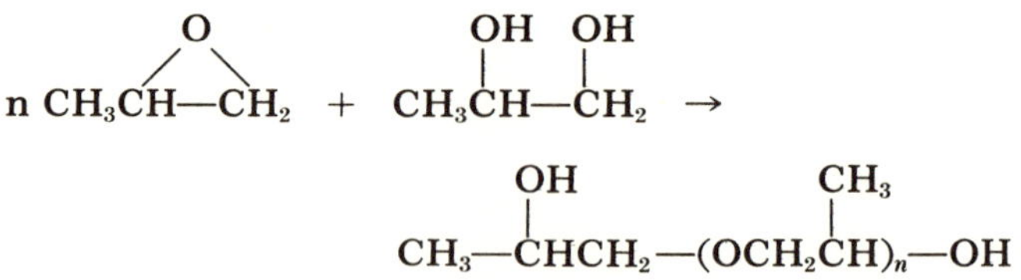

The polypropoxy ethers have more highly branched structures, and are similarly produced by reacting a polyol, such as glycerol, with propylene oxide:

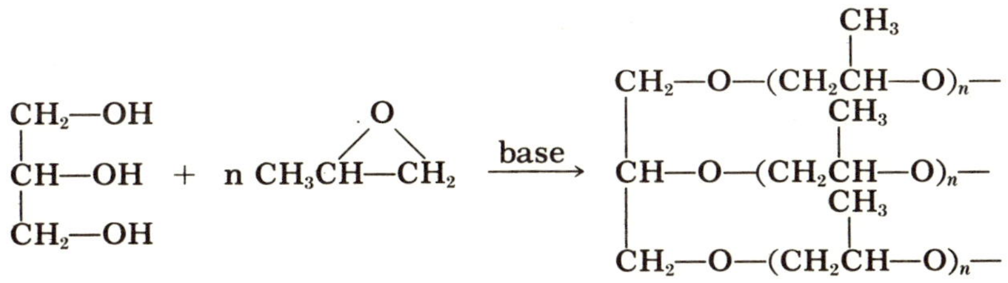

ACRYLONITRILE

Acrylonitrile was originally produced solely from acetylene (Chapter 9). With the introduction of Sohio's ammoxidation route to acrylonitrile, propylene-based processes rapidly supplanted acetylene technology. By year end 1971, all US acrylonitrile processes were based on propylene (Fig. 11.5).

The economic advantage of propylene ammoxidation over acetylene-HCN addition was delineated in Chapter 9 (Table 9.8) where it was seen that a 73% advantage in raw material costs is possible with propylene technology. A more complete, economic picture for a new 200-MM-lb plant is given in Table 11.33. The acrylonitrile production

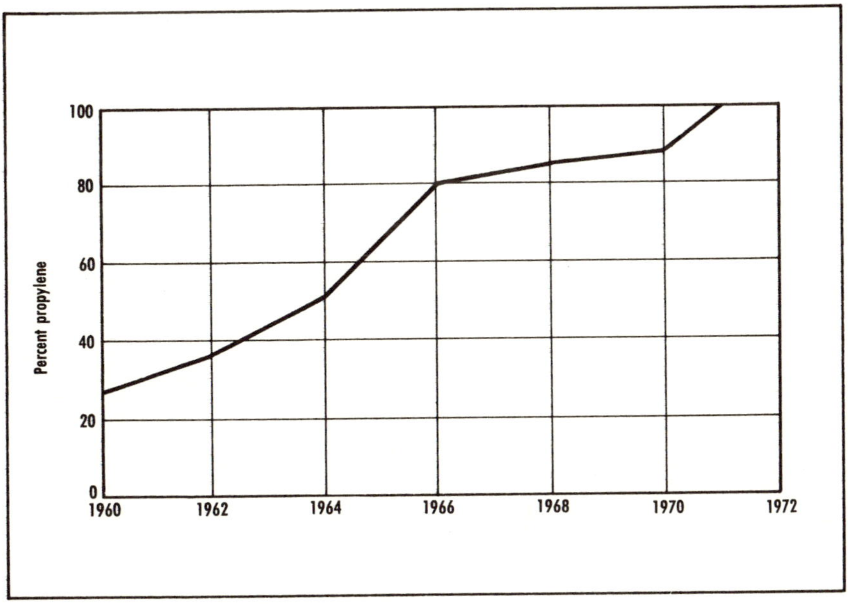

FIG. 11.5 Propylene-based acrylonitrile capacity

cost of 7.96¢/lb (5.85¢/lb excluding depreciation) is more compatible with recent prices of 12–14¢/lb, than could be any acetylene process (Table 11.34).

TABLE 11.33
Estimated economics for acrylonitrile
(200-MM-lb plant; 1972 construction)

Total fixed capital	$42.2 MM
Production cost	
	¢/lb
Raw material[1]	4.22
Utilities	0.79
Labor	0.21
Maintenance (6% ISBL + 3% OSBL)	1.01
Overhead (45% labor + maintenance)	0.55
Taxes & insurance (1.5% of investment)	0.32
Depreciation (10 years)	2.11
Total	9.21
Byproducts credit[2]	1.25
Net cost	7.96

[1] 1.16 lb propylene at 3.0¢/lb and 0.37 lb ammonia at 2.0¢/lb.

[2] HCN, 0.05 lb at 10¢/lb; 0.05 lb acetonitrile at 24.0¢/lb.

TABLE 11.34
Acrylonitrile[1]
(MM lb)

Year	*Production*	*Sales*	*Unit value, ¢/lb*
1961	250	157	18
1962	360	203	15
1963	455	212	14
1964	594	311	16
1965	772	303	16
1966	716	318	13
1967	671	270	12
1968	1,021	na	na
1969	1,157	562	12
1970	1,037	na	14.5[2]
1971	965	na	14.5[2]

[1] US Tariff Commission; 1971 based on 8 months' data.

[2] Oil, Paint, & Drug Reporter posted price (tanks, works).

Producers of acrylonitrile are summarized in Table 9.12 of Chapter 9.

As cited previously, the ammoxidation route employs a fluid bed catalyst over which a mixture of propylene, air and ammonia (1 to 13 to 1 v/v) are passed at 425° to 470°C. Steam is often used as a diluent to improve product selectivity. Principal byproducts are HCN and acetonitrile, which are credited as shown in Table 11.33.

$$CH_2{=}CH{-}CH_3 + NH_3 \xrightarrow[\text{catalyst}\ 425\text{–}470°C.]{\text{air}} CH_2{=}CH{-}CN + HCN + CH_3CN + H_2O + CO_2$$

The reactor effluent goes to an absorption column where it is scrubbed to separate fixed gases and unreacted propylene as shown in Fig. 11.6.

More than 50% of acrylonitrile production is used in the manufacture of acrylic and modacrylic fibers as shown in Table 11.35. The acrylic fibers are polymers containing at least 85% acrylonitrile, while the modacrylic fibers contain 35 to 85% acrylonitrile and usually 20% or more of vinyl chloride or vinylidene chloride. Typical co-monomers for acrylic fibers are vinyl acetate and methyl acrylate.

The acrylic fibers are relatively insensitive to moisture and have good resistance to stains. They are used broadly in carpeting and knit apparel, and are commonly sold under the trade names of Orlon, Acrilan, and Dynel. Table 11.36 summarizes the recent history of US acrylic and modacrylic fiber production.

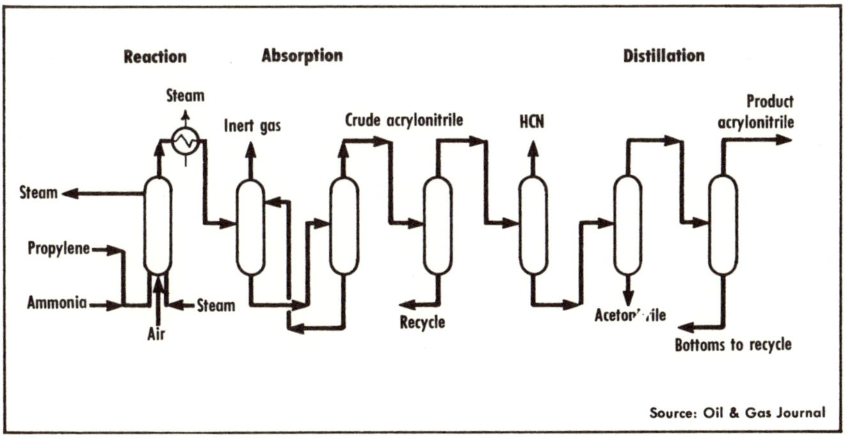

FIG. 11.6 Acrylonitrile process flow diagram

TABLE 11.35
1971 acrylonitrile end use pattern[1]

Item	*MM lb*
Acrylic and modacrylic fibers	590
ABS and SAN resins	164
Nitrile rubber	60
Misc. and exports	151
Total	965

[1] Based on Oil, Paint & Drug Reporter, Aug. 9, 1971.

TABLE 11.36
Acrylic and modacrylic fiber production[1]

Year	*Domestic use*	*For export*	*Total*
1963	199	11	210
1964	266	22	288
1965	348	20	368
1966	328	23	351
1967	371	28	399
1968	467	40	507
1969	474	48	522
1970	431	54	485
1971	470	81	551

[1] Textile Organon; 1971 based on 6 months' data.

Recently, substantial interest has evolved in the use of acrylonitrile as an intermediate for adiponitrile. The latter is useful for adipic acid and hexamethylene diamine synthesis, both co-monomers for nylon-6,6. Only Monsanto produces adiponitrile from acrylonitrile commercially in the US.

The Monsanto process is electrolytic hydrodimerization which proceeds at the cathode as indicated:

$$2CH_2{=}CH{-}CN + 2H^+ + 2e \xrightarrow[H_2O]{25°C.} NC{-}CH_2CH_2CH_2CH_2{-}CN$$

Small amounts of propionitrile and bis(cyanoethyl)ether are obtained as coproducts. Acrylonitrile has limited solubility in water so quaternary ammonium salts are used to overcome this problem. At 50 to 90% salt concentrations, acrylonitrile solubility can be increased from 7% to 50%. Other firms, notably ICI, UCB (Belgium), and Asahi have developed similar processes.

In late 1971, ICI had announced a novel process for acrylonitrile production directly from propane.

ISOPRENE

This represents a relatively small volume use for propylene. Other than Goodyear's 165-MM-lb plant at Beaumont, Texas, no other company uses propylene-based technology (Table 11.37). Most US capacity is based on isoamylene dehydrogenation, the feedstock being obtained as ethylene plant byproduct or the light gasoline stream from cat crackers.

TABLE 11.37
Isoprene capacity (1972)[1]

Company	*MM lb*	*Process*
Enjay (Baton Rouge, La.)	10	Ethylene byproduct
(Baytown, Tx.)	10	Ethylene byproduct
Goodrich-Gulf (Port Neches, Tx.)	100	Isoamylene dehydrogenation
Goodyear (Beaumont, Tx.)	165	Propylene dimerization & ethylene byproduct
Shell (Marietta, Ohio)	80	Isoamylene dehydrogenation
(Torrance, Calif.)	40	Isoamylene dehydrogenation
Total	405	

[1] Based on *Chemical Week*, March 24, 1971.

The dehydrogenation proceeds at 600°C. over an Fe_2O_3–K_2CO_3–Cr_2O_3 catalyst system in 85% selectivity on 35% conversion.

$$CH_3C(CH_3){=}CH{-}CH_3 + CH_2{=}C(CH_3){-}CH_2CH_3 \rightarrow CH_2{=}C(CH_3){-}CH{=}CH_2 + H_2$$

The Goodyear process was developed jointly with Scientific Design (Halcon, Int.) and comprises dimerization of refinery-grade propylene, principally to 2,3-dimethyl-1-pentene by use of a tripropyl aluminum catalyst. This is subsequently isomerized over a fixed bed SiO_2–Al_2O_3 catalyst to 2,3-dimethyl-2-pentene. Finally, pyrolysis in the presence of superheated steam leads to isoprene plus methane. Selectivity for this last step is 47% on 35% conversion.

$$2CH_3CH{=}CH_2 \xrightarrow[\substack{Al\ Pr_3 \\ 3000\ psig}]{200^\circ C.} CH_2{=}\overset{\displaystyle CH_3}{\overset{|}{C}}{-}CH_2CH_2CH_3 \quad (90\%)$$

$$CH_2{=}\overset{\displaystyle CH_3}{\overset{|}{C}}{-}CH_2CH_2CH_3 \xrightarrow[100^\circ C.]{SiO_2\text{-}Al_2O_3} CH_3\overset{\displaystyle CH_3}{\overset{|}{C}}{=}CH{-}CH_2CH_3 \quad (85\%)$$

$$CH_3\overset{\displaystyle CH_3}{\overset{|}{C}}{=}CH{-}CH_2CH_3 \xrightarrow[steam]{660^\circ C.} CH_2{=}\overset{\displaystyle CH_3}{\overset{|}{C}}{-}CH{=}CH_2 + CH_4 \quad (47\%)$$

The process flow scheme is illustrated in Fig. 11.7.

In recent years, a process developed independently in the USSR and by Institut Francais du Petrole has gained widespread interest.

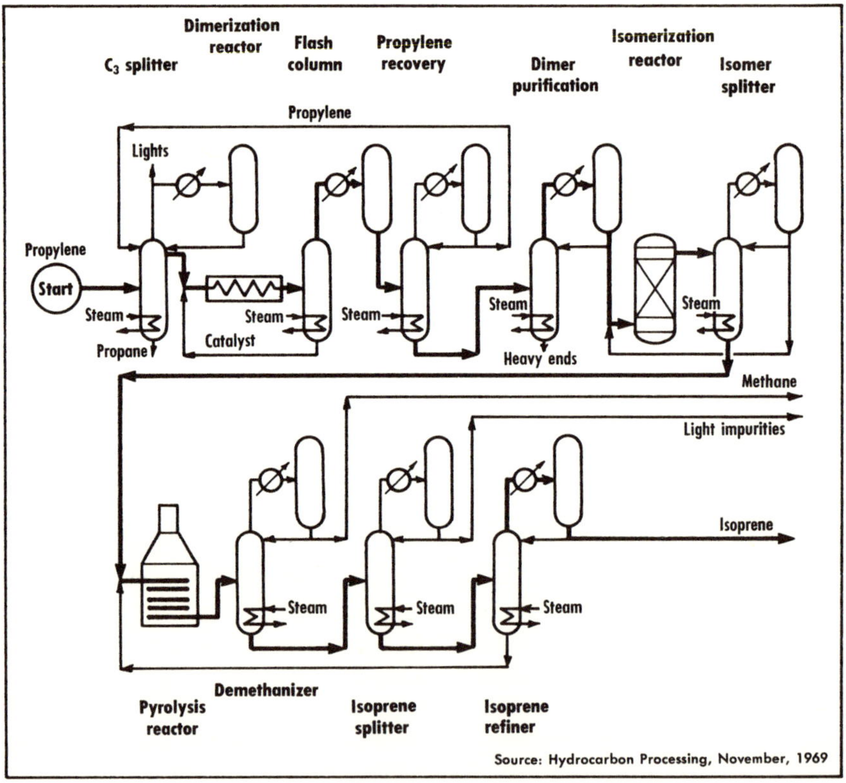

FIG. 11.7 Goodyear isoprene process flow scheme

This is an isobutylene-formaldehyde system and proceeds as indicated in Fig. 11.8.

Isobutylene is reacted with 2 moles of formaldehyde at 80°C. in the presence of H_2SO_4 to form dimethyldioxane (DMD). Yield is about 90% on both isobutylene and formaldehyde. This is pyrolyzed at 260°C. over $Ca_3(PO_4)_2$ to yield isoprene, water, and one mole of recyclable formaldehyde. Selectivity on pyrolysis is 95%. Conversions are held to 60% to reduce resinification between isoprene and formaldehyde.

Of these three synthetic routes, the propylene-based technology is generally regarded as the least economical. The isobutylene process may be the most attractive.

Although the amount of isoprene currently recovered as ethylene plant byproduct in the US is quite small, there is significant international activity which indicates that this may become quite important in the future (Table 11.38). Of 265,000 metric tons of planned capacity, at least 50% will be based on byproduct recovery.

Europe and Japan produce ethylene via naphtha cracking and thus have C_5 streams containing isoprene. As the US shifts away from

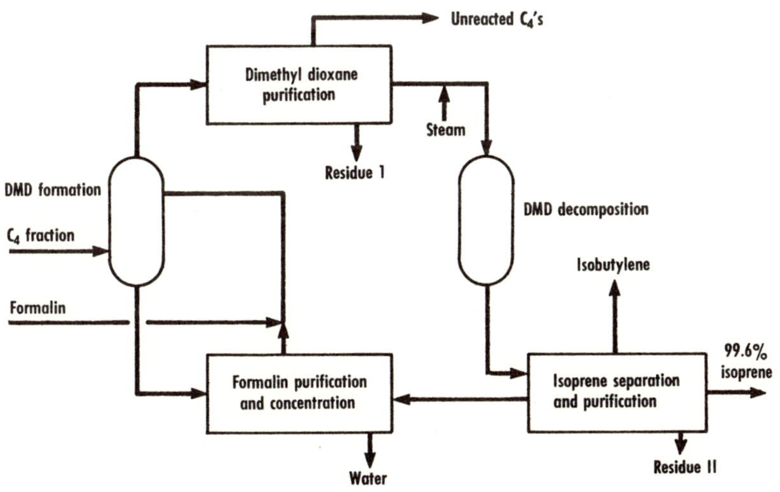

Source: Chemical & Process Eng., March, 1971

FIG. 11.8 Isobutylene-based isoprene route

TABLE 11.38

Planned isoprene capacity external to the US[1]

Producer	*Process*	*Metric tons/yr*
Venezuela	na	40,000
ANIC (Italy)	Acetylene	30,000
Erdolchemie (Germany)	Ethylene byproduct	80,000
Asahi (Japan)	Ethylene byproduct	30,000
Japanese Geon	Ethylene byproduct	25,000
Japan Synthetic rubber	Isobutylene or ethylene byproduct	30,000
Kurashiki Rayon (Japan)	Isobutylene	30,000
	Total	265,000

[1] *Chemical Week*, March 24, 1971.

natural-gas liquids to naphtha and cat gas oil, inexpensive isoprene recovery from this source may become significant in the US. The principal solvents used to extract isoprene from these streams are n-methylpyrrolidone and dimethylformamide. It has been estimated by D. O. Swan of Esso Chemical[1] that isoprene recovery from ethylene plants will increase from about 10% in 1970 to 75% of total demand by 1980. A 1-billion-lb ethylene plant using naphtha or cat gas oil can produce up to 60 MM lb of isoprene.

Production of isoprene via acetylene (Table 11.38) is an unusual route requiring a uniquely inexpensive starting material. Apparently, ANIC finds itself in such a situation. The process can be summarized by these reactions:

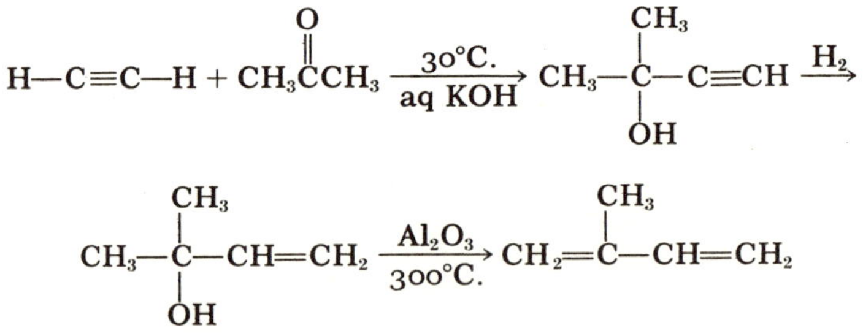

Isoprene is used captively for the production of synthetic rubbers, notably cis-polyisoprene (Table 11.39). Polymerization to cis-poly-

[1] 1971 Annual meeting of International Synthetic Rubber Producers (Australia).

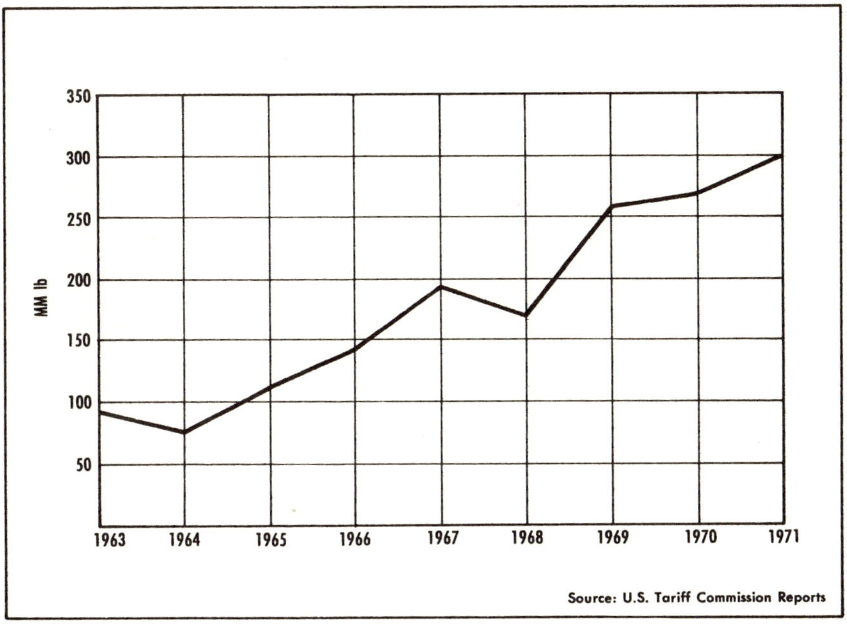

FIG. 11.9 US isoprene production

isoprene proceeds at 50°C. in a pentane solvent with a $TiCl_4$–$Al(iBu)_3$ catalyst. In some instances, butyl lithium is the catalyst:

$$CH_2{=}\underset{\displaystyle |}{\overset{\displaystyle CH_3}{C}}{-}CH{=}CH_2 \rightarrow {-}(CH_2{-}\overset{\displaystyle CH_3}{\overset{|}{C}}{=}CH{-}CH_2)_n{-}$$

Cis-polyisoprene offers more uniformity, better price stability, and

TABLE 11.39
1971 isoprene market pattern[1]

	(MM lb)
Cis-polyisoprene	280
Butyl rubber	11
Other synthetic rubber	5
Plastics	1
Chemicals	3
Total	300

[1] Based on R. A. Arnold (Goodrich-Gulf), Am. Chem. Soc. Nat'l. Meeting; N.Y., N.Y. (1969).

is as complete a replacement for natural rubber that has yet been developed. It is free from problems created by political uncertainties, and the only deterrent to more universal applications is the high cost of the monomer (ca., 10–12¢/lb).

Statistics on isoprene production are given in Fig. 11.9. Consumption of polyisoprene is expected to climb to 660 MM lb annually by 1979 as noted in Table 10.32 of Chapter 10.

Index

Index